The authors formulate and explore a new axiom of set theory, CPA, the Covering Property Axiom. CPA is consistent with the usual ZFC axioms; indeed, it is true in the iterated Sacks model and actually captures the combinatorial core of this model. A plethora of results known to be true in the Sacks model easily follow from CPA. Replacing iterated forcing arguments with deductions from CPA simplifies proofs, provides deeper insight, and leads to new results. One may say that CPA is similar in nature to Martin's axiom, as both capture the essence of the models of ZFC in which they hold.

The exposition is self-contained, and there are natural applications to real analysis and topology. Researchers who use set theory in their work will find much of interest in this book.

Krzysztof Ciesielski is Professor of Mathematics at West Virginia University.

Janusz Pawlikowksi is Professor of Mathematics at Wrocław University.

CAMBRIDGE TRACTS IN MATHEMATICS

General Editors

B. BOLLOBÁS, W. FULTON, A. KATOK, F. KIRWAN,
P. SARNAK, B. SIMON

164 The Covering Property Axiom, CPA: A Combinatorial Core of the Iterated Perfect Set Model

Krzysztof Ciesielski
West Virginia University

Janusz Pawlikowski
Wrocław University

The Covering Property Axiom, CPA

A Combinatorial Core of the Iterated Perfect Set Model

CAMBRIDGE
UNIVERSITY PRESS

PUBLISHED BY THE PRESS SYNDICATE OF THE UNIVERSITY OF CAMBRIDGE
The Pitt Building, Trumpington Street, Cambridge, United Kingdom

CAMBRIDGE UNIVERSITY PRESS
The Edinburgh Building, Cambridge CB2 2RU, UK
40 West 20th Street, New York, NY 10011-4211, USA
477 Williamstown Road, Port Melbourne, VIC 3207, Australia
Ruiz de Alarcón 13, 28014 Madrid, Spain
Dock House, The Waterfront, Cape Town 8001, South Africa

http://www.cambridge.org

First published 2004

Printed in the United States of America

Typeface Computer Modern 10/13 pt.   *System* LaTeX 2_ε [AU]

A catalog record for this book is available from the British Library.

Library of Congress Cataloging in Publication Data

Ciesielski, Krzysztof, 1957–
The covering property axiom, CPA : a combinatorial core of the iterated perfect
set model / Krzysztof Ciesielski, Janusz Pawlikowski.
p. cm. — (Cambridge tracts in mathematics ; 164)
Includes bibliographical references and index.
ISBN 0-521-83920-3
1. Axiomatic set theory. I. Pawlikowski, Janusz, 1957– II. Title. III. Series.
QA248.C473 2004
511.3'22—dc22   2004040788

ISBN 0 521 83920 3 hardback

To Monika and Joanna

Contents

Overview

Many interesting mathematical properties, especially those concerning real analysis, are known to be true in the iterated perfect set (Sacks) model, while they are false under the continuum hypothesis. However, the proofs that these facts are indeed true in this model are usually very technical and involve heavy forcing machinery. In this book we extract a combinatorial principle, an axiom similar to Martin's axiom, that is true in the model and show that this axiom implies the above-mentioned properties in a simple "mathematical" way. The proofs are essentially simpler than the original arguments.

It is also important that our axiom, which we call the *Covering Property Axiom* and denote by CPA, captures the essence of the Sacks model at least if it concerns most cardinal characteristics of continuum. This follows from a recent result of J. Zapletal [131], who proved that for a "nice" cardinal invariant κ, if $\kappa < \mathfrak{c}$ holds in any forcing extension, then $\kappa < \mathfrak{c}$ follows already from CPA. (In fact, $\kappa < \mathfrak{c}$ follows already from its weaker form, which we denote $\text{CPA}_{\text{prism}}^{\text{game}}$.)

To follow all but the last chapter of this book only a moderate knowledge of set theory is required. No forcing knowledge is necessary.

The iterated perfect set model, also known as the iterated Sacks model, is a model of the set theory ZFC in which the continuum $\mathfrak{c} = \omega_2$ and many of the consequences of the continuum hypothesis (CH) fail. In this book we describe a combinatorial axiom of the form similar to Martin's axiom, which holds in the iterated perfect set model and represents a combinatorial core of this model – it implies all the "general mathematical statements" that are known (to us) to be true in this model.

It should be mentioned here that our axiom is more an axiom schema with the perfect set forcing being a "built-in" parameter. Similar axioms also hold for several other forcings (like iterated Miller and iterated Laver forcings; see, e.g., [131, sec. 5.1]). In this book, however, we concentrate only on the axiom associated with the iterated perfect set model. This is dictated by two reasons: The axiom has the simplest form in this particular model, and the iterated perfect set model is the most studied from the class of forcing models we are interested in — we have a good supply of statements against which we can test the power of our axiom. In particular, we use for this purpose the statements listed below as (A)–(H). The citations in the parentheses refer to the proofs that a given property holds in the iterated perfect set model. For the definitions see the end of the next section.

(A) For every subset S of $\mathbb{R}$ of cardinality $\mathfrak{c}$ there exists a (uniformly) continuous function $f\colon \mathbb{R} \to [0,1]$ such that $f[S] = [0,1]$. (A. Miller [95])

(B) Every perfectly meager set $S \subset \mathbb{R}$ has cardinality less than $\mathfrak{c}$. (A. Miller [95])

(C) Every universally null set $S \subset \mathbb{R}$ has cardinality less than $\mathfrak{c}$. (R. Laver [84])

(D) The cofinality of the ideal $\mathcal{N}$ of null (i.e., Lebesgue measure zero) sets is less than $\mathfrak{c}$. (Folklore, see, e.g., [97] or [4, p. 339])

(E) There exist selective ultrafilters on ω, and any such ultrafilter is generated by less than $\mathfrak{c}$ many sets. (J. Baumgartner and R. Laver [7])

(F) There is no Darboux Sierpiński-Zygmund function $f\colon \mathbb{R} \to \mathbb{R}$; that is, for every Darboux function $f\colon \mathbb{R} \to \mathbb{R}$ there is a subset Y of $\mathbb{R}$ of cardinality $\mathfrak{c}$ such that $f \restriction Y$ is continuous. (M. Balcerzak, K. Ciesielski, and T. Natkaniec [2])

(G) For every Darboux function $g\colon \mathbb{R} \to \mathbb{R}$ there is a continuous nowhere constant function $f\colon \mathbb{R} \to \mathbb{R}$ such that $f + g$ is Darboux. (J. Steprāns [123])

(H) The plane $\mathbb{R}^2$ can be covered by less than $\mathfrak{c}$ many sets, each of which is a graph of a differentiable function (allowing infinite derivatives) of either a horizontal or vertical axis. (J. Steprāns [124])

The counterexamples under CH for (B) and (C) are Luzin and Sierpiński sets. They have been constructed in [87][1] and [116], respectively. The negation of (A) is witnessed by either a Luzin or a Sierpiński set, as noticed in [116, 117]. The counterexamples under CH for (F) and (G) can be found in [2] and [78], respectively. The fact that (D), (E), and (H) are false under CH is obvious.

---

[1] Constructed also a year earlier by Mahlo [88].

The book is organized as follows. Since our main axiom, which we call the *Covering Property Axiom* and denote by CPA, requires some extra definitions that are unnecessary for most of the applications, we will introduce the axiom in several approximations, from the easiest to state and use to the most powerful but more complicated. All the versions of the axiom will be formulated and discussed in the main body of the chapters. The sections that follow contain only the consequences of the axioms. In particular, most of the sections can be omitted in the first reading without causing any difficulty in following the rest of the material.

Thus, we start in Chapter 1 with a formulation of the simplest form of our axiom, CPA_{cube}, which is based on a natural notion of a cube in a Polish space. In Section 1.1 we show that CPA_{cube} implies properties (A)–(C), while in Section 1.2 we present A. Nowik's proof [107] that CPA_{cube} implies that

(I) Every uniformly completely Ramsey null $S \subset \mathbb{R}$ has cardinality less than $\mathfrak{c}$.

In Section 1.3 we prove that CPA_{cube} implies property (D), that is, $\text{cof}(\mathcal{N}) = \omega_1$, and Section 1.4 is devoted to the proof that CPA_{cube} implies the following fact, known as *the total failure of Martin's axiom*:

(J) $\mathfrak{c} > \omega_1$ and for every nontrivial forcing $\mathbb{P}$ satisfying the countable chain condition (ccc), there exists ω_1 many dense sets in $\mathbb{P}$ such that no filter intersects all of them.

Recall that a forcing $\mathbb{P}$ is ccc provided it has no uncountable antichains, where $A \subset \mathbb{P}$ is an antichain in $\mathbb{P}$ provided no distinct elements of A have a common extension in $\mathbb{P}$. The consistency of (J) was first proved by J. Baumgartner [6] in a model obtained by adding Sacks reals side by side.

In Section 1.5 we show that CPA_{cube} implies that every selective ultrafilter is generated by ω_1 sets (i.e., the second part of property (E)) and that

(K) $\mathfrak{r} = \omega_1$,

where $\mathfrak{r}$ is the *reaping* (or *refinement*) number, that is,

$$\mathfrak{r} = \min\{|\mathcal{W}| : \mathcal{W} \subset [\omega]^\omega \ \& \ \forall A \in [\omega]^\omega \ \exists W \in \mathcal{W} \ (W \subset A \text{ or } W \subset \omega \setminus A)\}.$$

In Section 1.6 we prove that CPA_{cube} implies the following version of a theorem of S. Mazurkiewicz [91]:

(L) For each Polish space X and for every uniformly bounded sequence $\langle f_n \colon X \to \mathbb{R} \rangle_{n < \omega}$ of Borel measurable functions there are the sequences: $\langle P_\xi \colon \xi < \omega_1 \rangle$ of compact subsets of X and $\langle W_\xi \in [\omega]^\omega \colon \xi < \omega_1 \rangle$ such that $X = \bigcup_{\xi < \omega_1} P_\xi$ and for every $\xi < \omega_1$:

$\langle f_n \restriction P_\xi \rangle_{n \in W_\xi}$ is a monotone uniformly convergent sequence of uniformly continuous functions.

We also show that $\mathrm{CPA}_{\mathrm{cube}} +$ "$\exists$ selective ultrafilter on ω" implies the following variant of (L):

(L*) Let X be an arbitrary set and let $f_n \colon X \to \mathbb{R}$ be a sequence of functions such that the set $\{ f_n(x) \colon n < \omega \}$ is bounded for every $x \in X$. Then there are the sequences: $\langle P_\xi \colon \xi < \omega_1 \rangle$ of subsets of X and $\langle W_\xi \in \mathcal{F} \colon \xi < \omega_1 \rangle$ such that $X = \bigcup_{\xi < \omega_1} P_\xi$ and for every $\xi < \omega_1$:

$\langle f_n \restriction P_\xi \rangle_{n \in W_\xi}$ is monotone and uniformly convergent.

It should be noted here that a result essentially due to W. Sierpiński (see Example 1.6.2) implies that (L*) is false under Martin's axiom.

In Section 1.7 we present some consequences of $\mathrm{cof}(\mathcal{N}) = \omega_1$ that seem to be related to the iterated perfect set model. In particular, we prove that $\mathrm{cof}(\mathcal{N}) = \omega_1$ implies that

(M) $\mathfrak{c} > \omega_1$ and there exists a Boolean algebra B of cardinality ω_1 that is not a union of a strictly increasing ω-sequence of subalgebras of B.

The consistency of (M) was first proved by W. Just and P. Koszmider [72] in a model obtained by adding Sacks reals side by side, while S. Koppelberg [79] showed that Martin's axiom contradicts (M).

The last section of Chapter 1 consists of remarks on a form and consistency of $\mathrm{CPA}_{\mathrm{cube}}$. In particular, we note that $\mathrm{CPA}_{\mathrm{cube}}$ is false in a model obtained by adding Sacks reals side by side.

In Chapter 2 we revise slightly the notion of a cube and introduce a cube-game $\mathrm{GAME}_{\mathrm{cube}}$ — a covering game of length ω_1 that is a foundation for our next (stronger) variant of the axiom, $\mathrm{CPA}_{\mathrm{cube}}^{\mathrm{game}}$. In Section 2.1, as its application, we show that $\mathrm{CPA}_{\mathrm{cube}}^{\mathrm{game}}$ implies that:

(N) $\mathfrak{c} > \omega_1$ and for every Polish space there exists a partition of X into ω_1 disjoint closed nowhere dense measure zero sets.

In Section 2.2 we show that $\mathrm{CPA}_{\mathrm{cube}}^{\mathrm{game}}$ implies that:

(O) $\mathfrak{c} > \omega_1$ and there exists a family $\mathcal{F} \subset [\omega]^\omega$ of cardinality ω_1 that is simultaneously maximal almost disjoint (MAD) and reaping.

Section 2.3 is devoted to the proof that, under $\text{CPA}^{\text{game}}_{\text{cube}}$,

(P) there exists an uncountable γ-set.

Chapter 3 begins with a definition of a prism, which is a generalization of a notion of cube in a Polish space. This notion, perhaps the most important notion of this text, is then used in our next generation of the axioms, $\text{CPA}^{\text{game}}_{\text{prism}}$ and $\text{CPA}_{\text{prism}}$, which are prism (stronger) counterparts of axioms $\text{CPA}^{\text{game}}_{\text{cube}}$ and CPA_{cube}. Since the notion of a prism is rather unknown, in the first two sections of Chapter 3 we develop the tools that will help us to deal with them (Section 3.1) and prove for them the main duality property that distinguishes them from cubes (Section 3.2). In the remaining sections of the chapter we discuss some applications of $\text{CPA}_{\text{prism}}$. In particular, we prove that $\text{CPA}_{\text{prism}}$ implies the following generalization of property (A):

(A*) There exists a family $\mathcal{G}$ of uniformly continuous functions from $\mathbb{R}$ to $[0,1]$ such that $|\mathcal{G}| = \omega_1$ and for every $S \in [\mathbb{R}]^{\mathfrak{c}}$ there exists a $g \in \mathcal{G}$ with $g[S] = [0,1]$.

We also show that $\text{CPA}_{\text{prism}}$ implies that:

(Q) $\text{add}(s_0)$, the additivity of the Marczewski's ideal s_0, is equal to $\omega_1 < \mathfrak{c}$.

In Section 3.4 we prove that:

(N*) If $G \in G_{\omega_1}$, where G_{ω_1} is the family of the intersections of ω_1 many open subsets of a given Polish space X, and $|G| = \mathfrak{c}$, then G contains a perfect set; however, there exists a $G \in G_{\omega_1}$ that is not a union of ω_1 many closed subsets of X.

Thus, under $\text{CPA}_{\text{prism}}$, G_{ω_1} sets act to some extent as Polish spaces, but they fall short of having property (N). The fact that the first part of (N*) holds in the iterated perfect set model was originally proved by J. Brendle, P. Larson, and S. Todorcevic [12, thm. 5.10]. The second part of (N*) refutes their conjecture [12, conj. 5.11]. We finish Chapter 3 with several remarks on $\text{CPA}^{\text{game}}_{\text{prism}}$. In particular, we prove that $\text{CPA}^{\text{game}}_{\text{prism}}$ implies axiom $\text{CPA}^{\text{game}}_{\text{prism}}(\mathcal{X})$, in which the game is played simultaneously over ω_1 Polish spaces.

Chapters 4 and 5 deal with the applications of the axioms $\text{CPA}_{\text{prism}}$ and $\text{CPA}^{\text{game}}_{\text{prism}}$, respectively. Chapter 4 contains a deep discussion of a problem of covering $\mathbb{R}^2$ and Borel functions from $\mathbb{R}$ to $\mathbb{R}$ by continuous functions of different smoothness levels. In particular, we show that $\text{CPA}_{\text{prism}}$ implies the following strengthening of property (H):

(H*) There exists a family $\mathcal{F}$ of less than continuum many $\mathcal{C}^1$ functions from $\mathbb{R}$ to $\mathbb{R}$ (i.e., differentiable functions with continuous derivatives) such that $\mathbb{R}^2$ is covered by functions from $\mathcal{F}$ and their inverses.

We also show the following covering property for the Borel functions:

(R) For every Borel function $f\colon \mathbb{R} \to \mathbb{R}$ there exists a family $\mathcal{F}$ of less than continuum many "$\mathcal{C}^1$" functions (i.e., differentiable functions with continuous derivatives, where the derivative can be infinite) whose graphs cover the graph of f.

We also examine which functions can be covered by less than $\mathfrak{c}$ many $\mathcal{C}^n$ functions for $n > 1$ and give examples showing that all of the covering theorems discussed are the best possible.

Chapter 5 concentrates on several specific applications of $\mathrm{CPA}_{\mathrm{prism}}^{\mathrm{game}}$. Thus, in Section 5.1 we show that $\mathrm{CPA}_{\mathrm{prism}}^{\mathrm{game}}$ implies that:

(S) There is a family $\mathcal{H}$ of ω_1 pairwise disjoint perfect subsets of $\mathbb{R}$ such that $H = \bigcup \mathcal{H}$ is a Hamel basis, that is, a linear basis of $\mathbb{R}$ over $\mathbb{Q}$.

We also show that the following two properties are the consequences of (S):

(T) There exists a nonmeasurable subset X of $\mathbb{R}$ without the Baire property that is $\mathcal{N} \cap \mathcal{M}$-rigid, that is, such that $X \triangle (r + X) \in \mathcal{N} \cap \mathcal{M}$ for every $r \in \mathbb{R}$.

(U) There exists a function $f\colon \mathbb{R} \to \mathbb{R}$ such that for every $h \in \mathbb{R}$ the difference function $\Delta_h(x) = f(x + h) - f(x)$ is Borel; however, for every $\alpha < \omega_1$ there is an $h \in \mathbb{R}$ such that Δ_h is not of Borel class α.

The implication $\mathrm{CPA}_{\mathrm{prism}}^{\mathrm{game}} \Longrightarrow$ (T) answers a question related to the work of J. Cichoń, A. Jasiński, A. Kamburelis, and P. Szczepaniak [23]. The implication $\mathrm{CPA}_{\mathrm{prism}}^{\mathrm{game}} \Longrightarrow$ (U) shows that a recent construction of such a function from CH due to R. Filipów and I. Recław [57] (and answering a question of M. Laczkovich from [82]) can also be repeated with the help of our axiom. In Section 5.2 we show that $\mathrm{CPA}_{\mathrm{prism}}^{\mathrm{game}}$ implies that:

(V) There exists a discontinuous, almost continuous, and additive function $f\colon \mathbb{R} \to \mathbb{R}$ whose graph is of measure zero.

The first construction of such a function, under Martin's axiom, was given by K. Ciesielski in [27]. It is unknown whether it can be constructed in ZFC. We also prove there that, under $\mathrm{CPA}_{\mathrm{prism}}^{\mathrm{game}}$:

(W) There exists a Hamel basis H such that $E^+(H)$ has measure zero.

Here $E^+(A)$ is a linear combination of $A \subset \mathbb{R}$ with nonnegative rational coefficients. This relates to the work of P. Erdős [54], H. Miller [98], and K. Muthuvel [100], who constructed such and similar Hamel bases under different set theoretical assumptions. It is unknown whether (W) holds in ZFC. In Section 5.3 we deduce from $\mathrm{CPA}_{\mathrm{prism}}^{\mathrm{game}}$ that every selective ideal on ω can be extended to a maximal selective ideal. In particular, the first part of condition (E) holds and $\mathfrak{u} = \mathfrak{r}_\sigma = \omega_1$, where $\mathfrak{u}$ is the smallest cardinality of the base for a nonprincipal ultrafilter on ω. In Section 5.4 we prove that $\mathrm{CPA}_{\mathrm{prism}}^{\mathrm{game}}$ implies that:

(X) There exist many nonselective P-points as well as a family $\mathcal{F} \subset [\omega]^\omega$ of cardinality ω_1 that is simultaneously independent and splitting.

In particular, $\mathfrak{i} = \omega_1$, where $\mathfrak{i}$ is the smallest cardinality of an infinite maximal independent family. We finish the chapter with the proof that $\mathrm{CPA}_{\mathrm{prism}}^{\mathrm{game}}$ implies that:

(Y) There exists a nonprincipal ultrafilter on $\mathbb{Q}$ that is crowded.

In Chapter 6 we formulate the most general form of our axiom, CPA, and show that it implies all the other versions of the axiom. In Section 6.1 we conclude from CPA that

(Z) $\mathrm{cov}(s_0) = \mathfrak{c}$.

In Section 6.2 we show that CPA implies the following two generalizations of property (F):

(F*) For an arbitrary function h from a subset S of a Polish space X onto a Polish space Y there exists a uniformly continuous function f from a subset of X into Y such that $|f \cap h| = \mathfrak{c}$.

(F′) For any function h from a subset S of $\mathbb{R}$ onto a perfect subset of $\mathbb{R}$ there exists a function $f \in$ "$\mathcal{C}_{\mathrm{perf}}^\infty$" such that $|f \cap h| = \mathfrak{c}$, and f can be extended to a function $\bar{f} \in$ "$\mathcal{C}^1(\mathbb{R})$" such that either $\bar{f} \in \mathcal{C}^1$ or $\bar{f}$ is an autohomeomorphism of $\mathbb{R}$ with $\bar{f}^{-1} \in \mathcal{C}^1$.

In Section 6.3 we show that (A)&(F*)$\Longrightarrow$(G). In particular, (G) follows from CPA.

Finally, in Chapter 7 we show that CPA holds in the iterated perfect set model.

Preliminaries

Our set theoretic terminology is standard and follows that of [4], [25], and [81]. The sets of real, rational, and integer numbers are denoted by $\mathbb{R}$, $\mathbb{Q}$, and $\mathbb{Z}$, respectively. If $a, b \in \mathbb{R}$ and $a < b$, then $(b, a) = (a, b)$ will stand for the open interval $\{x \in \mathbb{R}: a < x < b\}$. Similarly, $[b, a] = [a, b]$ is an appropriate closed interval. The Cantor set 2^ω will be denoted by the symbol $\mathfrak{C}$. In this text we use the term *Polish space* for a complete separable metric space **without isolated points**. A subset of a Polish space is *perfect* if it is closed and contains no isolated points. For a Polish space X, the symbol $\mathrm{Perf}(X)$ will denote the collection of all subsets of X homeomorphic to $\mathfrak{C}$; the closure of an $A \subset X$ will be denoted by $\mathrm{cl}(A)$; and, as usual, $\mathcal{C}(X)$ will stand for the family of all continuous functions from X into $\mathbb{R}$.

A function $f: \mathbb{R} \to \mathbb{R}$ is *Darboux* if a conclusion of the intermediate value theorem holds for f or, equivalently, when f maps every interval onto an interval; f is a *Sierpiński-Zygmund function* if its restriction $f \restriction Y$ is discontinuous for every subset Y of $\mathbb{R}$ of cardinality $\mathfrak{c}$; and f is *nowhere constant* if it is not constant on any nontrivial interval.

A set $S \subset \mathbb{R}$ is *perfectly meager* if $S \cap P$ is meager in P for every perfect set $P \subset \mathbb{R}$, and S is *universally null* provided for every perfect set $P \subset \mathbb{R}$ the set $S \cap P$ has measure zero with respect to every countably additive probability measure on P vanishing on singletons.

For an ideal $\mathcal{I}$ on a set X, its *cofinality* is defined by

$$\mathrm{cof}(\mathcal{I}) = \min\{|\mathcal{B}|: \mathcal{B} \text{ generates } \mathcal{I}\}$$

and its *covering* as

$$\mathrm{cov}(\mathcal{I}) = \min\left\{|\mathcal{B}|: \mathcal{B} \subset \mathcal{I} \ \& \ \bigcup \mathcal{B} = X\right\}.$$

The symbol $\mathcal{N}$ will stand for the ideal of Lebesgue measure zero subsets of $\mathbb{R}$. For a fixed Polish space X. the ideal of its meager subsets will be denoted by $\mathcal{M}$, and we will use the symbol s_0 (or $s_0(X)$) to denote the σ-ideal of Marczewski's s_0-sets, that is,

$$s_0 = \{S \subset X : (\forall P \in \mathrm{Perf}(X))(\exists Q \in \mathrm{Perf}(X)) \ Q \subset P \setminus S\}.$$

For an ideal $\mathcal{I}$ on a set X we use the symbol $\mathcal{I}^+$ to denote its coideal, that is, $\mathcal{I}^+ = \mathcal{P}(X) \setminus \mathcal{I}$.

For an ideal $\mathcal{I}$ on ω containing all finite subsets of ω we use the following generalized selectivity terminology. We say (see I. Farah [55]) that an ideal $\mathcal{I}$ is *selective* provided for every sequence $F_0 \supset F_1 \supset \cdots$ of sets from $\mathcal{I}^+$ there exists an $F_\infty \in \mathcal{I}^+$ (called a *diagonalization* of this sequence) with the property that $F_\infty \setminus \{0, \dots, n\} \subset F_n$ for all $n \in F_\infty$. Notice that this definition agrees with the definition of selectivity given by S. Grigorieff in [65, p. 365]. (The ideals selective in the above sense Grigorieff calls *inductive*, but he also proves [65, cor. 1.15] that the inductive ideals and the ideals selective in his sense are the same notions.)

For $A, B \subset \omega$ we write $A \subseteq^* B$ when $|A \setminus B| < \omega$. A set $\mathcal{D} \subset \mathcal{I}^+$ is *dense* in $\mathcal{I}^+$ provided for every $B \in \mathcal{I}^+$ there exists an $A \in \mathcal{D}$ such that $A \subseteq^* B$, and the set $\mathcal{D}$ is *open* in $\mathcal{I}^+$ if $B \in \mathcal{D}$ provided there is an $A \in \mathcal{D}$ such that $B \subseteq^* A$. For $\bar{\mathcal{D}} = \langle \mathcal{D}_n \subset \mathcal{I}^+ : n < \omega \rangle$ we say that $F_\infty \in \mathcal{I}^+$ is a *diagonalization* of $\bar{\mathcal{D}}$ provided $F_\infty \setminus \{0, \dots, n\} \in \mathcal{D}_n$ for every $n < \omega$. Following I. Farah [55] we say that an ideal $\mathcal{I}$ on ω is *semiselective* provided for every sequence $\bar{\mathcal{D}} = \langle \mathcal{D}_n \subset \mathcal{I}^+ : n < \omega \rangle$ of dense and open subsets of $\mathcal{I}^+$ the family of all diagonalizations of $\bar{\mathcal{D}}$ is dense in $\mathcal{I}^+$.

Following S. Grigorieff [65, p. 390] we say that $\mathcal{I}$ is *weakly selective* (or *weak selective*) provided for every $A \in \mathcal{I}^+$ and $f : A \to \omega$ there exists a $B \in \mathcal{I}^+$ such that $f \restriction B$ is either one to one or constant. (I. Farah, in [55, sec. 2], terms such ideals as *having the Q^+-property*. Note also that J. Baumgartner and R. Laver, in [7], call such ideals selective, despite the fact that they claim to use Grigorieff's terminology from [65].)

We have the following implications between these notions (see I. Farah [55, sec. 2]):

$$\mathcal{I} \text{ is selective} \implies \mathcal{I} \text{ is semiselective} \implies \mathcal{I} \text{ is weakly selective}$$

All these notions represent different generalizations of the properties of the ideal $[\omega]^{<\omega}$. In particular, it is easy to see that $[\omega]^{<\omega}$ is selective.

We say that an ideal $\mathcal{I}$ on a countable set X is selective (weakly selective) provided it is such upon an identification of X with ω via an arbitrary bijection. A filter $\mathcal{F}$ on a countable set X is selective (semiselective, weakly selective) provided the same property has its dual ideal $\mathcal{I} = \{X \setminus F : F \in \mathcal{F}\}$.

It is important to note that a maximal ideal (or an ultrafilter) is selective if and only if it is weakly selective. This follows, for example, directly from the definitions of these notions as in S. Grigorieff [65]. Recall also that the existence of selective ultrafilters cannot be proved in ZFC. (K. Kunen [80] proved that there are no selective ultrafilters in the random real model. This also follows from the fact that every selective ultrafilter is a P-point, while S. Shelah proved that there are models with no P-points; see, e.g., [4, thm. 4.4.7].)

Acknowledgments. Janusz Pawlikowski wishes to thank West Virginia University for its hospitality in the years 1998–2001, where most of the results presented in this text were obtained. The authors also thank Dr. Elliott Pearl for proofreading this monograph.

1

Axiom CPA$_{\text{cube}}$ and its consequences: properties (A)–(E)

For a Polish space X we will consider $\text{Perf}(X)$, the family of all subsets of X homeomorphic to the Cantor set $\mathfrak{C}$, as ordered by inclusion. Thus, a family $\mathcal{E} \subset \text{Perf}(X)$ is *dense in* $\text{Perf}(X)$ provided for every $P \in \text{Perf}(X)$ there exists a $Q \in \mathcal{E}$ such that $Q \subset P$.

All different versions of our axiom will be more or less of the form:

If $\mathcal{E} \subset \text{Perf}(X)$ is *appropriately* dense in $\text{Perf}(X)$, then some portion $\mathcal{E}_0$ of $\mathcal{E}$ covers almost all of X in a sense that $|X \setminus \bigcup \mathcal{E}_0| < \mathfrak{c}$.

If the word "appropriately" in the above is ignored, then it implies the following statement:

Naïve-CPA: If $\mathcal{E}$ is dense in $\text{Perf}(X)$, then $|X \setminus \bigcup \mathcal{E}| < \mathfrak{c}$.

It is a very good candidate for our axiom in the sense that it implies all the properties we are interested in. It has, however, one major flaw — *it is false!* This is the case since $S \subset X \setminus \bigcup \mathcal{E}$ for some dense set $\mathcal{E}$ in $\text{Perf}(X)$ provided:

For each $P \in \text{Perf}(X)$ there is a $Q \in \text{Perf}(X)$ such that $Q \subset P \setminus S$.

This means that the family $\mathcal{G}$ of all sets of the form $X \setminus \bigcup \mathcal{E}$, where $\mathcal{E}$ is dense in $\text{Perf}(X)$, coincides with the σ-ideal s_0 of Marczewski's sets, since $\mathcal{G}$ is clearly hereditary. Thus we have

$$s_0 = \left\{ X \setminus \bigcup \mathcal{E} \colon \mathcal{E} \text{ is dense in } \text{Perf}(X) \right\}. \tag{1.1}$$

However, it is well known (see, e.g., [96, thm. 5.10]) that there are s_0-sets of cardinality $\mathfrak{c}$. Thus, our Naïve-CPA "axiom" cannot be consistent with ZFC.

In order to formulate the real axiom CPA$_{\text{cube}}$, we need the following terminology and notation. A subset C of a product $\mathfrak{C}^\eta$ of the Cantor set is

1

said to be a *perfect cube* if $C = \prod_{n \in \eta} C_n$, where $C_n \in \text{Perf}(\mathfrak{C})$ for each n. For a fixed Polish space X let $\mathcal{F}_{\text{cube}}$ stand for the family of all continuous injections from perfect cubes $C \subset \mathfrak{C}^\omega$ onto perfect subsets of X. Each such injection f is called a *cube* in X and is considered as a coordinate system imposed on $P = \text{range}(f)$.[1] We will usually abuse this terminology and refer to P itself as a *cube* (in X) and to f as a *witness function* for P. A function $g \in \mathcal{F}_{\text{cube}}$ is a *subcube* of f provided $g \subset f$. In the above spirit we call $Q = \text{range}(g)$ a *subcube of a cube* P. Thus, when we say that Q *is a subcube of a cube* $P \in \text{Perf}(X)$ we mean that $Q = f[C]$, where f is a witness function for P and $C \subset \text{dom}(f) \subset \mathfrak{C}^\omega$ is a perfect cube. Here and in what follows, the symbol $\text{dom}(f)$ stands for the domain of f.

We say that a family $\mathcal{E} \subset \text{Perf}(X)$ is $\mathcal{F}_{\text{cube}}$-*dense* (or *cube-dense*) in $\text{Perf}(X)$ provided every cube $P \in \text{Perf}(X)$ contains a subcube $Q \in \mathcal{E}$. More formally, $\mathcal{E} \subset \text{Perf}(X)$ is $\mathcal{F}_{\text{cube}}$-dense provided

$$\forall f \in \mathcal{F}_{\text{cube}} \, \exists g \in \mathcal{F}_{\text{cube}} \, (g \subset f \ \& \ \text{range}(g) \in \mathcal{E}). \tag{1.2}$$

It is easy to see that the notion of $\mathcal{F}_{\text{cube}}$-density is a generalization of the notion of density as defined in the first paragraph of this chapter:

If $\mathcal{E}$ is $\mathcal{F}_{\text{cube}}$-dense in $\text{Perf}(X)$, then $\mathcal{E}$ is dense in $\text{Perf}(X)$. $\qquad$ (1.3)

On the other hand, the converse implication is not true, as shown by the following simple example.

Example 1.0.1 *Let* $X = \mathfrak{C} \times \mathfrak{C}$ *and let* $\mathcal{E}$ *be the family of all* $P \in \text{Perf}(X)$ *such that either*

- *all vertical sections* $P_x = \{y \in \mathfrak{C} \colon \langle x, y \rangle \in P\}$ *of* P *are countable, or*
- *all horizontal sections* $P^y = \{x \in \mathfrak{C} \colon \langle x, y \rangle \in P\}$ *of* P *are countable.*

Then $\mathcal{E}$ *is dense in* $\text{Perf}(X)$, *but it is not* $\mathcal{F}_{\text{cube}}$-*dense in* $\text{Perf}(X)$.

PROOF. To see that $\mathcal{E}$ is dense in $\text{Perf}(X)$, let $R \in \text{Perf}(X)$. We need to find a $P \subset R$ with $P \in \mathcal{E}$. If all vertical sections of R are countable, then $P = R \in \mathcal{E}$. Otherwise, there exists an x such that R_x is uncountable. Then there exists a perfect subset P of $\{x\} \times R_x \subset R$ and clearly $P \in \mathcal{E}$.

To see that $\mathcal{E}$ is not $\mathcal{F}_{\text{cube}}$-dense in $\text{Perf}(X)$, it is enough to notice that $P = X = \mathfrak{C} \times \mathfrak{C}$ considered as a cube, where the second coordinate is identified with $\mathfrak{C}^\omega \setminus \{0\}$, has no subcube in $\mathcal{E}$. More formally, let h be a homeomorphism from $\mathfrak{C}$ onto $\mathfrak{C}^\omega \setminus \{0\}$, let $g \colon \mathfrak{C} \times \mathfrak{C} \to \mathfrak{C}^\omega = \mathfrak{C} \times \mathfrak{C}^\omega \setminus \{0\}$ be given by $g(x, y) = \langle x, h(y) \rangle$, and let $f = g^{-1} \colon \mathfrak{C}^\omega \to \mathfrak{C} \times \mathfrak{C}$ be the coordinate

---

[1] In a language of forcing, a coordinate function f is simply a nice name for an element from X.

function making $\mathfrak{C} \times \mathfrak{C} = \text{range}(f)$ a cube. Then range(f) does not contain a subcube from $\mathcal{E}$.   ∎

With these notions in hand we are ready to formulate our axiom CPA$_{cube}$. For a Polish space X let

CPA$_{cube}[X]$: $\mathfrak{c} = \omega_2$, and for every $\mathcal{F}_{cube}$-dense family $\mathcal{E} \subset \text{Perf}(X)$ there is an $\mathcal{E}_0 \subset \mathcal{E}$ such that $|\mathcal{E}_0| \leq \omega_1$ and $|X \setminus \bigcup \mathcal{E}_0| \leq \omega_1$.

Then

CPA$_{cube}$: CPA$_{cube}[X]$ for every Polish space X.

We will show in Remark 1.8.3 that both these versions of the axiom are equivalent, that is, that CPA$_{cube}[X]$ is equivalent to CPA$_{cube}[Y]$ for arbitrary Polish spaces X and Y.

The proof that CPA$_{cube}$ is consistent with ZFC (it holds in the iterated perfect set model) will be presented in the next chapters. In the remainder of this chapter we will take a closer look at CPA$_{cube}$ and its consequences.

It is also worth noticing that, in order to check that $\mathcal{E}$ is $\mathcal{F}_{cube}$-dense, it is enough to consider in condition (1.2) only functions f defined on the entire space $\mathfrak{C}^\omega$, that is:

Fact 1.0.2 $\mathcal{E} \subset \text{Perf}(X)$ *is* $\mathcal{F}_{cube}$-*dense if and only if*

$$\forall f \in \mathcal{F}_{cube},\ \text{dom}(f) = \mathfrak{C}^\omega,\ \exists g \in \mathcal{F}_{cube}\ (g \subset f\ \&\ \text{range}(g) \in \mathcal{E}).\quad (1.4)$$

PROOF. To see this, let Φ be the family of all bijections $h = \langle h_n \rangle_{n < \omega}$ between perfect subcubes $\prod_{n \in \omega} D_n$ and $\prod_{n \in \omega} C_n$ of $\mathfrak{C}^\omega$ such that each h_n is a homeomorphism between D_n and C_n. Then

$$f \circ h \in \mathcal{F}_{cube}\quad \text{for every } f \in \mathcal{F}_{cube} \text{ and } h \in \Phi \text{ with range}(h) \subset \text{dom}(f).$$

Now take an arbitrary $f \colon C \to X$ from $\mathcal{F}_{cube}$ and choose an $h \in \Phi$ mapping $\mathfrak{C}^\omega$ onto C. Then $\hat{f} = f \circ h \in \mathcal{F}_{cube}$ maps $\mathfrak{C}^\omega$ into X, and, using (1.4), we can find a $\hat{g} \in \mathcal{F}_{cube}$ such that $\hat{g} \subset \hat{f}$ and range($\hat{g}$) $\in \mathcal{E}$. Then $g = f \restriction h[\text{dom}(\hat{g})]$ satisfies (1.2).   ∎

Next, let us consider[1]

$$s_0^{cube} = \left\{ X \setminus \bigcup \mathcal{E} \colon \mathcal{E} \text{ is } \mathcal{F}_{cube}\text{-dense in } \text{Perf}(X) \right\}\quad (1.5)$$

$$= \{ S \subset X \colon \forall \text{ cube } P \in \text{Perf}(X)\ \exists \text{ subcube } Q \subset P \setminus S \}.$$

---

[1] The second equation follows immediately from the fact that if $\mathcal{E}$ is $\mathcal{F}_{cube}$-dense and $Y \subset X \setminus \bigcup \mathcal{E}$, then $Y = X \setminus \bigcup \mathcal{E}'$ for some $\mathcal{F}_{cube}$-dense $\mathcal{E}'$. To see this, for every $x \in X$ choose $T_x \in \text{Perf}(X)$ such that $T_x \subset \{x\} \cup \bigcup \mathcal{E}$ and note that $\mathcal{E}' = \mathcal{E} \cup \{T_x \colon x \in X \setminus Y\}$ is as desired.

It can be easily shown, in ZFC, that s_0^{cube} forms a σ-ideal. However, we will not use this fact in this text in that general form. This is the case, since we will usually assume that CPA$_{\text{cube}}$ holds while CPA$_{\text{cube}}$ implies the following stronger fact.

Proposition 1.0.3 *If* CPA$_{\text{cube}}$ *holds, then* $s_0^{\text{cube}} = [X]^{\leq \omega_1}$.

PROOF. It is obvious that CPA$_{\text{cube}}$ implies $s_0^{\text{cube}} \subset [X]^{<\mathfrak{c}}$. The other inclusion is always true, and it follows from the following simple fact. ∎

Fact 1.0.4 $[X]^{<\mathfrak{c}} \subset s_0^{\text{cube}} \subset s_0$ *for every Polish space* X.

PROOF. Choose $S \in [X]^{<\mathfrak{c}}$. In order to see that $S \in s_0^{\text{cube}}$, note that the family $\mathcal{E} = \{P \in \text{Perf}(X): P \cap S = \emptyset\}$ is $\mathcal{F}_{\text{cube}}$-dense in Perf$(X)$. Indeed, if function $f: \mathfrak{C}^\omega \to X$ is from $\mathcal{F}_{\text{cube}}$, then there is a perfect subset P_0 of $\mathfrak{C}$ that is disjoint with the projection $\pi_0(f^{-1}(S))$ of $f^{-1}(S)$ into the first coordinate. Then $f\left[\prod_{i<\omega} P_i\right] \cap S = \emptyset$, where $P_i = \mathfrak{C}$ for all $0 < i < \omega$. Therefore, $f\left[\prod_{i<\omega} P_i\right] \in \mathcal{E}$. Thus, $X \setminus \bigcup \mathcal{E} \in s_0^{\text{cube}}$. Since clearly $S \subset X \setminus \bigcup \mathcal{E}$, we get $S \in s_0^{\text{cube}}$.

The inclusion $s_0^{\text{cube}} \subset s_0$ follows immediately from (1.1), (1.5), and (1.3). ∎

1.1 Perfectly meager sets, universally null sets, and continuous images of sets of cardinality continuum

The results presented in this section come from K. Ciesielski and J. Pawlikowski [39]. An important quality of the ideal s_0^{cube}, and so the power of the assumption $s_0^{\text{cube}} = [X]^{<\mathfrak{c}}$, is well depicted by the following fact.

Proposition 1.1.1 *If* X *is a Polish space and* $S \subset X$ *does not belong to* s_0^{cube}, *then there exist a* $T \in [S]^{\mathfrak{c}}$ *and a uniformly continuous function* h *from* T *onto* $\mathfrak{C}$.

PROOF. Take an S as above and let $f: \mathfrak{C}^\omega \to X$ be a continuous injection such that $f[C] \cap S \neq \emptyset$ for every perfect cube C. Let $g: \mathfrak{C} \to \mathfrak{C}$ be a continuous function such that $g^{-1}(y)$ is perfect for every $y \in \mathfrak{C}$. Then clearly $h_0 = g \circ \pi_0 \circ f^{-1}: f[\mathfrak{C}^\omega] \to \mathfrak{C}$ is uniformly continuous. Moreover, if $T = S \cap f[\mathfrak{C}^\omega]$, then $h_0[T] = \mathfrak{C}$ since

$$T \cap h_0^{-1}(y) = T \cap f[\pi_0^{-1}(g^{-1}(y))] = S \cap f[g^{-1}(y) \times \mathfrak{C} \times \mathfrak{C} \times \cdots] \neq \emptyset$$

for every $y \in \mathfrak{C}$. ∎

Corollary 1.1.2 *Assume* $s_0^{\text{cube}} = [X]^{<\mathfrak{c}}$ *for a Polish space* X. *If* $S \subset X$ *has cardinality* $\mathfrak{c}$, *then there is a uniformly continuous function* $f \colon X \to [0,1]$ *such that* $f[S] = [0,1]$. *In particular,* CPA_{cube} *implies property* (A).

PROOF. If S is as above, then, by CPA_{cube}, $S \notin s_0^{\text{cube}}$. Thus, by Proposition 1.1.1 there exists a uniformly continuous function h from a subset of S onto $\mathfrak{C}$. Consider $\mathfrak{C}$ as a subset of $[0,1]$ and let $\hat{h} \colon X \to [0,1]$ be a uniformly continuous extension of h. If $g \colon [0,1] \to [0,1]$ is continuous and such that $g[\mathfrak{C}] = [0,1]$, then $f = g \circ \hat{h}$ is as desired. ∎

For more on property (A) see also Corollary 3.3.5.

It is worth noticing here that the function f in Corollary 1.1.2 cannot be required to be either monotone or in the class "D^1" of all functions having a finite or infinite derivative at every point. This follows immediately from the following proposition, since each function that is either monotone or "D^1" belongs to the Banach class

$$(T_2) = \{f \in \mathcal{C}(\mathbb{R}) \colon \{y \in \mathbb{R} \colon |f^{-1}(y)| > \omega\} \in \mathcal{N}\}.$$

(See [58] or [114, p. 278].)

Proposition 1.1.3 *There is, in ZFC, an* $S \in [\mathbb{R}]^{\mathfrak{c}}$ *such that* $[0,1] \not\subset f[S]$ *for every* $f \in (T_2)$.

PROOF. Let $\{f_\xi \colon \xi < \mathfrak{c}\}$ be an enumeration of all functions from (T_2) whose range contains $[0,1]$. Construct by induction a sequence $\langle \langle s_\xi, y_\xi \rangle \colon \xi < \mathfrak{c} \rangle$ such that, for every $\xi < \mathfrak{c}$,

(i) $y_\xi \in [0,1] \setminus f_\xi[\{s_\zeta \colon \zeta < \xi\}]$ and $|f_\xi^{-1}(y_\xi)| \le \omega$.
(ii) $s_\xi \in \mathbb{R} \setminus \left(\{s_\zeta \colon \zeta < \xi\} \cup \bigcup_{\zeta \le \xi} f_\zeta^{-1}(y_\zeta) \right)$.

Then the set $S = \{s_\xi \colon \xi < \mathfrak{c}\}$ is as required since $y_\xi \in [0,1] \setminus f_\xi[S]$ for every $\xi < \mathfrak{c}$. ∎

Theorem 1.1.4 *If* $S \subset \mathbb{R}$ *is either perfectly meager or universally null, then* $S \in s_0^{\text{cube}}$. *In particular,*

$$\text{CPA}_{\text{cube}} \implies \text{"}s_0^{\text{cube}} = [\mathbb{R}]^{<\mathfrak{c}}\text{"} \implies \text{"(B) \& (C)."}$$

PROOF. Take an $S \subset \mathbb{R}$ that is either perfectly meager or universally null and let $f \colon \mathfrak{C}^\omega \to \mathbb{R}$ be a continuous injection. Then $S \cap f[\mathfrak{C}^\omega]$ is either meager or null in $f[\mathfrak{C}^\omega]$. Thus $G = \mathfrak{C}^\omega \setminus f^{-1}(S)$ is either comeager or of full measure in $\mathfrak{C}^\omega$. Hence the theorem follows immediately from the following claim, which will be used many times in the sequel. ∎

Claim 1.1.5 *Consider* $\mathfrak{C}^\omega$ *with its usual topology and its usual product measure. If G is a Borel subset of $\mathfrak{C}^\omega$ that is either of the second category or of positive measure, then G contains a perfect cube $\prod_{i<\omega} P_i$.*

In particular, if $\mathcal{G}$ is a countable cover of $\mathfrak{C}^\omega$ formed by either measurable sets or by sets with the Baire property, then there is a $G \in \mathcal{G}$ that contains a perfect cube.

The measure version of the claim is a variant the following theorem:

(m) For every full measure subset H of $[0,1] \times [0,1]$ there are a perfect set $P \subset [0,1]$ and a positive inner measure subset $\hat{H}$ of $[0,1]$ such that $P \times \hat{H} \subset H$.

This was proved by H.G. Eggleston [52] and, independently, by M.L. Brodskiĭ [13]. The category version of the claim is a consequence of the category version of (m):

(c) For every Polish space X and every comeager subset G of $X \times X$ there are a perfect set $P \subset X$ and a comeager subset $\hat{G}$ of X such that $P \times \hat{G} \subset G$.

This well-known result can be found in [74, exercise 19.3]. (Its version for $\mathbb{R}^2$ is also proved, for example, in [45, condition $(\star)$, p. 416].) For completeness, we will show here in detail how to deduce the claim from (m) and (c).

We will start the argument with a simple fact, in which we will use the following notations. If X is a Polish space endowed with a Borel measure, then $\psi_0(X)$ will stand for the sentence

$\psi_0(X)$: For every full measure subset H of $X \times X$ there are a perfect set $P \subset X$ and a positive inner measure subset $\hat{H}$ of X such that $P \times \hat{H} \subset H$.

Thus $\psi_0([0,1])$ is a restatement of (m). We will also use the following seemingly stronger variants of $\psi_0(X)$.

$\psi_1(X)$: For every full measure subset H of $X \times X$ there are a perfect set $P \subset X$ and a subset $\hat{H}$ of X of full measure such that $P \times \hat{H} \subset H$.

$\psi_2(X)$: For a subset H of $X \times X$ of positive inner measure there are a perfect set $P \subset X$ and a positive inner measure subset $\hat{H}$ of X such that $P \times \hat{H} \subset H$.

Fact 1.1.6 *Let $n = 1, 2, 3, \ldots$.*

(i) *If E is a subset of $\mathbb{R}^n$ of a positive Lebesgue measure, then the set $\mathbb{Q}^n + E = \bigcup_{q \in \mathbb{Q}^n}(q + E)$ has a full measure.*

(ii) *$\psi_k(X)$ holds for all $k < 3$ and $X \in \{[0,1], (0,1), \mathbb{R}, \mathfrak{C}\}$.*

PROOF. (i) Let λ be the Lebesgue measure on $\mathbb{R}^n$, and for $\varepsilon > 0$ and $x \in \mathbb{R}^n$ let $B(x, \varepsilon)$ be an open ball in $\mathbb{R}^n$ of radius ε centered at x. By way of contradiction, assume that there exists a positive measure set $A \subset \mathbb{R}^n$ disjoint with $\mathbb{Q}^n + E$. Let $a \in A$ and $x \in E$ be the Lebesgue density points of A and X, respectively. Take an $\varepsilon > 0$ such that $\lambda(A \cap B(a, \varepsilon)) > (1 - 4^{-n})\lambda(B(a, \varepsilon))$ and $\lambda(E \cap B(x, \varepsilon)) > (1 - 4^{-n})\lambda(B(x, \varepsilon))$. Now, if $q \in \mathbb{Q}^n$ is such that $q + x \in B(a, \varepsilon/2)$, then $A \cap (q + E) \cap B(a, \varepsilon/2) \neq \emptyset$ since $B(a, \varepsilon/2) \subset B(a, \varepsilon) \cap B(q+x, \varepsilon)$, and thus $\lambda(A \cap (q+E) \cap B(a, \varepsilon/2)) > \lambda(B(a, \varepsilon/2)) - 2 \cdot 4^{-n}\lambda(B(a, \varepsilon)) \geq 0$. Hence $A \cap (\mathbb{Q}^n + E) \neq \emptyset$, contradicting the choice of A.

(ii) First note that $\psi_k(\mathbb{R}) \Leftrightarrow \psi_k((0,1)) \Leftrightarrow \psi_k([0,1]) \Leftrightarrow \psi_k(\mathfrak{C})$ for every $k < 3$. This is justified by the fact that, for the mappings $f \colon (0,1) \to \mathbb{R}$ given by $f(x) = \cot(x\pi)$, the identity mapping $id \colon (0,1) \to [0,1]$, and a function $d \colon \mathfrak{C} \to [0,1]$ given by $d(x) = \sum_{i < \omega} \frac{x(i)}{2^{i+1}}$, the image and the preimage of a measure zero (respectively, full measure) set is of measure zero (respectively, of full measure).

Since, by (m), $\psi_0([0,1])$ is true, we also have that $\psi_0(X)$ also holds for $X \in \{(0,1), \mathbb{R}, \mathfrak{C}\}$. To finish the proof it is enough to show that $\psi_0(\mathbb{R})$ implies $\psi_1(\mathbb{R})$ and $\psi_2(\mathbb{R})$.

To prove $\psi_1(\mathbb{R})$, let H be a full measure subset of $\mathbb{R} \times \mathbb{R}$ and let us define $H_0 = \bigcap_{q \in \mathbb{Q}}(\langle 0, q \rangle + H)$. Then H_0 is still of full measure, so, by $\psi_0(\mathbb{R})$, there are perfect set $P \subset \mathbb{R}$ and a positive inner measure subset $\hat{H}_0$ of $\mathbb{R}$ such that $P \times \hat{H}_0 \subset H_0$. Thus, $P \times (q + \hat{H}_0) \subset \langle 0, q \rangle + H_0 = H_0$ for every $q \in \mathbb{Q}$. Let $\hat{H} = \bigcup_{q \in \mathbb{Q}}(q + \hat{H}_0)$. Then $P \times \hat{H} \subset H_0 \subset H$, and, by (i), $\hat{H}$ has full measure. So, $\psi_1(\mathbb{R})$ is proved.

To prove $\psi_2(\mathbb{R})$, let $H \subset \mathbb{R} \times \mathbb{R}$ be of positive inner measure. Decreasing H, if necessary, we can assume that H is compact. Let $H_0 = \mathbb{Q}^2 + H$. Then, by (i), H_0 is of full measure, and so, by $\psi_0(\mathbb{R})$, there are a perfect set $P_0 \subset \mathbb{R}$ and a positive inner measure subset $\hat{H}_0$ of $\mathbb{R}$ such that $P_0 \times \hat{H}_0 \subset H_0$. Once again, decreasing P_0 and $\hat{H}_0$ if necessary, we can assume that they are homeomorphic to $\mathfrak{C}$ and that no relatively open subset of $\hat{H}_0$ has measure zero. Since $P_0 \times \hat{H}_0 \subset \bigcup_{q \in \mathbb{Q}^2}(q + H)$ is covered by countably many compact sets $(P_0 \times \hat{H}_0) \cap (q + H)$ with $q \in \mathbb{Q}^2$, there is a $q = \langle q_0, q_1 \rangle \in \mathbb{Q}^2$ such that $(P_0 \times \hat{H}_0) \cap (q + H)$ has a nonempty interior in $P_0 \times \hat{H}_0$. Let U and V be nonempty clopen (i.e., simultaneously closed and open) subsets of P_0 and

$\hat{H}_0$, respectively, such that $U \times V \subset (P_0 \times \hat{H}_0) \cap (q + H) \subset \langle q_0, q_1 \rangle + H$. Then U and V are perfect and V has positive measure. Let $P = -q_0 + U$ and $\hat{H} = -q_1 + V$. Then $P \times \hat{H} = (-q_0 + U) \times (-q_1 + V) = -\langle q_0, q_1 \rangle + (U \times V) \subset H$, and so $\psi_2(\mathbb{R})$ holds. ∎

PROOF OF CLAIM 1.1.5. Since the natural homeomorphism between $\mathfrak{C}$ and $\mathfrak{C}^{\omega \setminus \{0\}}$ preserves product measure, we can identify $\mathfrak{C}^\omega = \mathfrak{C} \times \mathfrak{C}^{\omega \setminus \{0\}}$ with $\mathfrak{C} \times \mathfrak{C}$ considered with its usual topology and its usual product measure. With this identification, the result follows easily, by induction on coordinates, from the following fact:

(•) For every Borel subset H of $\mathfrak{C} \times \mathfrak{C}$ that is of the second category (of positive measure) there are a perfect set $P \subset \mathfrak{C}$ and a second category (positive measure) subset $\hat{H}$ of $\mathfrak{C}$ such that $P \times \hat{H} \subset H$.

The measure version of (•) is a restatement of $\psi_2(\mathfrak{C})$, which was proved in Fact 1.1.6(ii). To see the category version of (•), let H be a Borel subset of $\mathfrak{C} \times \mathfrak{C}$ of the second category. Then there are clopen subsets U and V of $\mathfrak{C}$ such that $H_0 = H \cap (U \times V)$ is comeager in $U \times V$. Since U and V are homeomorphic to $\mathfrak{C}$, we can apply (c) to H_0 and $U \times V$ to find a perfect set $P \subset U$ and a comeager Borel subset $\hat{H}$ of V such that $P \times \hat{H} \subset H_0 \subset H$, finishing the proof. ∎

We will finish this section with the following consequence of CPA$_{\text{cube}}$ that follows easily from Claim 1.1.5. In what follows we will use the following notation: Σ_1^1 will stand for the class of analytic sets, that is, continuous images of Borel sets; Π_1^1 will stand for the class of coanalytic sets, the complements of analytic sets; and Σ_2^1 will stand for continuous images of coanalytic sets, and Π_2^1 for the class of all complements of Σ_2^1 sets. For the argument that follows we also need to recall a theorem of W. Sierpiński that every Σ_2^1 set is the union of ω_1 Borel sets. (See, e.g., [74, p. 324].)

Fact 1.1.7 If CPA$_{\text{cube}}$ holds, then for every Σ_2^1 subset B of a Polish space X there exists a family $\mathcal{P}$ of ω_1 many compact sets such that $B = \bigcup \mathcal{P}$.

PROOF. Since every Σ_2^1 set is a union of ω_1 Borel sets, we can assume that B is Borel. Let $\mathcal{E}$ be the family of all $P \in \text{Perf}(X)$ such that either $P \subset B$ or $P \cap B = \emptyset$. We claim that $\mathcal{E}$ is $\mathcal{F}_{\text{cube}}$-dense. Indeed, if $f \colon \mathfrak{C}^\omega \to X$ is a continuous injection, then $f^{-1}(B)$ is Borel in $\mathfrak{C}^\omega$. Thus, there exists a basic open set U in $\mathfrak{C}^\omega$, which is homeomorphic to $\mathfrak{C}^\omega$, such that either $U \cap f^{-1}(B)$ or $U \setminus f^{-1}(B)$ is comeager in U. Apply Claim 1.1.5 to this comeager set to find a perfect cube P contained in it. Then $f[P] \in \mathcal{E}$ is a subcube of range(f). So, $\mathcal{E}$ is $\mathcal{F}_{\text{cube}}$-dense.

By CPA$_{\text{cube}}$, there is an $\mathcal{E}_0 \subset \mathcal{E}$ such that $|\mathcal{E}_0| \leq \omega_1$ and $|X \setminus \bigcup \mathcal{E}_0| \leq \omega_1$. Let $\mathcal{P}_0 = \{P \in \mathcal{E}_0 : P \subset B\}$ and $\mathcal{P} = \mathcal{P}_0 \cup \{\{x\} : x \in B \setminus \bigcup \mathcal{E}_0\}$. Then $\mathcal{P}$ is as desired. ∎

1.2 Uniformly completely Ramsey null sets

Uniformly completely Ramsey null sets are small subsets of $[\omega]^\omega$ that are related to the Ramsey property. The notion has been formally defined by U. Darji [47], though it was already studied by F. Galvin and K. Prikry in [63]. Instead of using the original definition for this class, we will use its characterization due to A. Nowik [107], in which we consider $\mathcal{P}(\omega)$ as a Polish space by identifying it with 2^ω via the characteristic functions.

Proposition 1.2.1 (A. Nowik [107]) *A subset X of $[\omega]^\omega$ is uniformly completely Ramsey null if and only if for every continuous function $G \colon \mathcal{P}(\omega) \to \mathcal{P}(\omega)$ there exists an $A \in [\omega]^\omega$ such that $|G[\mathcal{P}(A)] \cap X| \leq \omega$.*

Recently A. Nowik [108] proved that under CPA$_{\text{cube}}$ every uniformly completely Ramsey null set has cardinality less than continuum. This answered a question of U. Darji, who asked whether there is a ZFC example of a uniformly completely Ramsey null set of cardinality continuum. Since Nowik's argument is typical for the use of CPA$_{\text{cube}}$, we reproduce it here, with the author's approval.

Theorem 1.2.2 (A. Nowik [108]) *If $X \in [\omega]^\omega$ is uniformly completely Ramsey null, then $X \in s_0^{\text{cube}}$.*

PROOF. Let $f \colon \mathfrak{C}^\omega \to \mathcal{P}(\omega)$ be a continuous injection. We need to find a perfect cube $C \subset \mathfrak{C}^\omega$ such that $f[C] \cap X = \emptyset$.

Let $\langle \cdot, \cdot \rangle \colon \omega \times \omega \to \omega$ be a bijection and define a function $F \colon \mathcal{P}(\omega) \to \mathfrak{C}^\omega$ by $F(A)(n) = \chi_{\{a_{\langle k,n \rangle} : k < \omega\}}$, where $\{a_0, a_1, \dots\}$ is an increasing enumeration of A. It is easy to see that F is a continuous injection. Therefore, the function $G = f \circ F \colon \mathcal{P}(\omega) \to \mathcal{P}(\omega)$ is continuous and so, by Proposition 1.2.1, there exists an $A \in [\omega]^\omega$ such that $|(f \circ F)[\mathcal{P}(A)] \cap X| \leq \omega$.

Let $A = \{a_0, a_1, a_2, \dots\}$ be an increasing enumeration of elements of A and define a function $\Xi \colon \mathfrak{C}^\omega \to \mathcal{P}(A)$ by

$$\Xi(x) = \{a_{2\langle k,n \rangle} : x(n)(k) = 0\} \cup \{a_{2\langle k,n \rangle + 1} : x(n)(k) = 1\}.$$

We claim that:

(∗) $F[\Xi[\mathfrak{C}^\omega]]$ is a perfect cube in $\mathfrak{C}^\omega$.

To see this, for every $k, n < \omega$, let $E_{k,n} = \{a_{2\langle k,n\rangle}, a_{2\langle k,n\rangle+1}\}$ and put

$$D_n = \{x \in \mathfrak{C} : x^{-1}(1) \subseteq \bigcup_{k\in\omega} E_{k,n} \ \& \ (\forall k \in \omega)\,|E_{k,n} \cap x^{-1}(1)| = 1\}.$$

It is easy to see that each D_n is perfect in $\mathfrak{C}$. We will show that

$$F[\Xi[\mathfrak{C}^\omega]] = \prod_{n\in\omega} D_n.$$

So, let $x \in \mathfrak{C}^\omega$. To see that $F(\Xi(x)) \in \prod_{n\in\omega} D_n$, first notice that, if $\{b_0, b_1, \ldots\}$ is an increasing enumeration of $\Xi(x)$, then $b_i \in \{a_{2i}, a_{2i+1}\}$ for every $i < \omega$. Therefore $b_{\langle k,n\rangle} \in E_{k,n}$ for every $k, n < \omega$. In particular, $F(\Xi(x))(n)^{-1}(1) = \{b_{\langle k,n\rangle} : k < \omega\} \in D_n$ for every $n < \omega$.

To see the other inclusion, take $\langle x_n : n < \omega\rangle \in \prod_{n\in\omega} D_n$ and define $B = \bigcup_{n<\omega}(x_n)^{-1}(1)$. Then $F(B) = \langle x_n : n < \omega\rangle$ and $|B \cap E_{k,n}| = 1$ for every $k, n < \omega$. Let $x \in \mathfrak{C}^\omega$ be such that $x(n)(k) = 0$ if and only if $a_{2\langle k,n\rangle} \in B$. Then $\Xi(x) = B$ and so $\langle x_n : n < \omega\rangle = F(\Xi(x)) \in \prod_{n\in\omega} D_n$, finishing the proof of (∗).

Now, $D = F[\Xi[\mathfrak{C}^\omega]] \subset \mathfrak{C}^\omega$ is a perfect cube and $|f[D] \cap X| \leq \omega$, since $f[D] = f[F[\Xi[\mathfrak{C}^\omega]]] \subset f[F[\mathcal{P}(A)]] = (f \circ F)[\mathcal{P}(A)]$. Since D can be partitioned into continuum many disjoint perfect cubes, for some member of the partition, say C, we will have $f[C] \cap X = \emptyset$.   ∎

Corollary 1.2.3 (A. Nowik [108]) CPA$_{\text{cube}}$ *implies that every uniformly completely Ramsey null set has cardinality less than continuum.*

To discuss another application of CPA$_{\text{cube}}$, let us consider the following covering number connected to a theorem of H. Blumberg (see Section 1.7) and studied by F. Jordan in [70]. Here $\mathcal{B}_1$ stands for the class of all Baire class 1 functions $f : \mathbb{R} \to \mathbb{R}$.

- $\operatorname{cov}(\mathcal{B}_1, \operatorname{Perf}(\mathbb{R}))$ is the smallest cardinality of $F \subset \mathcal{B}_1$ such that for each $P \in \operatorname{Perf}(\mathbb{R})$ there is an $f \in F$ with $f \upharpoonright P$ not continuous.

Jordan also proves [70, thm. 7(a)] that $\operatorname{cov}(\mathcal{B}_1, \operatorname{Perf}(\mathbb{R}))$ is equal to the covering number of the space $\operatorname{Perf}(\mathbb{R})$ (considered with the Hausdorff metric) by the elements of some σ-ideal $\mathcal{Z}_p$ and notices [70, thm. 17(a)] that every compact set $C \in \operatorname{Perf}(\mathbb{R})$ contains a dense G_δ subset that belongs to $\mathcal{Z}_p$. So, by Claim 1.1.5, the elements of $\operatorname{Perf}(\mathbb{R}) \cap \mathcal{Z}_p$ are $\mathcal{F}_{\text{cube}}$-dense in $\operatorname{Perf}(\mathbb{R})$. Thus

Corollary 1.2.4 CPA$_{\text{cube}}$ *implies that* $\operatorname{cov}(\mathcal{B}_1, \operatorname{Perf}(\mathbb{R})) = \operatorname{cov}(\mathcal{Z}_p) = \omega_1$.

1.3 cof($\mathcal{N}$) = ω_1

Next, following the argument of K. Ciesielski and J. Pawlikowski from [39], we show that $\mathrm{CPA_{cube}}$ implies that cof($\mathcal{N}$) = ω_1. So, under $\mathrm{CPA_{cube}}$, all cardinals from Cichoń's diagram (see, e.g., [4]) are equal to ω_1.

Let $\mathcal{C}_H$ be the family of all subsets $\prod_{n<\omega} T_n$ of ω^ω such that $T_n \in [\omega]^{\leq n+1}$ for all $n < \omega$. We will use the following characterization.

Proposition 1.3.1 (T. Bartoszyński [4, thm. 2.3.9])

$$\mathrm{cof}(\mathcal{N}) = \min\left\{|\mathcal{F}|: \mathcal{F} \subset \mathcal{C}_H \ \& \ \bigcup \mathcal{F} = \omega^\omega\right\}.$$

Lemma 1.3.2 *The family* $\mathcal{C}_H^* = \{X \subset \omega^\omega: X \subset T$ *for some* $T \in \mathcal{C}_H\}$ *is* $\mathcal{F}_{\mathrm{cube}}$-*dense in* $\mathrm{Perf}(\omega^\omega)$.

PROOF. Let $f: \mathfrak{C}^\omega \to \omega^\omega$ be a continuous function. By (1.4) it is enough to find a perfect cube C in $\mathfrak{C}^\omega$ such that $f[C] \in \mathcal{C}_H^*$.

Construct, by induction on $n < \omega$, the families $\{E_s^i: s \in 2^n \ \& \ i < \omega\}$ of nonempty clopen subsets of $\mathfrak{C}$ such that, for every $n < \omega$ and $s, t \in 2^n$,

(i) $E_s^i = E_t^i$ for every $n \leq i < \omega$;
(ii) $E_{s^\frown 0}^i$ and $E_{s^\frown 1}^i$ are disjoint subsets of E_s^i for every $i < n+1$;
(iii) for every $\langle s_i \in 2^n: i < \omega \rangle$

$$f(x_0) \restriction 2^{(n+1)^2} = f(x_1) \restriction 2^{(n+1)^2} \quad \text{for every} \quad x_0, x_1 \in \prod_{i<\omega} E_{s_i}.$$

For each $i < \omega$ the fusion of $\{E_s^i: s \in 2^{<\omega}\}$ will give us the i-th coordinate set of the desired perfect cube C.

Condition (iii) can be ensured by the uniform continuity of f. Indeed, let $\delta > 0$ be such that $f(x_0) \restriction 2^{(n+1)^2} = f(x_1) \restriction 2^{(n+1)^2}$ for every $x_0, x_1 \in \mathfrak{C}^\omega$ of distance less than δ. Then it is enough to choose $\{E_s^i: s \in 2^n \ \& \ i < \omega\}$ such that (i) and (ii) are satisfied and every set $\prod_{i<\omega} E_{s_i}$ from (iii) has diameter less than δ. This finishes the construction.

Next, for every $i, n < \omega$, let $E_n^i = \bigcup\{E_s^i: s \in 2^n\}$ and $E_n = \prod_{i<\omega} E_n^i$. Then $C = \bigcap_{n<\omega} E_n = \prod_{i<\omega}\left(\bigcap_{n<\omega} E_n^i\right)$ is a perfect cube in $\mathfrak{C}^\omega$, since $\bigcap_{n<\omega} E_n^i \in \mathrm{Perf}(\mathfrak{C})$ for every $i < \omega$. Thus, to finish the proof it is enough to show that $f[C] \in \mathcal{C}_H^*$.

So, for every $k < \omega$, let $n < \omega$ be such that $2^{n^2} \leq k+1 < 2^{(n+1)^2}$, put

$$T_k = \{f(x)(k): x \in E_n\} = \left\{f(x)(k): x \in \prod_{i<\omega} E_{s_i} \text{ for some } \langle s_i \in 2^n: i < \omega\rangle\right\},$$

and notice that T_k has at most $2^{n^2} \leq k+1$ elements. Indeed, by (iii), the

set $\{f(x)(k)\colon x \in \prod_{i<\omega} E_{s_i}\}$ is a singleton for every $\langle s_i \in 2^n\colon i<\omega\rangle$ while
(i) implies that $\{\prod_{i<\omega} E_{s_i}\colon \langle s_i \in 2^n\colon i<\omega\rangle\}$ has 2^{n^2} elements. Therefore
$\prod_{k<\omega} T_k \in \mathcal{C}_H$.

To finish the proof it is enough to notice that $f[C] \subset \prod_{k<\omega} T_k$. ∎

Corollary 1.3.3 *If* CPA$_{\mathrm{cube}}$ *holds, then* $\mathrm{cof}(\mathcal{N}) = \omega_1$.

PROOF. By CPA$_{\mathrm{cube}}$ and Lemma 1.3.2, there exists an $\mathcal{F} \in [\mathcal{C}_H]^{\leq \omega_1}$ such
that $|\omega^\omega \setminus \bigcup \mathcal{F}| \leq \omega_1$. This and Proposition 1.3.1 imply $\mathrm{cof}(\mathcal{N}) = \omega_1$. ∎

1.4 Total failure of Martin's axiom

In this section we prove that CPA$_{\mathrm{cube}}$ implies the total failure of Martin's
axiom, that is, the property that:

For every nontrivial ccc forcing $\mathbb{P}$ there exists ω_1 many dense sets in $\mathbb{P}$ such
that no filter intersects all of them.

The consistency of this fact with $\mathfrak{c} > \omega_1$ was first proved by J. Baumgartner [6] in a model obtained by adding Sacks reals side by side. The
topological and Boolean algebraic formulations of the theorem follow immediately from the following proposition. The proof presented below comes
from K. Ciesielski and J. Pawlikowski [39].

Proposition 1.4.1 *The following conditions are equivalent.*

(a) *For every nontrivial ccc forcing $\mathbb{P}$ there exists ω_1 many dense sets in $\mathbb{P}$
such that no filter intersects all of them.*

(b) *Every compact ccc topological space without isolated points is a union
of ω_1 nowhere dense sets.*

(c) *For every atomless ccc complete Boolean algebra B there exist ω_1 many
dense sets in B such that no filter intersects all of them.*

(d) *For every atomless ccc complete Boolean algebra B there exist ω_1 many
maximal antichains in B such that no filter intersects all of them.*

(e) *For every countably generated atomless ccc complete Boolean algebra
B there exists ω_1 many maximal antichains in B such that no filter
intersects all of them.*

PROOF. The equivalence of conditions (a), (b), (c), and (d) is well known.
In particular, equivalences (a)–(c) are explicitly given in [6, thm. 0.1].

Clearly (d) implies (e). The remaining implication, (e)$\Longrightarrow$(d), is a version of the theorem from [89, p. 158]. However, it is expressed there in slightly different language, so we include its proof here.

Let $\langle B, \vee, \wedge, \mathbf{0}, \mathbf{1} \rangle$ be an atomless ccc complete Boolean algebra. For every $\sigma \in 2^{<\omega_1}$ define, by induction on the length $\mathrm{dom}(\sigma)$ of a sequence σ, a $b_\sigma \in B$ such that the following conditions are satisfied:

- $b_\emptyset = \mathbf{1}$.
- b_σ is a disjoint union of $b_{\sigma^\frown 0}$ and $b_{\sigma^\frown 1}$.
- If $b_\sigma > \mathbf{0}$, then $b_{\sigma^\frown 0} > \mathbf{0}$ and $b_{\sigma^\frown 1} > \mathbf{0}$.
- If $\lambda = \mathrm{dom}(\sigma)$ is a limit ordinal, then $b_\sigma = \bigwedge_{\xi < \lambda} b_{\sigma \restriction \xi}$.

Let $T = \{s \in 2^{<\omega_1} : b_s > \mathbf{0}\}$. Then T is a subtree of $2^{<\omega_1}$; its levels determine antichains in B, so they are countable.

First assume that T has a countable height. Then T itself is countable. Let B_0 be the smallest complete subalgebra of B containing $\{b_\sigma : \sigma \in T\}$ and notice that B_0 is atomless. Indeed, if there were an atom a in B_0, then $S = \{\sigma \in T : a \le b_\sigma\}$ would be a branch in T so that $\delta = \bigcup S$ would belong to $2^{<\omega_1}$. Since $b_\delta \ge a > \mathbf{0}$, we would also have $\delta \in T$. But then $a \le b_\delta = b_{\delta^\frown 0} \vee b_{\delta^\frown 1}$, so that either $\delta^\frown 0$ or $\delta^\frown 1$ belongs to S, which is impossible.

Thus, B_0 is a complete, countably generated, atomless subalgebra of B. So, by (e), there exists a family $\mathcal{A}$ of ω_1 many maximal antichains in B_0 with no filter in B_0 intersecting all of them. But then each $A \in \mathcal{A}$ is also a maximal antichain in B and no filter in B would intersect all of them. So, we have (d).

Next, assume that T has height ω_1 and for every $\alpha < \omega_1$ let

$$T_\alpha = \{\sigma \in T : \mathrm{dom}(\sigma) = \alpha\}$$

be the α-th level of T. Also let $b_\alpha = \bigvee_{\sigma \in T_\alpha} b_\sigma$. Notice that $b_\alpha = b_{\alpha+1}$ for every $\alpha < \omega_1$. On the other hand, it may happen that $b_\lambda > \bigwedge_{\alpha < \lambda} b_\alpha$ for some limit $\lambda < \omega_1$; however, this may happen only countably many times, since B is ccc. Thus, there is an $\alpha < \omega_1$ such that $b_\beta = b_\alpha$ for every $\alpha < \beta < \omega_1$.

Now, let B_0 be the smallest complete subalgebra of B below $\mathbf{1} \setminus b_\alpha$ containing $\{b_\sigma \setminus b_\alpha : \sigma \in T\}$. Then B_0 is countably generated and, as before, it can be shown that B_0 is atomless. Thus, there exists a family $\mathcal{A}_0$ of ω_1 many maximal antichains in B_0 with no filter in B_0 intersecting all of them. Then no filter in B containing $\mathbf{1} \setminus b_\alpha$ intersects every $A \in \mathcal{A}_0$. But for every $\alpha < \beta < \omega_1$ the set $A^\beta = \{b_\sigma : \sigma \in T_\beta\}$ is a maximal antichain in B below b_α. Therefore, $\mathcal{A}_1 = \{A^\beta : \alpha < \beta < \omega_1\}$ is an uncountable family of maximal antichains in B below b_α with no filter in B containing b_α intersecting every $A \in \mathcal{A}_1$. Then it is easy to see that the family $\mathcal{A} = \{A_0 \cup A_1 : a_0 \in \mathcal{A}_0 \ \& \ A_1 \in \mathcal{A}_1\}$ is a family of ω_1 many maximal antichains in B with no filter in B intersecting all of them. This proves condition (d). ∎

Theorem 1.4.2 CPA$_{\text{cube}}$ *implies the total failure of Martin's axiom.*

PROOF. Let $\mathcal{A}$ be a countably generated, atomless, ccc complete Boolean algebra and let $\{A_n : n < \omega\}$ generate $\mathcal{A}$. By Proposition 1.4.1 it is enough to show that $\mathcal{A}$ contains ω_1 many maximal antichains such that no filter in $\mathcal{A}$ intersects all of them.

Next let $\mathcal{B}$ be the σ-algebra of Borel subsets of a space $\mathfrak{C} = 2^\omega$. Recall that it is a free countably generated σ-algebra, with the free generators $B_i = \{s \in \mathfrak{C} : s(i) = 0\}$. Define $h_0 : \{B_n : n < \omega\} \to \{A_n : n < \omega\}$ by $h_0(B_n) = A_n$ for all $n < \omega$. Then h_0 can be uniquely extended to a σ-homomorphism $h : \mathcal{B} \to \mathcal{A}$ between σ-algebras $\mathcal{B}$ and $\mathcal{A}$. (See, e.g., [121, 34.1 p. 117].) Let $\mathcal{I} = \{B \in \mathcal{B} : h[B] = \mathbf{0}\}$. Then $\mathcal{I}$ is a σ-ideal in $\mathcal{B}$ and the quotient algebra $\mathcal{B}/\mathcal{I}$ is isomorphic to $\mathcal{A}$. (Compare also Loomis and Sikorski's theorem in [121, p. 117] or [86].) In particular, $\mathcal{I}$ contains all singletons and is ccc, since $\mathcal{A}$ is atomless and ccc.

It follows that we need only to consider complete Boolean algebras of the form $\mathcal{B}/\mathcal{I}$, where $\mathcal{I}$ is some ccc σ-ideal of Borel sets containing all singletons. To prove that such an algebra has ω_1 maximal antichains as desired, it is enough to prove that:

(∗) $\mathfrak{C}$ is a union of ω_1 perfect sets $\{N_\xi : \xi < \omega_1\}$ that belong to $\mathcal{I}$.

Indeed, assume that (∗) holds and for every $\xi < \omega_1$ let $\mathcal{D}_\xi^*$ be a family of all $B \in \mathcal{B} \setminus \mathcal{I}$ with closures $\mathrm{cl}(B)$ disjoint from N_ξ. Then $\mathcal{D}_\xi = \{B/\mathcal{I} : B \in \mathcal{D}_\xi^*\}$ is dense in $\mathcal{B}/\mathcal{I}$, since $\mathfrak{C} \setminus N_\xi$ is σ-compact and $\mathcal{B}/\mathcal{I}$ is a σ-algebra. Let $\mathcal{A}_\xi^* \subset \mathcal{D}_\xi^*$ be such that $\mathcal{A}_\xi = \{B/\mathcal{I} : B \in \mathcal{A}_\xi^*\}$ is a maximal antichain in $\mathcal{B}/\mathcal{I}$. It is enough to show that no filter intersects all $\mathcal{A}_\xi$'s. But if there were a filter $\mathcal{F}$ in $\mathcal{B}/\mathcal{I}$ intersecting all $\mathcal{A}_\xi$'s, then for every $\xi < \omega_1$ there would exist a $B_\xi \in \mathcal{A}_\xi^*$ with $B_\xi/\mathcal{I} \in \mathcal{F} \cap \mathcal{A}_\xi$. Thus, the set $\bigcap_{\xi < \omega_1} \mathrm{cl}(B_\xi)$ would be nonempty, despite the fact that it is disjoint from $\bigcup_{\xi < \omega_1} N_\xi = \mathfrak{C}$.

To finish the proof it is enough to show that (∗) follows from CPA$_{\text{cube}}$. But this follows immediately from the fact that any cube P in $\mathfrak{C}$ contains a subcube $Q \in \mathcal{I}$ as any cube P can be partitioned into $\mathfrak{c}$ many disjoint subcubes, and, by the ccc property of $\mathcal{I}$, only countably many of them can be outside $\mathcal{I}$. ∎

1.5 Selective ultrafilters and the reaping numbers $\mathfrak{r}$ and $\mathfrak{r}_\sigma$

In this section, which is based in part on a paper by K. Ciesielski and J. Pawlikowski [37], we will show that $\mathrm{CPA_{cube}}$ implies that every selective ultrafilter is generated by ω_1 sets and that the reaping number $\mathfrak{r}$ is equal to ω_1. The actual construction of a selective ultrafilter will require a stronger version of the axiom and will be done in Theorem 5.3.3.

We will use here the terminology introduced in the Preliminaries chapter. In particular, recall that the ideal $[\omega]^{<\omega}$ of finite subsets of ω is semiselective.

The most important combinatorial fact for us concerning semiselective ideals is the following property. (See Theorem 2.1 and Remark 4.1 in [55].) This is a generalization of a theorem of R. Laver [85], who proved this fact for the ideal $\mathcal{I} = [\omega]^{<\omega}$.

Proposition 1.5.1 (I. Farah [55]) *Let* $\mathcal{I}$ *be a semiselective ideal on* ω. *For every analytic set* $S \subset \mathfrak{C}^\omega \times [\omega]^\omega$ *and every* $A \in \mathcal{I}^+$ *there exist a* $B \in \mathcal{I}^+ \cap \mathcal{P}(A)$ *and a perfect cube* C *in* $\mathfrak{C}^\omega$ *such that* $C \times [B]^\omega$ *is either contained in or disjoint with* S.

With this fact in hand we can prove the following theorem.

Theorem 1.5.2 *Assume that* $\mathrm{CPA_{cube}}$ *holds. If* $\mathcal{I}$ *is a semiselective ideal, then there is a family* $\mathcal{W} \subset \mathcal{I}^+$, $|\mathcal{W}| \leq \omega_1$, *such that for every analytic set* $A \subset [\omega]^\omega$ *there is a* $W \in \mathcal{W}$ *for which either* $[W]^\omega \subset A$ *or* $[W]^\omega \cap A = \emptyset$.

PROOF. Let $S \subset \mathfrak{C} \times [\omega]^\omega$ be a universal analytic set, that is, such that the family $\{S_x : x \in \mathfrak{C}\}$ (where $S_x = \{y \in [\omega]^\omega : \langle x, y \rangle \in S\}$) contains all analytic subsets of $[\omega]^\omega$. (See, e.g., [69, lem. 39.4].) In fact, we will take S such that, for any analytic set A in $[\omega]^\omega$,

$$|\{x \in \mathfrak{C} : S_x = A\}| = \mathfrak{c}. \tag{1.6}$$

(If $U \subset \mathfrak{C} \times [\omega]^\omega$ is a universal analytic set, then $S = \mathfrak{C} \times U \subset \mathfrak{C} \times \mathfrak{C} \times [\omega]^\omega$ satisfies (1.6), where we identify $\mathfrak{C} \times \mathfrak{C}$ with $\mathfrak{C}$.) For this particular set S consider the family $\mathcal{E}$ of all $Q \in \mathrm{Perf}(\mathfrak{C})$ for which there exists a $W_Q \in \mathcal{I}^+$ such that

$$Q \times [W_Q]^\omega \text{ is either contained in or disjoint from } S. \tag{1.7}$$

Note that, by Proposition 1.5.1, the family $\mathcal{E}$ is $\mathcal{F}_{cube}$-dense in $\mathrm{Perf}(\mathfrak{C})$. So, by $\mathrm{CPA_{cube}}$, there exists an $\mathcal{E}_0 \subset \mathcal{E}$, $|\mathcal{E}_0| \leq \omega_1$, such that $|\mathfrak{C} \setminus \bigcup \mathcal{E}_0| < \mathfrak{c}$. Let

$$\mathcal{W} = \{W_Q : Q \in \mathcal{E}_0\}.$$

It is enough to see that this $\mathcal{W}$ is as required.

Clearly $|\mathcal{W}| \leq \omega_1$. Also, by (1.6), for an analytic set $A \subset [\omega]^\omega$ there exist a $Q \in \mathcal{E}_0$ and an $x \in Q$ such that $A = S_x$. So, by (1.7), $\{x\} \times [W_Q]^\omega$ is either contained in or disjoint from $\{x\} \times S_x = \{x\} \times A$.   ∎

Recall (see, e.g., [4] or [128]) that a family $\mathcal{W} \subset [\omega]^\omega$ is a *reaping family* provided

$$\forall A \in [\omega]^\omega \ \exists W \in \mathcal{W} \ (W \subset A \ \text{ or } \ W \subset \omega \setminus A).$$

The *reaping* (or *refinement*) number $\mathfrak{r}$ is defined as the minimum cardinality of a reaping family. Also, a number $\mathfrak{r}_\sigma$ is defined as the smallest cardinality of a family $\mathcal{W} \subset [\omega]^\omega$ such that for every sequence $\langle A_n \in [\omega]^\omega : n < \omega \rangle$ there exists a $W \in \mathcal{W}$ such that for every $n < \omega$ either $W \subseteq^* A_n$ or $W \subseteq^* \omega \setminus A_n$. (See [18] or [128].) Clearly $\mathfrak{r} \leq \mathfrak{r}_\sigma$.

Corollary 1.5.3 *If CPA$_{\text{cube}}$ holds, then for every semiselective ideal $\mathcal{I}$ there exists a family $\mathcal{W} \subset \mathcal{I}^+$, $|\mathcal{W}| \leq \omega_1$, such that for every $A \in [\omega]^\omega$ there is a $W \in \mathcal{W}$ for which either $W \subseteq^* A$ or $W \subseteq^* \omega \setminus A$. In particular, CPA$_{\text{cube}}$ implies that $\mathfrak{r} = \omega_1 < \mathfrak{c}$.*

PROOF. The family $\mathcal{W}$ from Theorem 1.5.2 works: since $[A]^\omega$ is analytic in $[\omega]^\omega$, there exists a $W \in \mathcal{W}$ such that either $[W]^\omega \subset [A]^\omega$ or $[W]^\omega \cap [A]^\omega = \emptyset$. ∎

Note also that CPA$_{\text{cube}}$ implies the second part of property (E).

Corollary 1.5.4 *If CPA$_{\text{cube}}$ holds, then every selective ultrafilter $\mathcal{F}$ on ω is generated by a family of size $\omega_1 < \mathfrak{c}$.*

PROOF. If $\mathcal{F}$ is a selective ultrafilter on ω, then $\mathcal{I} = \mathcal{P}(\omega) \setminus \mathcal{F}$ is a selective ideal and $\mathcal{I}^+ = \mathcal{F}$. Let $\mathcal{W} \subset \mathcal{I}^+ = \mathcal{F}$ be as in Corollary 1.5.3. Then $\mathcal{W}$ generates $\mathcal{F}$.

Indeed, if $A \in \mathcal{F}$, then there exists a $W \in \mathcal{W}$ such that either $W \subset A$ or $W \subset \omega \setminus A$. But it is impossible that $W \subset \omega \setminus A$, since then we would have $\emptyset = A \cap W \in \mathcal{F}$.   ∎

As mentioned above, in Theorem 5.3.3 we will prove that some version of our axiom implies that there exists a selective ultrafilter on ω. In particular, the assumptions of the next corollary are implied by such a version of our axiom.

Corollary 1.5.5 *If CPA$_{\text{cube}}$ holds and there exists a selective ultrafilter $\mathcal{F}$ on ω, then $\mathfrak{r}_\sigma = \omega_1 < \mathfrak{c}$.*

PROOF. Let $\mathcal{W} \in [\mathcal{F}]^{\leq \omega_1}$ be a generating family for $\mathcal{F}$. We will show that it justifies $\mathfrak{r}_\sigma = \omega_1$. Indeed, take a sequence $\langle A_n \in [\omega]^\omega : n < \omega \rangle$. For every $n < \omega$, let A_n^* belong to $\mathcal{F} \cap \{A_n, \omega \setminus A_n\}$. Since $\mathcal{F}$ is selective, there exists an $A \in \mathcal{F}$ such that $A \subseteq^* A_n^*$ for every $n < \omega$. Let $W \in \mathcal{W}$ be such that $W \subset A$. Then for every $n < \omega$ either $W \subseteq^* A_n$ or $W \subseteq^* \omega \setminus A_n$.   ∎

We are particularly interested in the number $\mathfrak{r}_\sigma$ since it is related to different variants of sets of uniqueness coming from harmonic analysis, as described in the survey paper [18]. In particular, from [18, thm. 12.6] it follows that an appropriate version of our axiom implies that all covering numbers described in the paper are equal to ω_1.

1.6 On the convergence of subsequences of real-valued functions

This section can be viewed as an extension of the discussion of Egorov's theorem presented in [77, chapter 9]. In 1932, S. Mazurkiewicz [91] proved the following variant of Egorov's theorem, where a sequence $\langle f_n \rangle_{n<\omega}$ of real-valued functions is *uniformly bounded* provided there exists an $r \in \mathbb{R}$ such that $\text{range}(f_n) \subset [-r, r]$ for every n.

Mazurkiewicz' Theorem *For every uniformly bounded sequence $\langle f_n \rangle_{n<\omega}$ of real-valued continuous functions defined on a Polish space X there exists a subsequence that is uniformly convergent on some perfect set P.*

The proof of the next theorem comes from the paper [39] of K. Ciesielski and J. Pawlikowski .

Theorem 1.6.1 *If* CPA_{cube} *holds, then for every Polish space X and every uniformly bounded sequence $\langle f_n : X \to \mathbb{R} \rangle_{n<\omega}$ of Borel measurable functions there are two sequences, $\langle P_\xi : \xi < \omega_1 \rangle$ of compact subsets of X and $\langle W_\xi \in [\omega]^\omega : \xi < \omega_1 \rangle$, such that $X = \bigcup_{\xi < \omega_1} P_\xi$ and for every $\xi < \omega_1$:*

$\langle f_n \restriction P_\xi \rangle_{n \in W_\xi}$ *is a monotone uniformly convergent sequence of uniformly continuous functions.*

PROOF. We first note that the family $\mathcal{E}$ of all $P \in \text{Perf}(X)$ for which there exists a $W \in [\omega]^\omega$ such that

the sequence $\langle f_n \restriction P \rangle_{n \in W}$ is monotone and uniformly convergent

is $\mathcal{F}_{\text{cube}}$-dense in $\text{Perf}(X)$.

Indeed, let $g \in \mathcal{F}_{\text{cube}}$, $g : \mathfrak{C}^\omega \to X$ and consider the functions $h_n = f_n \circ g$. Since $h = \langle h_n : n < \omega \rangle : \mathfrak{C}^\omega \to \mathbb{R}^\omega$ is Borel measurable, there is a dense G_δ subset G of $\mathfrak{C}^\omega$ such that $h \restriction G$ is continuous. So, we can find a perfect cube $C \subset G \subset \mathfrak{C}^\omega$, and for this C the function $h \restriction C$ is continuous. Thus, identifying the coordinate spaces of C with $\mathfrak{C}$, without loss of generality we can assume that $C = \mathfrak{C}^\omega$, that is, that each function $h_n \colon \mathfrak{C}^\omega \to \mathbb{R}$ is continuous. Now, by [126, thm. 6.9], there is a perfect cube C in $\mathfrak{C}^\omega$ and a $W \in [\omega]^\omega$ such that the sequence $\langle h_n \restriction C \rangle_{n \in W}$ is monotone and uniformly convergent.[1] So $P = g[C]$ is in $\mathcal{E}$.

Now, by CPA_{cube}, there exists an $\mathcal{E}_0 \in [\mathcal{E}]^{\leq \omega_1}$ such that $|X \setminus \bigcup \mathcal{E}_0| \leq \omega_1$. Then $\{P_\xi \colon \xi < \omega_1\} = \mathcal{E}_0 \cup \{\{x\} \colon x \in X \setminus \bigcup \mathcal{E}_0\}$ is as desired: If $P_\xi \in \mathcal{E}_0$, then the existence of an appropriate W_ξ follows from the definition of $\mathcal{E}$. If P_ξ is a singleton, then the existence of W_ξ follows from the fact that every sequence of reals has a monotone subsequence.   ∎

Of course neither Theorem 1.6.1 nor Mazurkiewicz' theorem can be proved if we do not assume some regularity of the functions f_n even if $X = \mathbb{R}$. But is it at least true that

(∗) for every uniformly bounded sequence $\langle f_n \colon \mathbb{R} \to \mathbb{R} \rangle_{n < \omega}$ the conclusion of Mazurkiewicz' theorem holds for some $P \subset \mathbb{R}$ of cardinality $\mathfrak{c}$?

The consistency of the negative result follows from the next example, which is essentially due to W. Sierpiński [118].[2] (See [77, pp. 193–194], where it is proved under the assumption of the existence of an ω_1-Luzin set. The same proof also works for our more general statement.)

Example 1.6.2 *Assume that there exists a κ-Luzin set.[3] Then for every X of cardinality κ there exists a sequence $\langle f_n \colon X \to \{0, 1\} \rangle_{n < \omega}$ with the property that for every $W \in [\omega]^\omega$ the subsequence $\langle f_n \rangle_{n \in W}$ converges pointwise for less than κ many points $x \in X$.*

In particular, under Martin's axiom, the above sequence exists for every Polish space X and $\kappa = \mathfrak{c}$.

Note also that under Martin's axiom the above example can hold only for $\kappa = \mathfrak{c}$, since Martin's axiom implies that:

For every set S of cardinality less than $\mathfrak{c}$, every uniformly bounded sequence $\langle f_n \colon S \to \mathbb{R} \rangle_{n < \omega}$ has a pointwise convergent subsequence.

---

[1] Actually, [126, thm. 6.9] is stated for functions defined on $[0, 1]^\omega$. However, the proof presented there also works for functions defined on $\mathfrak{C}^\omega$.

[2] Sierpiński constructed this example under the assumption of the continuum hypothesis.

[3] A set $L \subset \mathbb{R}$ is a κ-*Luzin set* if $|L| = \kappa$ but $|L \cap N| < \kappa$ for every nowhere dense subset N of $\mathbb{R}$. Recall that Martin's axiom implies the existence of a $\mathfrak{c}$-Luzin set.

(See [77, p. 195].) Sharper results concerning the above two facts were recently obtained by S. Fuchino and Sz. Plewik [59], in which they relate them to the splitting number $\mathfrak{s}$. (For the definition of $\mathfrak{s}$ see, e.g., [4]. For what follows it is only important that $\omega_1 \leq \mathfrak{s} \leq \mathfrak{d} \leq \mathrm{cof}(\mathcal{N})$; therefore our axiom implies that $\mathfrak{s} = \omega_1$.) More precisely, the authors show there that: *For any $X \in [\mathbb{R}]^{<\mathfrak{s}}$, any sequence $\langle f_n : X \to [-\infty, \infty] \rangle_{n<\omega}$ has a subsequence convergent pointwise on X; however, for any $X \in [\mathbb{R}]^{\mathfrak{s}}$ there exists a sequence $\langle f_n : X \to [0,1] \rangle_{n<\omega}$ with no pointwise convergent subsequence.*

Our main result of this section is the proof that $(*)$ is consistent with (and so, by the example, also independent from) the usual axioms of set theory ZFC. This follows from Corollaries 1.3.3, 1.5.5, and 1.6.4 and the fact that the existence of a selective ultrafilter on ω follows from a stronger version of our axiom. (See Theorem 5.3.3.) The proof presented below follows the argument given by K. Ciesielski and J. Pawlikowski in [36].

Theorem 1.6.3 *Assume that $\mathrm{cof}(\mathcal{N}) = \omega_1$ and that there exists a selective ω_1-generated ultrafilter on ω.*

Let X be an arbitrary set and $\langle f_n : X \to \mathbb{R} \rangle_{n<\omega}$ be a sequence of functions such that the set $\{f_n(x) : n < \omega\}$ is bounded for every $x \in X$. Then there are sequences $\langle P_\xi : \xi < \omega_1 \rangle$ of subsets of X and $\langle W_\xi \in \mathcal{F} : \xi < \omega_1 \rangle$ such that $X = \bigcup_{\xi<\omega_1} P_\xi$ and for every $\xi < \omega_1$:

the sequence $\langle f_n \restriction P_\xi \rangle_{n \in W_\xi}$ is monotone and uniformly convergent.

The conclusion of Theorem 1.6.3 is obvious for sets X with cardinality $\leq \omega_1$, since sets P_ξ can be chosen just as singletons. Thus, we will be interested in the theorem only for the sets X of cardinality greater than ω_1. In particular, we get the following corollary, which, under the additional set-theoretical assumptions, generalizes Mazurkiewicz' theorem and implies $(*)$.

Corollary 1.6.4 *Assume that $\mathrm{cof}(\mathcal{N}) = \omega_1 < \mathfrak{c}$ and that there exists a selective ω_1-generated ultrafilter on ω. Then for every Polish space X and every uniformly bounded sequence $\langle f_n : X \to \mathbb{R} \rangle_{n<\omega}$ there exist sequences $\langle P_\xi : \xi < \omega_1 \rangle$ of subsets of X and $\langle W_\xi \in [\omega]^\omega : \xi < \omega_1 \rangle$ such that $X = \bigcup_{\xi<\omega_1} P_\xi$ and for every $\xi < \omega_1$:*

the sequence $\langle f_n \restriction P_\xi \rangle_{n \in W_\xi}$ is monotone and uniformly convergent.

In particular, there exists a $\xi < \omega_1$ such that $|P_\xi| = \mathfrak{c}$.

Moreover, if the functions f_n are continuous, then we can additionally require all sets P_ξ be compact.

PROOF. The main part follows immediately from the discussion above and the Pigeon Hole Principle. To see the additional part it is enough to note that for continuous functions sets P_ξ can be replaced by their closures, since for any sequence $\langle f_n\colon P \to \mathbb{R}\rangle_{n<\omega}$ of continuous functions, if $\langle f_n \restriction D\rangle_{n<\omega}$ is monotone and uniformly convergent for some dense subset D of P, then so is $\langle f_n\rangle_{n<\omega}$.   ∎

It is worthwhile to notice here that a stronger version of our axiom implies that the last part of Corollary 1.6.4 also holds for Borel measurable functions f_n. This follows immediately from Theorem 4.1.1(a).

Recall that a nonprincipal filter $\mathcal{F}$ on ω is said to be *Ramsey* provided for every $B \in \mathcal{F}$ and $h\colon [B]^2 \to \{0,1\}$ there exist $i < 2$ and $A \in \mathcal{F}$ such that $A \subset B$ and $h\left[[A]^2\right] = \{i\}$. It is well known that an ultrafilter on ω is Ramsey if and only if it is selective. (See, e.g., [55] or [65].)

PROOF OF THEOREM 1.6.3. Let $\mathcal{F}$ be a selective ultrafilter on ω for which there exists a family $\mathcal{W} \in [\mathcal{F}]^{\omega_1}$ generating it. For every $x \in X$ define $h_x\colon [\omega]^2 \to \{0,1\}$ by putting for every $n < m < \omega$

$$h_x(n,m) = 1 \text{ if and only if } f_n(x) \leq f_m(x).$$

Since $\mathcal{F}$ is Ramsey and $\mathcal{W}$ generates $\mathcal{F}$, we can find a $W_x \in \mathcal{W}$ and an $i_x < 2$ such that $h_x[[W_x]^2] = \{i_x\}$. Thus, the sequence $S_x = \langle f_n(x)\rangle_{n\in W_x}$ is monotone. It is increasing when $i_x = 1$ and it is decreasing for $i_x = 0$.

For $W \in \mathcal{W}$ and $i < 2$, let $P_W^i = \{x \in X\colon W_x = W \;\&\; i_x = i\}$. Then $\{P_W^i\colon W \in \mathcal{W} \;\&\; i < 2\}$ is a partition of X and for every $W \in \mathcal{W}$ and $i < 2$ the sequence $\langle f_n \restriction P_W^i\rangle_{n\in W}$ is monotone and pointwise convergent to some function $f\colon P_W^i \to \mathbb{R}$.

To get uniform convergence note that for every $x \in P_W^i$ there exists an $s_x \in \omega^\omega$ such that

$$(\forall k < \omega)\,(\forall n \in W \setminus s_x(k))\,|f_n(x) - f(x)| < 2^{-k}.$$

Since $\text{cof}(\mathcal{N}) = \omega_1$ implies that the dominating number $\mathfrak{d}$ is equal to ω_1 (see, e.g., [4]), there exists a $T \in [\omega^\omega]^{\omega_1}$ dominating ω^ω. In particular, for every $x \in P_W^i$ there exists a $t_x \in T$ and an $n_x < \omega$ such that $s_x(n) \leq t_x(n)$ for all $n_x \leq n < \omega$. For $t \in T$ and $n < \omega$, let

$$P_W^i(t,n) = \{x \in P_W^i\colon t_x = t \;\&\; n_x = n\}.$$

Then the sets $\{P_W^i(t,n)\colon i < 2,\ W \in \mathcal{W},\ t \in T, n < \omega\}$ form a desired covering $\{P_\xi\colon \xi < \omega_1\}$ of X, since every sequence $\langle f_k \restriction P_W^i(t,n)\rangle_{k\in W}$ is monotone and uniformly convergent.   ∎

1.7 Some consequences of $\mathrm{cof}(\mathcal{N}) = \omega_1$**: Blumberg's theorem, strong measure zero sets, magic sets, and the cofinality of Boolean algebras**

In this section we will show some consequences of $\mathrm{cof}(\mathcal{N}) = \omega_1$ (so also of $\mathrm{CPA}_{\mathrm{cube}}$) that are related to the Sacks model.

In 1922, H. Blumberg [10] proved that *for every* $f: \mathbb{R} \to \mathbb{R}$ *there exists a dense subset* D *of* $\mathbb{R}$ *such that* $f \upharpoonright D$ *is continuous.* This theorem sparked much of discussion and generalizations; see, e.g., [26, pp. 147–150]. In particular, S. Shelah [115] showed that there is a model of ZFC in which *for every* $f: \mathbb{R} \to \mathbb{R}$ *there is a nowhere meager subset* D *of* $\mathbb{R}$ *such that* $f \upharpoonright D$ *is continuous.* The dual measure result, that is, the consistency of a statement *for every* $f: \mathbb{R} \to \mathbb{R}$ *there is a subset* D *of* $\mathbb{R}$ *of positive outer Lebesgue measure such that* $f \upharpoonright D$ *is continuous,* also has been recently established by A. Rosłanowski and S. Shelah [112]. Below we note that each of these properties contradicts $\mathrm{CPA}_{\mathrm{cube}}$. (See also K. Ciesielski and J. Pawlikowski [36].)

Theorem 1.7.1 *Let* $\mathcal{I} \in \{\mathcal{N}, \mathcal{M}\}$*. If* $\mathrm{cof}(\mathcal{I}) = \omega_1$*, then there exists an* $f: \mathbb{R} \to \mathbb{R}$ *such that* $f \upharpoonright D$ *is discontinuous for every* $D \in \mathcal{P}(\mathbb{R}) \setminus \mathcal{I}$*.*

PROOF. We will assume that $\mathcal{I} = \mathcal{N}$, the proof for $\mathcal{I} = \mathcal{M}$ being essentially identical.

Let $\{N_\xi \subset \mathbb{R}^2 : \xi < \omega_1\}$ be a family cofinal in the ideal of null subsets of $\mathbb{R}^2$ and for each $\xi < \omega_1$ let

$$N_\xi^* = \{x \in \mathbb{R} : (N_\xi)_x \notin \mathcal{N}\},$$

where $(N_\xi)_x = \{y \in \mathbb{R} : \langle x, y \rangle \in N_\xi\}$. Notice that every $x \in \mathbb{R}$ belongs to some N_ξ^*, since there is a $\xi < \omega_1$ for which $\{x\} \times \mathbb{R} \subset N_\xi$. By Fubini's theorem, each N_ξ^* is null. (In the case of $\mathcal{I} = \mathcal{M}$, we use here the Kuratowski-Ulam theorem and conclude that N_ξ^* is meager.) Thus, for each $\xi < \omega_1$ and $x \in N_\xi^* \setminus \bigcup_{\zeta < \xi} N_\zeta^*$ the set $\bigcup_{\zeta < \xi} (N_\zeta)_x$ is null, so we can choose

$$f(x) \in \mathbb{R} \setminus \bigcup_{\zeta < \xi} (N_\zeta)_x.$$

Then function f is as desired.

Indeed, if $f \upharpoonright D$ is continuous for some $D \subset \mathbb{R}$, then $f \upharpoonright D$ is null in $\mathbb{R}^2$. In particular, there exists a $\xi < \omega_1$ such that $f \upharpoonright D \subset N_\xi$. But this means that $D \subset \bigcup_{\zeta \leq \xi} N_\zeta^*$. ∎

Note that essentially the same proof works if we assume only that $\text{cof}(\mathcal{I})$ is equal to the additivity number $\text{add}(\mathcal{I})$ of $\mathcal{I}$.

Corollary 1.7.2 *Assume* CPA$_{\text{cube}}$. *Then there exists an* $f\colon\mathbb{R}\to\mathbb{R}$ *such that if* $f\restriction D$ *is continuous, then* $D\in\mathcal{N}\cap\mathcal{M}$.

PROOF. By CPA$_{\text{cube}}$ we have $\text{cof}(\mathcal{N})=\text{cof}(\mathcal{M})=\omega_1$. Let $f_{\mathcal{N}}$ and $f_{\mathcal{M}}$ be from Theorem 1.7.1, constructed for the ideals $\mathcal{N}$ and $\mathcal{M}$, respectively. Let $G\subset\mathbb{R}$ be a dense G_δ of measure zero and put $f=[f_{\mathcal{M}}\restriction G]\cup[f_{\mathcal{N}}\restriction(\mathbb{R}\setminus G)]$. Then this f is as desired. ∎

Although CPA$_{\text{cube}}$ implies that all cardinals from Cichoń's diagram are as small as possible, this does not extend to all possible cardinal invariants. For example, we note that CPA$_{\text{cube}}$ refutes the Borel conjecture.

Corollary 1.7.3 *If* CPA$_{\text{cube}}$ *holds, then there exists an uncountable strong measure zero set.*

PROOF. This follows from the fact (see, e.g., [4, thm. 8.2.8]) that $\text{cof}(\mathcal{N})=\omega_1$ implies the existence of an uncountable strong measure zero set. ∎

Recall that a set $M\subset\mathbb{R}$ is a *magic set* (or *set of range uniqueness*) if $f[M]\neq g[M]$ for every different nowhere constant functions $f,g\in C(\mathbb{R})$. It has been proved by A. Berarducci and D. Dikranjan [8, thm. 8.5] that a magic set exists under the continuum hypothesis, while K. Ciesielski and S. Shelah [44] constructed a model with no magic set. It is relatively easy to see that if M is magic, then M is not meager and $f[M]\neq[0,1]$ for every $f\in C(\mathbb{R})$. (See [20, cor. 5.15 and thm. 5.6(5)].) So CPA$_{\text{cube}}$ implies there is no magic set of cardinality $\mathfrak{c}$. On the other hand, it was noticed by S. Todorcevic (see [20, p. 1097]) that in the iterated perfect set model there is a magic set (clearly of cardinality ω_1). We note here that the same is implied by CPA$_{\text{cube}}$. (See also K. Ciesielski and J. Pawlikowski [36].)

Proposition 1.7.4 *If the cofinality* $\text{cof}(\mathcal{M})$ *of the ideal* $\mathcal{M}$ *of meager sets is equal* ω_1, *then there exists a magic set. In particular,* CPA$_{\text{cube}}$ *implies that there exists a magic set.*

PROOF. An uncountable set $L\subset\mathbb{R}$ is a *2-Luzin set* provided for every disjoint subsets $\{x_\xi\colon\xi<\omega_1\}$ and $\{y_\xi\colon\xi<\omega_1\}$ of L, where the enumerations are one to one, the set of pairs $\{\langle x_\xi,y_\xi\rangle\colon\xi<\omega_1\}$ is not a meager subset of $\mathbb{R}^2$. In [20, prop. 4.8] it was noticed that every ω_1-dense 2-Luzin set is a magic set. It is also a standard and easy diagonal argument that $\text{cof}(\mathcal{M})=$

ω_1 implies the existence of a ω_1-dense 2-Luzin set. (The proof presented in [125, prop. 6.0] also works under the assumption $\mathrm{cof}(\mathcal{M}) = \omega_1$.) So,

$$\mathrm{CPA_{cube}} \implies \mathrm{cof}(\mathcal{N}) = \omega_1 \implies \mathrm{cof}(\mathcal{M}) = \omega_1 \implies \text{``there is a magic set.''} \blacksquare$$

Recall also that the existence of a magic set for the class "D^1" can be proved in ZFC. This follows from [21, thm. 3.1], since every function from "D^1" belongs to the Banach class (T_2). (Compare also [21, cor. 3.3 and 3.4].)

For an infinite Boolean algebra B, its *cofinality* $\mathrm{cof}(B)$ is defined as the least infinite cardinal number κ such that B is a union of strictly increasing sequence of type κ of subalgebras of B. S. Koppelberg [79] proved that Martin's axiom implies that $\mathrm{cof}(B) = \omega$ for every Boolean algebra B with $|B| < \mathfrak{c}$ while W. Just and P. Koszmider [72] proved that in the model obtained by adding at least ω_2 Sacks reals side by side there exists a Boolean algebra B of cardinality ω_1 such that $\mathrm{cof}(B) = \omega_1 < \mathfrak{c}$. (See also [51].)

We would like to point out here that this also follows from $\mathrm{CPA_{cube}}$, since we have:

Theorem 1.7.5 $\mathrm{cof}(\mathcal{N}) = \omega_1$ *implies that there exists a Boolean algebra B of cardinality ω_1 such that $\mathrm{cof}(B) = \omega_1$.*

The proof of Theorem 1.7.5 presented below follows the argument given by K. Ciesielski and J. Pawlikowski in [35]. It will be based on the following lemma.

Lemma 1.7.6 *If* $\mathrm{cof}(\mathcal{N}) = \omega_1$, *then for every infinite countable Boolean algebra $\mathcal{A}$ there exists a family $\{a_n^\xi \in \mathcal{A} : n < \omega \ \& \ \xi < \omega_1\}$ with the following properties.*

(i) $a_n^\xi \wedge a_m^\xi = \mathbf{0}$ *for every $n < m < \omega$ and $\xi < \omega_1$.*

(ii) *For every increasing sequence $\langle \mathcal{A}_n : n < \omega \rangle$ of proper subalgebras of $\mathcal{A}$ with $\mathcal{A} = \bigcup_{n<\omega} \mathcal{A}_n$ there exists a $\xi < \omega_1$ such that $a_n^\xi \notin \mathcal{A}_n$ for all $n < \omega$.*

PROOF. In the argument that follows, every sequence $\bar{\mathcal{A}} = \langle \mathcal{A}_n : n < \omega \rangle$ as in (ii) will be identified with a function $f_{\bar{\mathcal{A}}} = f \in \omega^{\mathcal{A}}$ for which we have $f^{-1}(n) = \mathcal{A}_n \setminus \bigcup_{i<n} \mathcal{A}_i$. We will denote the set of all such functions by X. Also, let $\{b_n : n < \omega\}$ be an enumeration of $\mathcal{A}$, and for each $n < \omega$ let B_n be a finite algebra generated by $\{b_i : i < n\}$. Thus, $\mathcal{A} = \bigcup_{i<\omega} B_i$.

Since $\text{cof}(\mathcal{N}) = \omega_1$, the dominating number

$$\mathfrak{d} = \min\{|\mathcal{K}|: \mathcal{K} \subset \omega^\omega \,\&\, (\forall f \in \omega^\omega)(\exists g \in \mathcal{K})(\forall n < \omega)\, f(n) < g(n)\}$$

is equal to ω_1. (See, e.g., [4].) So, there exists a dominating family $\mathcal{K} \subset \omega^\omega$ of cardinality ω_1. We can also assume that the sequences in $\mathcal{K}$ are strictly increasing and that for every $g \in \mathcal{K}$ the function $\bar{g}$ defined by $\bar{g}(n) = \sum_{i \leq n} g(i)$ also belongs to $\mathcal{K}$.

Next notice that for every $f \in X$ there exist $\bar{d} = \langle d_k : k < \omega \rangle \in \mathcal{K}$ and $\bar{r} = \langle r_k : k < \omega \rangle \in \mathcal{K}$ such that, for every $k < \omega$,

(a) $f(b) < r_k$ for all $b \in B_k$, and

(b) there are disjoint $b_0, \dots, b_{2k} \in B_{d_k}$ with $r_{d_{k-1}} < f(b_0) < \cdots < f(b_{2k})$.

Indeed, the existence of $\bar{r}$ satisfying (a) follows directly from the definition of a dominating family. Moreover, since all algebras $\mathcal{A}_n = f^{-1}(\{0, \dots, n\})$ are proper, for every number $d < \omega$ there exist disjoint $b_0, \dots, b_{2d} \in \mathcal{A}$ such that $r_d < f(b_0) < \cdots < f(b_{2d})$. Let $h \in \omega^\omega$ be such that $b_0, \dots, b_{2d} \in B_{d+h(d)}$ for every $d < \omega$ and let $g \in \mathcal{K}$ be a function dominating h. Then $\bar{d} = \bar{g}$ is as required.

The above implies, in particular, that for every $f \in X$ there are $\bar{d}, \bar{r} \in \mathcal{K}$ such that f satisfies (b) and the sequence $f^{\bar{r}} = \langle f \restriction (B_{d_k} \setminus B_{d_{k-1}}) : k < \omega \rangle$ belongs to

$$X(\bar{d}, \bar{r}) = \prod_{k < \omega} (r_{d_k})^{B_{d_k} \setminus B_{d_{k-1}}}.$$

Now, since $\text{cof}(\mathcal{N}) = \omega_1$, by Proposition 1.3.1 (applied to $\prod_{k < \omega} \omega^{B_{d_k} \setminus B_{d_{k-1}}}$ in place of ω^ω) we can find an ω_1-covering of $X(\bar{d}, \bar{r})$ by sets T of the form $\prod_{k < \omega} T_k$, where $T_k \in \left[\omega^{B_{d_k} \setminus B_{d_{k-1}}}\right]^{\leq k+1}$ for all $k < \omega$. Since the total number of these sets T (for different $\bar{d}, \bar{r} \in \mathcal{K}$) is equal to ω_1, to finish the proof it is enough to show that for any such T there is one sequence $\langle a_n : n < \omega \rangle$ satisfying (i) and such that (ii) holds for every $\bar{A} = \langle \mathcal{A}_n : n < \omega \rangle$ for which $f^{\bar{r}}_{\bar{A}}$ belongs to T and $f_{\bar{A}}$ satisfies (b).

So, let T be as above and let T^* be the set of all functions $f_{\bar{A}}$ satisfying (b) for which $f^{\bar{r}}_{\bar{A}} \in T$. By induction on $k < \omega$ we will construct a sequence $\langle c_k \in B_{d_k} \setminus B_{d_{k-1}} : k < \omega \rangle$ such that:

(∗) $f(c_k) > r_{d_k} \geq k$ for every $k < \omega$ and $f \in T^*$.

So fix a $k < \omega$ and let $\{f_i : i < k\}$ be such that

$$\{f_i \restriction B_{d_k} \setminus B_{d_k-1} : i < k\} = \{f \restriction B_{d_k} \setminus B_{d_k-1} : f \in T^*\} \subset T_k.$$

We show inductively that for every $m < k$

$$\text{there is a } c \in B_{d_k} \text{ such that } f_j(c) > r_{d_k} \text{ for all } j \leq m. \qquad (1.8)$$

So, fix an $m < k$ and let $a \in B_{d_k}$ such that $f_j(a^c) = f_j(a) > r_{d_k}$ for all $j < m$. If $f_m(a) > r_{d_k}$, then $c = a$ satisfies property (1.8). Thus, assume that $f_m(a^c) = f_m(a) \leq r_{d_k}$. By (b) we can find $b_0, \ldots, b_{2k} \in B_{d_k}$ such that $r_{d_k-1} < f_m(b_0) < \cdots < f_m(b_{2k})$. By the Pigeon Hole Principle we can find an $I \in [\{0, \ldots, 2k\}]^{k+1}$ and a $b \in \{a, a^c\}$ such that $f_m(b \wedge b_i) = f_m(b_i)$ for all $i \in I$. Without loss of generality we can assume that $I = \{0, \ldots, k\}$ and $b \wedge b_i = b_i$ for all $i \leq k$. Then

$$f_m(b^c \vee b_i) > r_{d_k} \quad \text{for all } i \leq k.$$

Moreover, for every $j < m$ there is at most one $i_j \leq k$ for which

$$f_j(b^c \vee b_{i_j}) \leq r_{d_k},$$

since for different $i, i' \leq k$ we have $f_j((b^c \vee b_i) \wedge (b^c \vee b_{i'})) = f_j(b^c) > r_{d_k}$. Thus, by the Pigeon Hole Principle, there is an $i \leq k$ such that $c = b^c \vee b_i$ satisfies (1.8). This finishes the proof of $(*)$.

Clearly the sequence $\langle c_k \in B_{d_k} \setminus B_{d_k-1} : k < \omega \rangle$ satisfies (ii) for every $\bar{A}$ with $f_{\bar{A}} \in T^*$. Thus, we need only to modify it to also get condition (i).

To do this, use the fact that

$$r_{d_k} < f(c_k) < r_{d_{k+1}} \text{ for every } k < \omega \text{ and } f \in T^*$$

to construct the sequences $\omega = I_0 \supset I_1 \supset \cdots$ of infinite subsets of ω, increasing $\langle k_j \in I_j : j < \omega \rangle$, and $\langle c^*_{k_j} \in \{c_{k_j}, c^c_{k_j}\} : j < \omega \rangle$ such that for every $j < \omega$

- $f(\bar{a}_{k_j} \wedge c_l) > r_{d_l}$ for every $f \in T^*$ and $l > k_j$ with $l \in I_j$,

where $\bar{a}_{k_j} = c^*_{k_0} \wedge \cdots \wedge c^*_{k_j}$. Then the sequence $\langle \bar{a}_{k_j} : j < \omega \rangle$ is a strictly decreasing sequence satisfying (ii) and it is now easy to see that by putting $a_j = \bar{a}_{k_j} \wedge \bar{a}^c_{k_{j+1}}$ we obtain the desired sequence. $\blacksquare$

PROOF OF THEOREM 1.7.5. The algebra B we construct will be a sub-algebra of the algebra $\mathcal{P}(\omega)$ of all subsets of ω. First, let $\mathcal{K} \subset \omega^\omega$ be a dominating family with $|\mathcal{K}| = \omega_1$ and fix a partition $\{D_k : k < \omega\}$ of ω into infinite subsets.

For every sequence $\bar{a} = \langle a_n : n < \omega \rangle$ of pairwise disjoint subsets of ω and $k < \omega$ put $a_k^* = \bigcup\{a_n : n \in D_k\}$. In addition, for every $h \in \mathcal{K}$ we put

$$a^h = \bigcup\{a_{n_h(k)} : k < \omega\},$$

where $n_h(k) = \min\{n \in D_k : n > \max\{h(k), k\}\}$. We also put

$$F(\bar{a}) = \{a_k^* : k < \omega\} \cup \{a^h : h \in \mathcal{K}\} \in [\mathcal{P}(\omega)]^{\le \omega_1}.$$

Next, we will construct an increasing sequence $\langle B_\xi \in [\mathcal{P}(\omega)]^{\omega_1} : \xi \le \omega_1 \rangle$ of subalgebras of $\mathcal{P}(\omega)$ aiming for $B = B_{\omega_1}$. Thus, we choose B_0 as an arbitrary subalgebra of $\mathcal{P}(\omega)$ with $|B_0| = \omega_1$, and for limit ordinal numbers $\lambda \le \omega_1$ we put $B_\lambda = \bigcup_{\xi < \lambda} B_\xi$. The algebra $B_{\xi+1}$ is formed from B_ξ in the following way.

Let $\{b_\eta : \eta < \omega_1\}$ be an enumeration of B_ξ, and for $\eta < \omega$ let $\mathcal{A}_\eta^\xi$ be a subalgebra of B_ξ generated by $\{b_\zeta : \zeta < \eta\}$. For each such algebra we apply Lemma 1.7.6 to find the sequences $\bar{a}^\gamma = \langle a_n^\gamma : n < \omega \rangle$, $\gamma < \omega_1$, satisfying (i) and (ii), and let

$$G(\mathcal{A}_\eta^\xi) = \bigcup_{\gamma < \omega_1} F(\bar{a}^\gamma).$$

$B_{\xi+1}$ is defined as the algebra generated by $B_\xi \cup \bigcup_{\eta < \omega_1} G(\mathcal{A}_\eta^\xi)$. This finishes the construction of B.

Clearly, $|B| = \omega_1$. To prove that $\mathrm{cof}(B) = \omega_1$, it is enough to show that B is not a union of an increasing sequence $\bar{B} = \langle B_n : n < \omega \rangle$ of proper subalgebras. So, by way of contradiction, assume that such a sequence $\bar{B}$ exists. For every $n < \omega$ choose $b_n \in B \setminus B_n$ and find $\xi, \eta < \omega_1$ such that $\{b_n : n < \omega\} \subset \mathcal{A}_\eta^\xi$. Then the algebras $\mathcal{A}_n = B_n \cap \mathcal{A}_\eta^\xi$ form an increasing sequence of proper subalgebras of $\mathcal{A} = \mathcal{A}_\eta^\xi$. Thus, one of the sequences $\bar{a}^\gamma$ satisfies (ii) for $\bar{A}$. So, if we put $\bar{a}^\gamma = \bar{a} = \langle a_n : n < \omega \rangle$, we conclude that $\{a_k^* : k < \omega\} \cup \{a^h : h \in \mathcal{K}\} \subset B$. Let $f(k) = \min\{n < \omega : a_k^* \in B_n\}$ and let $h \in \mathcal{K}$ be such that $f(k) < h(k)$ for all $k < \omega$.

The final contradiction is obtained by noticing that a^h cannot belong to any B_k. Indeed, if $a^h \in B_k$ for some k, then $a^h \cap a_k^* = a_{n_h(k)}$ belongs to $B_{\max\{f(k),k\}}$, since $a_k^* \in B_{f(k)}$. But $\max\{f(k), k\} \le \max\{h(k), k\} < n_h(k)$, so we get $a_{n_h(k)} \in B_{n_h(k)}$, contradicting the fact that $a_{n_h(k)}$ belongs to $\bar{A} \setminus \mathcal{A}_{n_h(k)}$, which is disjoint with $B_{n_h(k)}$. ∎

In the context of CPA$_{\text{cube}}$, recently T. Natkaniec [106] also proved that $\mathrm{cof}(\mathcal{N}) = \omega_1 + \text{"}2^{\omega_1} = \omega_2\text{"}$ implies that the density topology is extra-resolvable.

1.8 Remarks on a form and consistency of the axiom $\mathrm{CPA}_{\mathrm{cube}}$

In what follows we consider $\mathcal{F}_{\mathrm{cube}}$ as ordered by inclusion. Thus, $\mathcal{F} \subset \mathcal{F}_{\mathrm{cube}}$ is dense in $\mathcal{F}_{\mathrm{cube}}$ provided for every $f \in \mathcal{F}_{\mathrm{cube}}$ there exists a $g \in \mathcal{F}$ with $g \subset f$.

Proposition 1.8.1 $\mathrm{CPA}_{\mathrm{cube}}$ *is equivalent to the following.*

$\mathrm{CPA}^0_{\mathrm{cube}}$**:** $\mathfrak{c} = \omega_2$, *and for every Polish space* X *and dense subfamily* $\mathcal{F}$ *of* $\mathcal{F}_{\mathrm{cube}}$ *there is an* $\mathcal{F}_0 \in [\mathcal{F}]^{\leq \omega_1}$ *such that* $|X \setminus \bigcup_{g \in \mathcal{F}_0} \mathrm{range}(g)| \leq \omega_1$.

PROOF. First notice that

- $\mathcal{E} \subset \mathrm{Perf}(X)$ is $\mathcal{F}_{\mathrm{cube}}$-dense if and only if $\mathcal{F}_{\mathrm{cube}}(\mathcal{E})$ is dense in $\mathcal{F}_{\mathrm{cube}}$,

where $\mathcal{F}_{\mathrm{cube}}(\mathcal{E}) = \{g \in \mathcal{F}_{\mathrm{cube}} : \mathrm{range}(g) \in \mathcal{E}\}$.

Now, to see that $\mathrm{CPA}^0_{\mathrm{cube}}$ implies $\mathrm{CPA}_{\mathrm{cube}}$, take an $\mathcal{F}_{\mathrm{cube}}$-dense family $\mathcal{E} \subset \mathrm{Perf}(X)$. Then, using $\mathrm{CPA}^0_{\mathrm{cube}}$ with $\mathcal{F} = \mathcal{F}_{\mathrm{cube}}(\mathcal{E})$, we can find an appropriate $\mathcal{F}_0 \in [\mathcal{F}]^{\leq \omega_1}$. Clearly $\mathcal{E}_0 = \{\mathrm{range}(g) : g \in \mathcal{F}_0\}$ satisfies the conclusion of $\mathrm{CPA}_{\mathrm{cube}}$.

To see the other implication, take a dense subfamily $\mathcal{F}$ of $\mathcal{F}_{\mathrm{cube}}$ and let $\mathcal{E} = \{\mathrm{range}(g) : g \in \mathcal{F}\}$. Then, by •, $\mathcal{E}$ is $\mathcal{F}_{\mathrm{cube}}$-dense in $\mathrm{Perf}(X)$, since $\mathcal{F} \subset \mathcal{F}_{\mathrm{cube}}(\mathcal{E})$. By $\mathrm{CPA}_{\mathrm{cube}}$ we can find an $\mathcal{E}_0 \in [\mathcal{E}]^{\leq \omega_1}$ such that $|X \setminus \bigcup \mathcal{E}_0| < \mathfrak{c}$. For each $E \in \mathcal{E}_0$ take an $f_E \in \mathcal{F}$ such that $\mathrm{range}(f_E) = E$. Then $\mathcal{F}_0 = \{f_E : E \in \mathcal{E}_0\}$ satisfies the conclusion of $\mathrm{CPA}^0_{\mathrm{cube}}$.   ∎

Condition $\mathrm{CPA}^0_{\mathrm{cube}}$ has a form that is closer to our main axiom CPA than $\mathrm{CPA}_{\mathrm{cube}}$. Also, $\mathrm{CPA}^0_{\mathrm{cube}}$ does not require a new notion of $\mathcal{F}_{\mathrm{cube}}$-density. So why bother with $\mathrm{CPA}_{\mathrm{cube}}$ at all? The reason is that the consequences presented in this chapter follow more naturally from $\mathrm{CPA}_{\mathrm{cube}}$ than from $\mathrm{CPA}^0_{\mathrm{cube}}$. Moreover, from $\mathrm{CPA}_{\mathrm{cube}}$ it is also clearer that the axiom describes, in fact, a property of perfect subsets of X, rather than of some coordinate functions.

Note that if $\mathcal{F}_{\mathrm{all}}$ stands for the family of all injections from perfect subsets of $\mathfrak{C}^\omega$ into X, then we can use schema (1.2) to define a notion of $\mathcal{F}$-density of $\mathcal{E} \subset \mathrm{Perf}(X)$ for any family $\mathcal{F} \subset \mathcal{F}_{\mathrm{all}}$: $\mathcal{E} \subset \mathrm{Perf}(X)$ is $\mathcal{F}$-*dense* provided

$$\forall f \in \mathcal{F} \; \exists g \in \mathcal{F} \; (g \subset f \; \& \; \mathrm{range}(g) \in \mathcal{E}). \tag{1.9}$$

In particular, it is easy to see that a family $\mathcal{E} \subset \mathrm{Perf}(X)$ is dense in $\mathrm{Perf}(X)$ if and only if $\mathcal{E}$ is $\mathcal{F}_{\mathrm{all}}$-dense in $\mathrm{Perf}(X)$.

Next notice that the equation $\mathfrak{c} = \omega_2$ is, in some sense, a consequence of the second part of the axiom $\mathrm{CPA}_{\mathrm{cube}}$.

Remark 1.8.2 The following statements are equivalent.

CPA$_{\text{cube}}$: $\mathfrak{c} = \omega_2$, and for every Polish space X and every $\mathcal{F}_{\text{cube}}$-dense family $\mathcal{E} \subset \text{Perf}(X)$ there is an $\mathcal{E}_0 \subset \mathcal{E}$ such that $|\mathcal{E}_0| \leq \omega_1$ and $|X \setminus \bigcup \mathcal{E}_0| \leq \omega_1$.

(i) $\mathfrak{c} \geq \omega_2$, and for every Polish space X and $\mathcal{F}_{\text{cube}}$-dense family $\mathcal{E} \subset \text{Perf}(X)$ there is an $\mathcal{E}_0 \subset \mathcal{E}$ such that $|\mathcal{E}_0| \leq \omega_1$ and $|X \setminus \bigcup \mathcal{E}_0| \leq \omega_1$.

(ii) $\mathfrak{c} \leq \omega_2$, and for every Polish space X and $\mathcal{F}_{\text{cube}}$-dense family $\mathcal{E} \subset \text{Perf}(X)$ there is an $\mathcal{E}_0 \subset \mathcal{E}$ such that $|\mathcal{E}_0| < \mathfrak{c}$ and $|X \setminus \bigcup \mathcal{E}_0| < \mathfrak{c}$.

PROOF.   Clearly (ii) implies that either CPA$_{\text{cube}}$ or CH hold. But (ii) implies, as in Proposition 1.0.3, that $s_0^{\text{cube}} = [X]^{<\mathfrak{c}}$; so, by Corollary 1.1.2, property (A) holds, which contradicts CH. So, (ii) implies CPA$_{\text{cube}}$. It is also obvious that CPA$_{\text{cube}}$ implies (i). To see that (i) $\Longrightarrow$ (ii) it is enough to notice that (i) implies that $\mathfrak{c} \leq \omega_2$. This follows from Proposition 1.1.1 and the fact that the conclusion of Proposition 1.0.3 already follows from (i). (These two facts imply that every subset of X of cardinality greater than ω_1 can be mapped onto $\mathfrak{C}$, so $\mathfrak{c} \leq \omega_2$.)   ∎

Recall that CPA$_{\text{cube}}[X]$ stands for CPA$_{\text{cube}}$ for a fixed Polish space X.

Remark 1.8.3 For any Polish space X, axiom CPA$_{\text{cube}}[X]$ implies the full axiom CPA$_{\text{cube}}$.

PROOF.   Let X be a Polish space. First notice the following two facts.

(F1) If Y is a Polish subspace of X, then CPA$_{\text{cube}}[X]$ implies CPA$_{\text{cube}}[Y]$.

Indeed, let $\mathcal{E}$ be an $\mathcal{F}_{\text{cube}}$-dense subset of $\text{Perf}(Y)$, and for every cube Q in Y let $\varphi(Q) \in \mathcal{E}$ be a subcube of Q. Next, for each cube P in X let Q_P be its subcube such that either $Q_P \cap Y = \emptyset$ or $Q_P \subset Y$. Such a subcube can be found by Claim 1.1.5, since Y is a G_δ subset of X. If $Q_P \subset Y$, put $\psi(P) = \varphi(Q_P) \in \mathcal{E}$; otherwise, we put $\psi(P) = Q_P$. Now $\bar{\mathcal{E}} = \{\psi(P) : P \text{ is a cube in } X\}$ is $\mathcal{F}_{\text{cube}}$-dense subset of $\text{Perf}(X)$. So, by CPA$_{\text{cube}}[X]$, there exists an $\bar{\mathcal{E}}_0 \in [\bar{\mathcal{E}}]^{\leq \omega_1}$ such that $|X \setminus \bigcup \bar{\mathcal{E}}_0| \leq \omega_1$. Then $\mathcal{E}_0 = \mathcal{E} \cap \bar{\mathcal{E}}_0 \in [\mathcal{E}]^{\leq \omega_1}$ and $|Y \setminus \bigcup \mathcal{E}_0| \leq \omega_1$. So (F1) is proved.

(F2) If a Polish space Y is a one to one continuous image of X, then CPA$_{\text{cube}}[X]$ implies CPA$_{\text{cube}}[Y]$.

Indeed, let f be a continuous bijection from X onto Y and let $\mathcal{E}$ be an $\mathcal{F}_{\text{cube}}$-dense subset of $\text{Perf}(Y)$. Put $\bar{\mathcal{E}} = \{f^{-1}(P) : P \in \mathcal{E}\}$ and notice that $\bar{\mathcal{E}}$ is an $\mathcal{F}_{\text{cube}}$-dense subset of $\text{Perf}(X)$. (If $g \in \mathcal{F}_{\text{cube}}$ maps $\mathfrak{C}^\omega$ onto $R \subset X$, then $f[R]$ is a cube in Y witnessed by $f \circ g \in \mathcal{F}_{\text{cube}}$. Now, if $S = f \circ g[C] \in \mathcal{E}$

is a subcube of $f[R]$, then $g[C] = f^{-1}(S) \in \bar{\mathcal{E}}$ is a subcube of R.) So, by CPA$_{\mathrm{cube}}[X]$, there is an $\bar{\mathcal{E}}_0 \in [\bar{\mathcal{E}}]^{\leq \omega_1}$ such that $|X \setminus \bigcup \bar{\mathcal{E}}_0| \leq \omega_1$. Then $\mathcal{E}_0 = \{f[Q] : Q \in \bar{\mathcal{E}}_0\} \in [\mathcal{E}]^{\leq \omega_1}$ and $|Y \setminus \bigcup \mathcal{E}_0| = |f[X \setminus \bar{\mathcal{E}}_0]| \leq \omega_1$. So (F2) is proved.

To finish the proof, take a Polish space X for which CPA$_{\mathrm{cube}}[X]$ holds and recall that the Baire space ω^ω is homeomorphic to a subspace of X (since X contains a copy of $\mathfrak{C}$ and $\mathfrak{C}$ contains a copy of ω^ω). Thus, by (F1), CPA$_{\mathrm{cube}}[Z]$ holds for an arbitrary Polish subspace Z of ω^ω. Now, if Y is an arbitrary Polish space, then there exists a closed subset F of ω^ω such that Y is a one to one continuous image of F. (See, e.g., [74, thm. 7.9].) So, by (F2), CPA$_{\mathrm{cube}}[Y]$ holds as well.   ∎

Remark 1.8.4 $s_0^{\mathrm{cube}} \neq [X]^{\leq \omega_1}$ in a model obtained by adding Sacks numbers side by side. In particular, CPA$_{\mathrm{cube}}$ and all other versions of the CPA axiom considered in this book are false in this model.

PROOF. This follows from the fact that $s_0^{\mathrm{cube}} = [X]^{\leq \omega_1}$ implies property (A) (see Corollary 1.1.2) while it is false in the model mentioned above, as noticed by A. Miller in [95, p. 581]. (In this model, the set X of all Sacks generic numbers cannot be mapped continuously onto $[0, 1]$.)   ∎

One might wonder whether, in the formulation of the notion of cube density, it is necessary to consider the coordinate functions from the *countable* product $\mathfrak{C}^\omega$ of $\mathfrak{C}$. Wouldn't it be enough to consider the coordinate functions from the finite product of $\mathfrak{C}$? To make this question more precise, let us introduce the following terminology, which comes from a paper [30] by K. Ciesielski and A. Millán.

For a Polish space X and $0 < n \leq \omega$ we say $\mathcal{F} \subset \mathrm{Perf}(X)$ is *n-cube dense* provided that for every injection $f : \mathfrak{C}^n \to X$ there is a perfect cube $C = \prod_{i<n} P_i \subset \mathfrak{C}^n$ such that $f[C] \in \mathcal{F}$. Clearly, ω-cube density is a cube density we use in the CPA$_{\mathrm{cube}}$ and 1-cube density is just the standard perfect set density. Thus, (1.3) says that every ω-cube dense family is also 1-cube dense. More generally, we have the the following implications.

Fact 1.8.5 *If $0 < m < n \leq \omega$, then every n-cube dense family is also m-cube dense.*

PROOF. Let $m = k + 1$, $g : \mathfrak{C}^{n \setminus k} \to \mathfrak{C}$ be a homeomorphism, and define $h : \mathfrak{C}^n \to \mathfrak{C}^m$ by $h(x)(i) = x(i)$ for every $i < k$ and $h(x)(k) = g(x \restriction n \setminus k)$. It is easy to see that h is a homeomorphism and that if $P \subset \mathfrak{C}^n$ is a perfect cube, then $h[P]$ is a perfect cube.

Now, let $\mathcal{F} \subset \text{Perf}(X)$ be n-cube dense in X. To see that $\mathcal{F}$ is m-cube dense, take a continuous injection $f\colon \mathfrak{C}^m \to X$. Then $f \circ h\colon \mathfrak{C}^n \to X$ is also a continuous injection. Since $\mathcal{F}$ is n-cube dense, there exists a $P \subset \mathfrak{C}^n$ such that P is a perfect cube and $f[h[P]] = (f \circ h)[P] \in \mathcal{F}$. But $h[P]$ is a perfect cube, so $\mathcal{F}$ is m-cube dense. ∎

Example 1.0.1 shows that 1-cube density does not imply 2-cube density. In general, none of the implications from Fact 1.8.5 can be reversed, as we will show in Theorem 3.5.1. Moreover, we will see in Theorem 3.5.2 that in CPA$_{\text{cube}}$ we cannot replace cube density with "n-cube density for every $0 < n < \omega$," since the resulting statement is false in ZFC.

2

Games and axiom $\mathrm{CPA}_{\mathrm{cube}}^{\mathrm{game}}$

Before we get to the formulation of our next version of the axiom, it would be good to note that in many applications we would prefer to have a full covering of a Polish space X rather that the almost covering as claimed by $\mathrm{CPA}_{\mathrm{cube}}$. To get better access to the missing singletons[1] we will extend the notion of a cube by also allowing the "constant cubes:" A family $\mathcal{C}_{\mathrm{cube}}(X)$ of *constant cubes* is defined as the family of all constant functions from a perfect cube $C \subset \mathfrak{C}^\omega$ into X. We also define $\mathcal{F}_{\mathrm{cube}}^*(X)$ as

$$\mathcal{F}_{\mathrm{cube}}^* = \mathcal{F}_{\mathrm{cube}} \cup \mathcal{C}_{\mathrm{cube}}. \tag{2.1}$$

Thus, $\mathcal{F}_{\mathrm{cube}}^*$ is the family of all continuous functions from a perfect cube $C \subset \mathfrak{C}^\omega$ into X that are either one to one or constant. Now the range of every $f \in \mathcal{F}_{\mathrm{cube}}^*$ belongs to the family $\mathrm{Perf}^*(X)$ of all sets P such that either $P \in \mathrm{Perf}(X)$ or P is a singleton. The terms "$P \in \mathrm{Perf}^*(X)$ is a cube" and "Q is a subcube of a cube $P \in \mathrm{Perf}^*(X)$" are defined in a natural way.

Consider also the following game $\mathrm{GAME}_{\mathrm{cube}}(X)$ of length ω_1. The game has two players, Player I and Player II. At each stage $\xi < \omega_1$ of the game Player I can play an arbitrary cube $P_\xi \in \mathrm{Perf}^*(X)$ and Player II must respond with a subcube Q_ξ of P_ξ. The game $\langle \langle P_\xi, Q_\xi \rangle \colon \xi < \omega_1 \rangle$ is won by Player I provided

$$\bigcup_{\xi < \omega_1} Q_\xi = X;$$

otherwise, the game is won by Player II.

By a strategy for Player II we will consider any function S such that $S(\langle \langle P_\eta, Q_\eta \rangle \colon \eta < \xi \rangle, P_\xi)$ is a subcube of P_ξ, where $\langle \langle P_\eta, Q_\eta \rangle \colon \eta < \xi \rangle$ is any

---

[1] The logic for accessing the singletons in such a strange is justified by the versions of the axiom that will be presented in Chapter 6.

partial game. (We abuse here slightly the notation, since the function S also depends on the implicitly given coordinate functions $f_\eta: \mathfrak{C}^\omega \to P_\eta$, making each P_η a cube.) A game $\langle\langle P_\xi, Q_\xi\rangle: \xi < \omega_1\rangle$ is played according to a strategy S for Player II provided $Q_\xi = S(\langle\langle P_\eta, Q_\eta\rangle: \eta < \xi\rangle, P_\xi)$ for every $\xi < \omega_1$. A strategy S for Player II is a *winning strategy* for Player II provided Player II wins any game played according to the strategy S.

Here is our new version of the axiom.

CPA$_{\text{cube}}^{\text{game}}$: $\mathfrak{c} = \omega_2$, and for any Polish space X Player II has no winning strategy in the game GAME$_{\text{cube}}(X)$.

Notice that

Proposition 2.0.1 *Axiom* CPA$_{\text{cube}}^{\text{game}}$ *implies* CPA$_{\text{cube}}$.

PROOF. Let $\mathcal{E} \subset \text{Perf}(X)$ be an $\mathcal{F}_{\text{cube}}$-dense family. Thus, for every cube $P \in \text{Perf}(X)$ there exists a subcube $s(P) \in \mathcal{E}$ of P. Now, for a singleton $P \in \text{Perf}^*(X)$, put $s(P) = P$ and consider the following strategy (in fact, it is a tactic) S for Player II:

$$S(\langle\langle P_\eta, Q_\eta\rangle: \eta < \xi\rangle, P_\xi) = s(P_\xi).$$

By CPA$_{\text{cube}}^{\text{game}}$ it is not a winning strategy for Player II. So there exists a game $\langle\langle P_\xi, Q_\xi\rangle: \xi < \omega_1\rangle$ in which $Q_\xi = s(P_\xi)$ for every $\xi < \omega_1$ and Player II loses, that is, $X = \bigcup_{\xi < \omega_1} Q_\xi$. Now, let $\mathcal{E}_0 = \{Q_\xi: \xi < \omega_1 \ \& \ Q_\xi \in \text{Perf}(X)\}$. Then $|X \setminus \bigcup \mathcal{E}_0| \leq \omega_1$, so CPA$_{\text{cube}}$ is justified. ∎

2.1 CPA$_{\text{cube}}^{\text{game}}$ and disjoint coverings

The results presented in this section come from a paper [39] of K. Ciesielski and J. Pawlikowski.

Theorem 2.1.1 *Assume that* CPA$_{\text{cube}}^{\text{game}}$ *holds and let X be a Polish space. If $\mathcal{D} \subset \text{Perf}(X)$ is $\mathcal{F}_{\text{cube}}$-dense and it is closed under perfect subsets, then there exists a partition of X into ω_1 disjoint sets from $\mathcal{D} \cup \{\{x\}: x \in X\}$.*

In the proof we will use the following easy lemma.

Lemma 2.1.2 *Let $\mathcal{P} = \{P_i: i < \omega\} \subset \text{Perf}^*(X)$, where X is a Polish space. For every cube $P \in \text{Perf}(X)$ there exists a subcube Q of P such that either $Q \cap \bigcup_{i<\omega} P_i = \emptyset$ or $Q \subset P_i$ for some $i < \omega$.*

PROOF. Let $f \in \mathcal{F}_{\text{cube}}$ be such that $f[\mathfrak{C}^\omega] = P$.

If $P \cap \bigcup_{i<\omega} P_i$ is meager in P, then, by Claim 1.1.5, we can find a subcube Q of P such that $Q \subset P \setminus \bigcup_{i<\omega} P_i$.

If $P \cap \bigcup_{i<\omega} P_i$ is not meager in P, then there exists an $i < \omega$ such that $P \cap P_i$ has a nonempty interior in P. Thus, there exists a basic clopen set C in $\mathfrak{C}^\omega$, which is a perfect cube, such that $f[C] \subset P_i$. So, $Q = f[C]$ is a desired subcube of P. ∎

PROOF OF THEOREM 2.1.1. For a cube $P \in \text{Perf}(X)$ and a countable family $\mathcal{P} \subset \text{Perf}^*(X)$, let $D(P) \in \mathcal{D}$ be a subcube of P and $Q(\mathcal{P}, P) \in \mathcal{D}$ be as in Lemma 2.1.2 used with $D(P)$ in place of P. For a singleton $P \in \text{Perf}^*(X)$ we just put $Q(\mathcal{P}, P) = P$.

Consider the following strategy S for Player II:

$$S(\langle\langle P_\eta, Q_\eta\rangle : \eta < \xi\rangle, P_\xi) = Q(\{Q_\eta : \eta < \xi\}, P_\xi).$$

By CPA$_{\text{cube}}^{\text{game}}$, strategy S is not a winning strategy for Player II. So there exists a game $\langle\langle P_\xi, Q_\xi\rangle : \xi < \omega_1\rangle$ played according to S in which Player II looses, that is, $X = \bigcup_{\xi<\omega_1} Q_\xi$.

Notice that for every $\xi < \omega_1$ either $Q_\xi \cap \bigcup_{\eta<\xi} Q_\eta = \emptyset$ or there is an $\eta < \omega_1$ such that $Q_\xi \subset Q_\eta$. Let

$$\mathcal{F} = \left\{ Q_\xi : \xi < \omega_1 \ \& \ Q_\xi \cap \bigcup_{\eta<\xi} Q_\eta = \emptyset \right\}.$$

Then $\mathcal{F}$ is as desired. ∎

Since a family of all measure zero perfect subsets of $\mathbb{R}^n$ is $\mathcal{F}_{\text{cube}}$-dense, we get the following corollary.

Corollary 2.1.3 CPA$_{\text{cube}}^{\text{game}}$ *implies that there exists a partition of $\mathbb{R}^n$ into ω_1 many disjoint closed nowhere dense measure zero sets.*

Note that the conclusion of Corollary 2.1.3 does not follow from the fact that $\mathbb{R}^n$ can be covered by ω_1 perfect measure zero subsets. (See A. Miller [94, thm. 6].)

The next corollary is a generalization of Fact 1.1.7.

Corollary 2.1.4 CPA$_{\text{cube}}^{\text{game}}$ *implies that for every Borel subset B of a Polish space X there exists a family $\mathcal{P}$ of ω_1 many disjoint compact sets such that $B = \bigcup \mathcal{P}$.*

PROOF. This follows immediately from Theorem 2.1.1 and the fact that the family $\mathcal{E} = \{P \in \text{Perf}(X) : P \subset B \text{ or } P \cap B = \emptyset\}$ is $\mathcal{F}_{\text{cube}}$-dense. ∎

2.2 MAD families and the numbers $\mathfrak{a}$ and $\mathfrak{r}$

Recall that a family $\mathcal{A} \subset [\omega]^\omega$ is *almost disjoint* provided $|A \cap B| < \omega$ and it is *maximal almost disjoint*, *MAD*, provided it is not a proper subfamily of any other almost disjoint family. The cardinal number $\mathfrak{a}$ is defined as follows:

$$\mathfrak{a} = \min\{|\mathcal{A}|: \mathcal{A} \text{ is infinite and MAD}\}.$$

The fact that $\mathfrak{a} = \omega_1$ holds in the iterated perfect set model was apparently first noticed by Otmar Spinas (see A. Blass [9, sec. 11.5]), though it seems that the proof of this result was never provided. The argument presented below comes from K. Ciesielski and J. Pawlikowski [37].

Theorem 2.2.1 CPA$_{\text{cube}}^{\text{game}}$ *implies that* $\mathfrak{a} = \omega_1$.

Our proof of Theorem 2.2.1 is based on the following lemma.

Lemma 2.2.2 *For every countable infinite family* $\mathcal{W} \subset [\omega]^\omega$ *of almost disjoint sets and a cube* $P \in \text{Perf}([\omega]^\omega)$ *there exist a* $W \in [\omega]^\omega$ *and a subcube* Q *of* P *such that* $\mathcal{W} \cup \{W\}$ *is almost disjoint but* $\mathcal{W} \cup \{W, x\}$ *is not almost disjoint for every* $x \in Q$.

PROOF. Let $\mathcal{W} = \{W_i: i < \omega\}$. For every $i < \omega$ choose sets $V_i \subset W_i$ such that the V_i's are pairwise disjoint and each $W_i \setminus V_i$ is finite, but $V_\omega = \omega \setminus \bigcup_{i < \omega} V_i$ is infinite. (For example, for every $i < \omega$, put $V_i^* = W_i \setminus \bigcup_{j < i} W_j$ and $V_i = V_i^* \setminus \{\min V_i^*\}$.) Let

$$B = \{x \in P: (\forall i \le \omega)\, |x \cap V_i| < \omega\}$$

and notice that $B = \bigcap_{i \le \omega} \bigcup_{a \in [V_i]^{<\omega}} \{x \in P: x \cap V_i = a\}$ is a Borel subset of P, since each set $\{x \in P: x \cap V_i = a\}$ is closed. Since either B or $P \setminus B$ must be of the second category in P, by Claim 1.1.5 there is a subcube P^* of P such that either $P^* \subset B$ or $P^* \cap B = \emptyset$.

If $P^* \cap B = \emptyset$, then $W = V_\omega$ and $Q = P^*$ satisfy the conclusion of the lemma. So, suppose that $P^* \subset B$. Let $h: \mathfrak{C}^\omega \to P^*$, $h \in \mathcal{F}_{\text{cube}}$, be a coordinate function making P^* a cube, let λ be the standard product probability measure on $\mathfrak{C}^\omega$, and define a Borel measure μ on P^* by a formula $\mu(B) = \lambda(h^{-1}(B))$.

For $i, n < \omega$ let

$$P_i^n = \{x \in P^*: x \cap V_i \subset n\}.$$

Then all the sets $P_i^n = \bigcup_{a \subset n} \{x \in P^*: x \cap V_i = a\}$ are Borel (since each of the sets $\{x \in P^*: x \cap V_i = a\}$ is closed) and $P^* = \bigcup_{n < \omega} P_i^n$ for every

$i < \omega$. Thus for each $i < \omega$ there exists an $n(i) < \omega$ such that

$$\mu\left(P_i^{n(i)}\right) > 1 - 2^{-i}.$$

Then the set $T = \bigcup_{j<\omega} \bigcap_{j<i<\omega} P_i^{n(i)}$ has a μ-measure 1 so, by Claim 1.1.5, there is a subcube Q of P^* that is a subset of T. Let

$$W = \bigcup_{i<\omega} [V_i \cap n(i)].$$

We claim that W and Q satisfy the lemma.

It is obvious that W is almost disjoint with each W_i. So, fix an $x \in Q$. To finish the proof it is enough to show that

$$x \subseteq^* W.$$

But $x \in Q \subset \bigcup_{j<\omega} \bigcap_{j<i<\omega} P_i^{n(i)}$. Thus, there exists a $j < \omega$ such that $x \in \bigcap_{j<i<\omega} P_i^{n(i)}$. So,

$$x \cap \bigcup_{j<i<\omega} V_i = \bigcup_{j<i<\omega} (x \cap V_i) \subset \bigcup_{j<i<\omega} (V_i \cap n(i)) \subset W$$

and the set $x \setminus W \subset x \cap \left(V_\omega \cup \bigcup_{i\leq j} V_i\right) = (x \cap V_\omega) \cup \bigcup_{i\leq j}(x \cap V_i)$ is finite, as $x \in Q \subset P^* \subset \cdot B$. ∎

PROOF OF THEOREM 2.2.1. For a countably infinite almost disjoint family $\mathcal{W} \subset [\omega]^\omega$ and a cube $P \in \mathrm{Perf}([\omega]^\omega)$, let $W(\mathcal{W}, P) \in [\omega]^\omega$ and a subcube $Q(\mathcal{W}, P)$ of P be as in Lemma 2.2.2. For $P = \{x\} \in \mathrm{Perf}^*([\omega]^\omega)$, we put $Q(\mathcal{W}, P) = P$ and define $W(\mathcal{W}, P)$ as some arbitrary W almost disjoint with each set from $\mathcal{W}$ and such that $A \cap x$ is infinite for some $A \in \mathcal{W} \cup \{W\}$. (If $|x \cap V| < \omega$ for every $V \in \mathcal{W}$, we put $W = x$; otherwise, W is chosen as an arbitrary set almost disjoint with each set from $\mathcal{W}$.)

Let $\mathcal{A}_0 \subset [\omega]^\omega$ be an arbitrary infinite almost disjoint family and consider the following strategy S for Player II:

$$S(\langle\langle P_\eta, Q_\eta\rangle: \eta < \xi\rangle, P_\xi) = Q(\mathcal{A}_0 \cup \{W_\eta: \eta < \xi\}, P_\xi),$$

where the W_η's are defined inductively by $W_\eta = W(\mathcal{A}_0 \cup \{W_\zeta: \zeta < \eta\}, P_\eta)$. In other words, Player II remembers (recovers) the sets W_η associated with the sets P_η played so far, and he uses them (and Lemma 2.2.2) to get the next answer, $Q_\xi = Q(\mathcal{A}_0 \cup \{W_\eta: \eta < \xi\}, P_\xi)$, while remembering (or recovering each time) the set $W_\xi = W(\mathcal{A}_0 \cup \{W_\eta: \eta < \xi\}, P_\xi)$.

By CPA$_{\mathrm{cube}}^{\mathrm{game}}$, strategy S is not a winning strategy for Player II. So there exists a game $\langle\langle P_\xi, Q_\xi\rangle: \xi < \omega_1\rangle$ played according to S in which Player II loses, that is, $[\omega]^\omega = \bigcup_{\xi<\omega_1} Q_\xi$.

Now, notice that the family $\mathcal{A} = \mathcal{A}_0 \cup \{W_\xi : \xi < \omega_1\}$ is a MAD family. It is clear that $\mathcal{A}$ is almost disjoint, since every set W_ξ was chosen as almost disjoint with every set from $\mathcal{A}_0 \cup \{W_\zeta : \zeta < \xi\}$. To see that $\mathcal{A}$ is maximal it is enough to note that every $x \in [\omega]^\omega$ belongs to a Q_ξ for some $\xi < \omega_1$, and so there is an $A \in \mathcal{A}_0 \cup \{W_\eta : \eta \leq \xi\}$ such that $A \cap x$ is infinite.   ∎

By Theorem 2.2.1 we see that CPA$_{\text{cube}}^{\text{game}}$ implies the existence of a MAD family of size ω_1. Next we will show that such a family can be simultaneously a reaping family. This result is similar in flavor to that from Theorem 5.4.9.

Theorem 2.2.3 CPA$_{\text{cube}}^{\text{game}}$ *implies that there exists a family* $\mathcal{F} \subset [\omega]^\omega$ *of cardinality* ω_1 *that is simultaneously MAD and reaping.*

PROOF. The proof is just a slight modification of that for Theorem 2.2.1.

For a countable infinite almost disjoint family $\mathcal{W} \subset [\omega]^\omega$ and a cube $P \in \text{Perf}([\omega]^\omega)$, let $W_0 \in [\omega]^\omega$ and a subcube Q_0 of P be as in Lemma 2.2.2. Let $A \in [\omega]^\omega$ be almost disjoint with every set from $\mathcal{W} \cup \{W_0\}$. By Laver's theorem [85] we can also find a subcube Q_1 of Q_0 and a $W_1 \in [A]^\omega$ such that

- either $W_1 \cap x = \emptyset$ for every $x \in Q_1$,
- or else $W_1 \subset x$ for every $x \in Q_1$.

Let $Q(\mathcal{W}, P) = Q_1$ and $\mathcal{W}(\mathcal{W}, P) = \{W_0, W_1\}$. If $P \in \text{Perf}^*([\omega]^\omega)$ is a singleton, then we put $Q(\mathcal{W}, P) = P$ and we can easily find W_0 and W_1 satisfying the above conditions.

Let $\mathcal{A}_0 \subset [\omega]^\omega$ be an arbitrary infinite almost disjoint family and consider the following strategy S for Player II:

$$S(\langle \langle P_\eta, Q_\eta \rangle : \eta < \xi \rangle, P_\xi) = Q\left(\mathcal{A}_0 \cup \bigcup \{\mathcal{W}_\eta : \eta < \xi\}, P_\xi\right),$$

where the $\mathcal{W}_\eta$'s are defined as $\mathcal{W}_\eta = \mathcal{W}(\mathcal{A}_0 \cup \bigcup \{\mathcal{W}_\eta : \eta < \xi\}, P_\eta)$.

By CPA$_{\text{cube}}^{\text{game}}$, strategy S is not a winning strategy for Player II. So there exists a game $\langle \langle P_\xi, Q_\xi \rangle : \xi < \omega_1 \rangle$ played according to the strategy S in which Player II loses, that is, $[\omega]^\omega = \bigcup_{\xi < \omega_1} Q_\xi$. Then the family $\mathcal{F} = \mathcal{A}_0 \cup \bigcup \{\mathcal{W}_\xi : \xi < \omega_1\}$ is MAD and reaping.   ∎

2.3 Uncountable γ-sets and strongly meager sets

In this section we will prove that CPA$_{\text{cube}}^{\text{game}}$ implies the existence of an uncountable γ-set. We will also show that such a set can be, but need

not be, strongly meager. The results presented here were proved in [31] by
K. Ciesielski, A. Millán, and J. Pawlikowski.

Recall that a subset T of a Polish space X is a γ-*set* provided for every open ω-cover $\mathcal{U}$ of T there is a sequence $\langle U_n \in \mathcal{U}: n < \omega \rangle$ such that $T \subset \bigcup_{n<\omega} \bigcap_{i>n} U_i$, where $\mathcal{U}$ is an ω-*cover* of T if for every finite set $A \subset T$ there a $U \in \mathcal{U}$ with $A \subset U$. γ-sets were introduced by J. Gerlits and Zs. Nagy [64]. They were studied by F. Galvin and A. Miller [62], I. Recław [110], T. Bartoszyński and I. Recław [5], and others. It is known that under Martin's axiom there are γ-sets of cardinality continuum [62]. On the other hand, every γ-set is strong measure zero [64], so it is consistent with ZFC (in a model for Borel conjecture) that every γ-set is countable. Moreover, $\text{CPA}_{\text{cube}}^{\text{game}}$ implies that every γ-set has cardinality at most $\omega_1 < \mathfrak{c}$, since every strong measure zero is universally null, and so, by Theorem 1.1.4, under CPA_{cube} every γ-set has cardinality at most ω_1.

In the proofs of the next two theorems we will use the characterization of γ-sets due to I. Recław [110]. To formulate it we need to fix some terminology. Thus, we will consider $\mathcal{P}(\omega)$ as a Polish space by identifying it with 2^ω via characteristic functions. We say that a family $\mathcal{A} \subset \mathcal{P}(\omega)$ is *centered* provided $\bigcap \mathcal{A}_0$ is infinite for every finite $\mathcal{A}_0 \subset \mathcal{A}$, and $\mathcal{A}$ has a *pseudointersection* provided there exists a $B \in [\omega]^\omega$ such that $B \subseteq^* A$ for every $A \in \mathcal{A}$. In addition, for the rest of this section, $\mathcal{K}$ will stand for the family of all continuous functions from $\mathcal{P}(\omega)$ to $\mathcal{P}(\omega)$ and for $A \in \mathcal{P}(\omega)$ we put $A^* = \{B \in \mathcal{P}(\omega): B \subseteq^* A\}$.

Proposition 2.3.1 (I. Recław [110]) *For $T \subset \mathcal{P}(\omega)$, the following conditions are equivalent:*

(i) *T is a γ-set.*

(ii) *For every $f \in \mathcal{K}$, if $f[T]$ is centered, then $f[T]$ has a pseudointersection.*

In the proof that follows we will apply axiom $\text{CPA}_{\text{cube}}^{\text{game}}$ to the cubes from the space $\mathcal{K}$. The fact that the subcubes given by the axiom cover $\mathcal{K}$ will allow us to use the above characterization to conclude that the constructed set is indeed a γ-set. It is also possible to construct an uncountable γ-set by applying axiom $\text{CPA}_{\text{cube}}^{\text{game}}$ to the space $\mathcal{Y}$ of all ω-covers of $\mathcal{P}(\omega)$,[1] similarly as in the proof of Theorem 2.3.10. However, we believe that greater diversification of spaces to which we apply $\text{CPA}_{\text{cube}}^{\text{game}}$ makes the the presentation more interesting.

---

[1] More precisely, if $\mathcal{B}_0$ is a countable base for $\mathcal{P}(\omega)$ and $\mathcal{B}$ is the collection of all finite unions of elements from $\mathcal{B}_0$, then we can define $\mathcal{Y}$ as $\mathcal{B}^\omega$ considered with the product topology, where $\mathcal{B}$ is taken with discrete topology.

In what follows we will need the following two lemmas.

Lemma 2.3.2 *For every countable set* $Y \subset \mathcal{P}(\omega)$ *the set*

$$\mathcal{K}_Y = \{f \in \mathcal{K}: f[Y] \text{ is centered}\}$$

is Borel in $\mathcal{K}$.

PROOF. Let $Y = \{y_i: i < \omega\}$ and note that

$$\mathcal{K}_Y = \bigcap_{n,k<\omega} \bigcup_{m \geq k} \bigcap_{i<n} \{f \in \mathcal{K}: m \in f(y_i)\}.$$

So, $\mathcal{K}_Y$ is a G_δ set, since each set $\{f \in \mathcal{K}: m \in f(y_i)\}$ is open in $\mathcal{K}$. ∎

Lemma 2.3.3 *Let* $Y \subset \mathcal{P}(\omega)$ *be countable and such that* $[\omega]^{<\omega} \subset Y$. *For every* $W \in [\omega]^\omega$ *and a compact set* $Q \subset \mathcal{K}_Y$ *there exist* $V \in [W]^\omega$ *and a continuous function* $\varphi: Q \to [\omega]^\omega$ *such that* $\varphi(f)$ *is a pseudointersection of* $f[Y] \cup f[V^*]$ *for every* $f \in Q$.

Moreover, if $\mathcal{J}$ *is an infinite family of nonempty pairwise disjoint finite subsets of* W, *then we can choose* V *such that it contains infinitely many* J's *from* $\mathcal{J}$.

PROOF. First notice that there exists a continuous $\psi: Q \to [\omega]^\omega$ such that $\psi(f)$ is a pseudointersection of $f[Y]$ for every $f \in Q$.

Indeed, let $Y = \{y_i: i < \omega\}$ and for every $f \in Q$ let $\psi(f) = \left\{n_i^f: i < \omega\right\}$, where $n_0^f = \min f(y_0)$ and $n_{i+1}^f = \min\left\{n \in \bigcap_{j \leq i} f(y_j): n > n_i^f\right\}$. The set in the definition of a number n_{i+1}^f is nonempty, since $f[Y]$ is centered, as $f \in Q \subset \mathcal{K}_Y$. It is easy to see that ψ is continuous and that $\psi(f)$ is as desired.

We will define a sequence $\langle J_i \in \mathcal{J}: i < \omega \rangle$ such that $\max J_i < \min J_{i+1}$ for every $i < \omega$. We are aiming for $V = \bigcup_{i<\omega} J_i$.

A set $J_0 \in \mathcal{J}$ is chosen arbitrarily. Now, if J_i is already defined for some $i < \omega$, we define J_{i+1} as follows. Let $w_i = 1 + \max J_i$. Thus $J_i \subset w_i$. For every $f \in Q$ define

$$m_i^f = \min\left(\psi(f) \cap \bigcap f[\mathcal{P}(w_i)]\right).$$

The set $\psi(f) \cap \bigcap f[\mathcal{P}(w_i)]$ is infinite, since $\psi(f)$ is a pseudointersection of $f[Y]$ while $\mathcal{P}(w_i) \subset Y$. Let $k_i^f = \min K_i^f$, where

$$K_i^f = \left\{k \geq w_i: m_i^f \in f(a) \text{ for all } a \subset \omega \text{ with } a \cap k \subset w_i\right\}.$$

The fact that $K_i^f \neq \emptyset$ follows from the continuity of f since $m_i^f \in f(a)$ for all $a \subset w_i$. Notice that, by the continuity of ψ and the assignment of k_i^f, for every $p < \omega$ the set $U_p = \{f \in Q : k_i^f < p\}$ is open in Q. Since the sets $\{U_p : p < \omega\}$ form an increasing cover of Q, compactness of Q implies the existence of $p_i < \omega$ such that $Q \subset U_{p_i}$. Thus, $w_i \leq k_i^f < p_i$ for every $f \in Q$. We define J_{i+1} as an arbitrary element of $\mathcal{J}$ disjoint with p_i and notice that

$$m_i^f \in f(a) \text{ for every } f \in Q \text{ and } a \subset \omega \text{ with } a \cap \min J_{i+1} \subset w_i.$$

This finishes the inductive construction.

Let $V = \bigcup_{i<\omega} J_i \subset W$ and $\varphi(f) = \{m_i^f : i < \omega\}$. It is easy to see that φ is continuous (though we will not use this fact). To finish the proof it is enough to show that $\varphi(f)$ is a pseudointersection of $f[Y] \cup f[V^*]$ for every $f \in Q$.

So, fix an $f \in Q$. Clearly $\varphi(f) \subset \psi(f)$ is a pseudointersection of $f[Y]$ since so was $\psi(f)$. To see that $\varphi(f)$ is a pseudointersection of $f[V^*]$, take an $a \subseteq^* V$. Then, for almost all $i < \omega$ we have $a \cap \min J_{i+1} \subset w_i$, so that $m_i^f \in f(a)$. Thus $\varphi(f) \subseteq^* f(a)$. ∎

Theorem 2.3.4 CPA$_{\text{cube}}^{\text{game}}$ *implies that there is an uncountable γ-set in* $\mathcal{P}(\omega)$.

PROOF. For $\alpha < \omega_1$ and an $\subseteq^*$-decreasing sequence $\mathcal{V} = \{V_\xi \in [\omega]^\omega : \xi < \alpha\}$, let $W(\mathcal{V}) \in [\omega]^\omega$ be such that $W(\mathcal{V}) \subsetneq^* V_\xi$ for all $\xi < \alpha$. Moreover, if $P \in \text{Perf}^*(\mathcal{K})$ is a cube, then we define a subcube $Q = Q(\mathcal{V}, P)$ of P and an infinite subset $V = V(\mathcal{V}, P)$ of $W = W(\mathcal{V})$ as follows. Let $Y = \mathcal{V} \cup [\omega]^{<\omega}$ and choose a subcube Q of P such that either $Q \cap \mathcal{K}_Y = \emptyset$ or $Q \subset \mathcal{K}_Y$. This can be done by Claim 1.1.5 since $\mathcal{K}_Y$ is Borel. If $Q \cap \mathcal{K}_Y = \emptyset$, we put $V = W$; otherwise, we apply Lemma 2.3.3 to find V.

Consider the following strategy S for Player II:

$$S(\langle\langle P_\eta, Q_\eta\rangle : \eta < \xi\rangle, P_\xi) = Q(\{V_\eta : \eta < \xi\}, P_\xi),$$

where sets V_η are defined inductively by $V_\eta = V(\{V_\zeta : \zeta < \eta\}, P_\eta)$. In other words, Player II remembers (recovers) sets V_η associated with the cubes P_η played so far, and he uses them (and Lemma 2.3.3) to get the next answer, $Q_\xi = Q(\{V_\eta : \eta < \xi\}, P_\xi)$, while remembering (or recovering each time) the set $V_\xi = V(\{V_\eta : \eta < \xi\}, P_\xi)$.

By CPA$_{\text{cube}}^{\text{game}}$, strategy S is not a winning strategy for Player II. So there exists a game $\langle\langle P_\xi, Q_\xi\rangle : \xi < \omega_1\rangle$ played according to S in which Player II loses, that is, $\mathcal{K} = \bigcup_{\xi<\omega_1} Q_\xi$. Let $\mathcal{V} = \{V_\xi : \xi < \omega_1\}$ be a sequence associated with this game, which is strictly $\subseteq^*$-decreasing, and let $T = \mathcal{V} \cup [\omega]^{<\omega}$. We claim that T is a γ-set.

In the proof we use Lemma 2.3.2. So, let $f \in \mathcal{K}$ be such that $f[T]$ is centered. Then there exists an $\alpha < \omega_1$ such that $f \in Q_\alpha$. Since $f[\{V_\xi : \xi < \alpha\} \cup [\omega]^{<\omega}] \subset f[T]$ we must have applied Lemma 2.3.3 in the choice of Q_α and V_α. Therefore, the family $f[\{V_\xi : \xi < \alpha\} \cup [\omega]^{<\omega} \cup V_\alpha^*]$ has a pseudointersection. So, $f[T]$ has a pseudointersection too, since $T \subset \{V_\xi : \xi < \alpha\} \cup [\omega]^{<\omega} \cup V_\alpha^*$.   ■

Since $\mathcal{P}(\omega)$ embeds into any Polish space, we conclude that, under the axiom $\text{CPA}^{\text{game}}_{\text{cube}}$, any Polish space contains an uncountable γ-set. In particular, there exists an uncountable γ-set $T \subset \mathbb{R}$.

Recall (see, e.g., [4, p. 437]) that a subset X of $\mathbb{R}$ is *strongly meager* provided $X + G \neq \mathbb{R}$ for every measure zero subset G of $\mathbb{R}$. This is a notion that is dual to a strong measure zero subset of $\mathbb{R}$, since $X \subset \mathbb{R}$ is strong measure zero if and only if $X + M \neq \mathbb{R}$ for every meager subset M of $\mathbb{R}$.

Now, although every γ-set is strong measure zero, T. Bartoszyński and I. Recław [5] constructed, under Martin's axiom, a γ-set T in $\mathbb{R}$ that is not strongly meager. Next we will show that the existence of such a set follows also from $\text{CPA}^{\text{game}}_{\text{cube}}$. The construction is a generalization of that used in the proof of Theorem 2.3.4.

In the proof we will use the following notation. For $A, B \in \mathcal{P}(\omega)$ we define $A + B$ as the symmetric difference between A and B. Upon identification of a set $A \in \mathcal{P}(\omega)$ with its characteristic function $\chi_A \in 2^\omega$, this definition is motivated by the fact that $\chi_{A+B}(n) = \chi_A(n) +_2 \chi_B(n)$, where $+_2$ is the addition modulo 2. Also, let $\bar{\mathcal{J}} = \{J_n \in [\omega]^{2^n} : n < \omega\}$ be a family of pairwise disjoint sets and let $\tilde{G}$ be the family of all $W \subset \omega$ that are disjoint with infinitely many $J \in \bar{\mathcal{J}}$. Notice that $\tilde{G}$ has measure zero with respect to the standard measure on $\mathcal{P}(\omega)$ induced by the product measure on 2^ω.

Lemma 2.3.5 *If $\mathcal{J} \in [\bar{\mathcal{J}}]^\omega$ and P is a cube in $\mathcal{P}(\omega)$, then there is a subcube Q of P and a set $V \subset \bigcup \mathcal{J}$ containing infinitely many $J \in \mathcal{J}$ such that $V + Q \subset \tilde{G}$.*

PROOF. Let $D = \bigcup \mathcal{J}$ and

$$
\begin{aligned}
H &= \{\langle U, W \rangle \in \mathcal{P}(D) \times \mathcal{P}(\omega) : (U + W) \cap J = \emptyset \text{ for } \omega \text{ many } J \in \mathcal{J}\} \\
&\subseteq \{\langle U, W \rangle \in \mathcal{P}(D) \times \mathcal{P}(\omega) : U + W \in \tilde{G}\}.
\end{aligned}
$$

Notice that H is a G_δ subset of the product $\mathcal{P}(D) \times \mathcal{P}(\omega)$ since

$H_J = \{\langle U, W \rangle : (U + W) \cap J = \emptyset\}$ is open for every $J \in \mathcal{J}$. Moreover, the horizontal sections of H are dense in $\mathcal{P}(D)$. So, $\bar{H} = H \cap (\mathcal{P}(D) \times P)$ is a dense G_δ subset of $\mathcal{P}(D) \times P$, as all its horizontal sections are dense. Thus, by the Kuratowski-Ulam theorem, there is a dense G_δ subset $\mathcal{K}_0$ of $\mathcal{P}(D)$ such that for every $U \in \mathcal{K}_0$ the vertical section $\bar{H}_U$ of $\bar{H}$ is dense in P. Now, since

$$\mathcal{K}_1 = \{U \in \mathcal{P}(D) : J \subset U \text{ for infinitely many } J \in \mathcal{J}\}$$

is a dense G_δ, there exists a $V \in \mathcal{K}_0 \cap \mathcal{K}_1$. In particular, V contains infinitely many $J \in \mathcal{J}$ and $\bar{H}_V$ is a dense G_δ subset of P. Thus, by Claim 1.1.5, there exists a subcube Q of P contained in $\bar{H}_V$. Therefore, $Q \subset \bar{H}_V \subset \{W \in P : V + W \in \tilde{G}\}$, and so $V + Q \subset \tilde{G}$. $\blacksquare$

Theorem 2.3.6 $\mathrm{CPA}_{\mathrm{cube}}^{\mathrm{game}}$ *implies that there exists a γ-set $T \subset \mathcal{P}(\omega)$ such that $T + \tilde{G} = \mathcal{P}(\omega)$.*

PROOF. We will use $\mathrm{CPA}_{\mathrm{cube}}^{\mathrm{game}}$ for the space $X = \mathcal{K} \cup \mathcal{P}(\omega)$, a direct sum of $\mathcal{K}$ and $\mathcal{P}(\omega)$.

For $\alpha < \omega_1$ and an $\subseteq^*$-decreasing sequence $\mathcal{V} = \{V_\xi \in [\omega]^\omega : \xi < \alpha\}$ such that each V_ξ contains infinitely many $J \in \bar{\mathcal{J}}$, let $W(\mathcal{V}) \in [\omega]^\omega$ be such that $\mathcal{J} = \{J \in \bar{\mathcal{J}} : J \subset W(\mathcal{V})\}$ is infinite and $W(\mathcal{V}) \subsetneq^* V_\xi$ for all $\xi < \alpha$. For a cube $P \in \mathrm{Perf}^*(\mathcal{K})$ we define a subcube $Q = Q(\mathcal{V}, P)$ of P and an infinite subset $V = V(\mathcal{V}, P)$ of $W = W(\mathcal{V})$ as follows. By Claim 1.1.5 we can find subcube P' of P such that either $P' \subset \mathcal{K}$ or $P' \subset \mathcal{P}(\omega)$.

If $P' \subset \mathcal{K}$, we proceed as in the proof of Theorem 2.3.4. We define $Y = \mathcal{V} \cup [\omega]^{<\omega}$ and use Claim 1.1.5 to find a subcube Q of P' such that either $Q \cap \mathcal{K}_Y = \emptyset$ or $Q \subset \mathcal{K}_Y$. If $Q \cap \mathcal{K}_Y = \emptyset$, we put $V = W$; otherwise, we apply Lemma 2.3.3 to find V. If $P' \subset \mathcal{P}(\omega)$, we use Lemma 2.3.5 to find Q and V.

Consider the following strategy S for Player II:

$$S(\langle \langle P_\eta, Q_\eta \rangle : \eta < \xi \rangle, P_\xi) = Q(\{V_\eta : \eta < \xi\}, P_\xi),$$

where sets V_η are defined inductively by $V_\eta = V(\{V_\zeta : \zeta < \eta\}, P_\eta)$. By $\mathrm{CPA}_{\mathrm{cube}}^{\mathrm{game}}$, strategy S is not a winning strategy for Player II. So there exists a game $\langle \langle P_\xi, Q_\xi \rangle : \xi < \omega_1 \rangle$ played according to S in which Player II loses, that is, $X = \bigcup_{\xi < \omega_1} Q_\xi$. Let $\mathcal{V} = \{V_\xi : \xi < \omega_1\}$ be a sequence associated with this game, which is strictly $\subseteq^*$-decreasing, and let $T = \mathcal{V} \cup [\omega]^{<\omega}$. We claim that T is as desired.

The argument that T is a γ-set is the same as in the proof of Theorem 2.3.4. To see that $\mathcal{P}(\omega) \subset T + \tilde{G}$, notice that for every $A \in \mathcal{P}(\omega)$ there is an $\alpha < \omega_1$ such that $A \in Q_\alpha$. But then at step α we used Lemma 2.3.5 to find Q_α and V_α. In particular, $V_\alpha + Q_\alpha \subset \tilde{G}$. So, $A \in Q_\alpha \subset V_\alpha + \tilde{G} \subset T + \tilde{G}$. ∎

Corollary 2.3.7 $\text{CPA}_{\text{cube}}^{\text{game}}$ *implies that there exists a γ-set $X \subset \mathbb{R}$ that is not strongly meager.*

PROOF. This is the argument from [5]. Let T be as in Theorem 2.3.6 and let $f: \mathcal{P}(\omega) \to [0, 1]$, $f(A) = \sum_{i<\omega} 2^{-(i+1)} \chi_A(i)$. Then f is continuous, so $X = f[T]$ is a γ-set. Let $H = \bigcap_{m<\omega} \bigcup_{n>m} f[J_n]$. Then H has measure zero and it is easy to see that $[0, 1] = f[\mathcal{P}(\omega)] \subset f[T] + H = X + H$. Then $\bar{G} = H + \mathbb{Q}$ has measure zero and $X + \bar{G} = \mathbb{R}$. ∎

In the remainder of this section we will show that $\text{CPA}_{\text{cube}}^{\text{game}}$ also implies the existence of a γ-set $X \subset \mathbb{R}$ that is strongly meager.

Let X be a Polish space with topology τ. We say that $\mathcal{U} \subset \tau$ is a cover of $Z \subset [X]^{<\omega}$ provided for every $A \in Z$ there is a $U \in \mathcal{U}$ with $A \subset U$. Following [62], we say that a subset S of X is a *strong γ-set* provided there exists an increasing sequence $\langle k_n < \omega: n < \omega \rangle$ such that for every sequence $\langle J_n \subset \tau: n < \omega \rangle$, where each J_n is a cover of $[X]^{k_n}$, there exists a sequence $\langle D_n \in J_n: n < \omega \rangle$ with $X \subset \bigcup_{n<\omega} \bigcap_{m>n} D_m$. It is proved in [62] that every strong γ-set $X \subset \mathbb{R}$ is strongly meager. The goal of this section is to construct, under $\text{CPA}_{\text{cube}}^{\text{game}}$, an uncountable strong γ-set in $\mathcal{P}(\omega)$. So, after identifying $\mathcal{P}(\omega)$ with its homeomorphic copy in $\mathbb{R}$, this will become an uncountable γ-set in $\mathbb{R}$ that is strongly meager. Under Martin's axiom, a strong γ-set in $\mathcal{P}(\omega)$ of cardinality continuum exists; see [62].

Let $\mathcal{B}_0$ be a countable basis for the topology of $\mathcal{P}(\omega)$ and let $\mathcal{B}$ be the collection of all finite unions of elements from $\mathcal{B}_0$. Since every open cover of $[\mathcal{P}(\omega)]^k$, $k < \omega$, contains a refinement from $\mathcal{B}$, in the definition of strong γ-set it is enough to consider only sequences $\langle J_n: n < \omega \rangle$ with $J_n \subset \mathcal{B}$.

Now, consider $\mathcal{B}$ with the discrete topology. Since $\mathcal{B}$ is countable, the space $\mathcal{B}^\omega$, considered with the product topology, is a Polish space and so is $\mathcal{X} = (\mathcal{B}^\omega)^\omega$. For $J \in \mathcal{X}$ we will write J_n in place of $J(n)$. It is easy to see that a sub-basis for the topology of $\mathcal{X}$ is given for the clopen sets

$$\{J \in \mathcal{X}: J_n(m) = B\},$$

where $n, m < \omega$ and $B \in \mathcal{B}$.

For the remainder of this section we will fix a strictly increasing sequence $\langle k_n < \omega: n < \omega \rangle$ such that $k_n \geq n\, 2^n + n$ for every $n < \omega$. Then we have the following lemma.

Lemma 2.3.8 Let $X \in [\omega]^\omega$ and let F be a countable subset of $\mathcal{P}(\omega)$ such that $[\omega]^{<\omega} \subset F$. Assume that P is a compact subset of $\mathcal{X}$ such that for every $J \in P$ and $n < \omega$ the family $J_n[\omega] = \{J_n(m): m < \omega\}$ covers $[F]^{k_n}$. Then there exists a set $Y \in [X]^\omega$ and for each $J \in P$ a sequence $\langle D_n^J \in J_n: n < \omega \rangle$ such that $F \cup Y^* \subset \bigcup_{n<\omega} \bigcap_{m>n} D_m^J$.

PROOF. Let $\{F_n: n < \omega\}$ be an enumeration of $[\omega]^{<\omega}$ such that $F_n \subset n$ for all $n < \omega$ and let $F = \{f_n: n < \omega\}$. We will construct inductively the sequences $\langle s_n \in X: n < \omega \rangle$ and $\langle \{D_n^J \in J_n[\omega]: J \in P\}: n < \omega \rangle$ such that for every $n < \omega$, $J \in P$, and $A \subset \omega$ we have

(i) $\{f_i: i < n\} \subset D_n^J$ and $s_n < s_{n+1}$;
(ii) if $i < j \le n+1$ and $(A \cap s_{n+1}) \setminus \{s_0, \ldots, s_n\} = F_i$, then $A \in D_j^J$.

We chose $s_0 \in X$ and $\{D_n^J \in J_n[\omega]: J \in P\}$ arbitrarily. Then conditions (i) and (ii) are trivially satisfied. Next, assume that the sequence $\{s_i: i \le n\}$ is already constructed. We will construct s_{n+1} and sets D_{n+1}^J as follows. Let

$$Q = \{q \in [\omega]^{<\omega}: q \setminus \{s_0, \ldots, s_n\} = F_i \text{ for some } i \le n\}.$$

Then $|Q| \le (n+1)\, 2^{n+1}$ and $|Q \cup \{f_0, \ldots, f_n\}| \le k_{n+1}$.

Fix $J \in P$. Since $J_{n+1}[\omega]$ covers $[F]^{\le k_{n+1}}$, there exists a $\bar{D}_{n+1}^J \in J_{n+1}[\omega]$ containing $Q \cup \{f_0, \ldots, f_n\}$. Since $\bar{D}_{n+1}^J$ is open and covers a finite set Q, there is an $s_{n+1}^J > s_n$ in X such that for every $q \in Q$

$$\{x \subset \omega: x \cap s_{n+1}^J = q \cap s_{n+1}^J\} \subset \bar{D}_{n+1}^J.$$

Notice that:

(∗) For every $A \subset \omega$ and $\bar{s}_{n+1} \ge s_{n+1}^J$ condition (ii) holds.

Indeed, assume that $(A \cap \bar{s}_{n+1}) \setminus \{s_0, \ldots, s_n\} = F_i$ for some $i < j \le n+1$. If $j \le n$, then $n \ge 1$ and since $F_i \subset i \subset s_{n-1}$ we have

$$(A \cap s_n) \setminus \{s_0, \ldots, s_{n-1}\} = (A \cap \bar{s}_{n+1}) \setminus \{s_0, \ldots, s_n\} = F_i.$$

Therefore, by the inductive assumption, $A \in D_j^J$. Now, if $j = n+1$, then $q = A \cap \bar{s}_{n+1} \in Q$. Thus $A \in \{x \subset \omega: x \cap \bar{s}_{n+1} = q \cap \bar{s}_{n+1}\} \subset \{x \subset \omega: x \cap s_{n+1}^J = q \cap s_{n+1}^J\} \subset \bar{D}_{n+1}^J$, finishing the proof of (∗).

For each $J \in P$ let $m^J < \omega$ be such that $J_{n+1}(m^J) = \bar{D}_{n+1}^J$ and define $U_J = \{K \in \mathcal{X}: K_{n+1}(m^J) = \bar{D}_{n+1}^J\}$. Then U_J is an open neighborhood of J. In particular, $\{U_J: J \in P\}$ is an open cover of a compact set P, so there exists a finite $P_0 \subset P$ such that $P \subset \bigcup\{U_{\bar{J}}: \bar{J} \in P_0\}$. Choose $s_{n+1} \in X$ such that $s_{n+1} \ge \max\{s_{n+1}^{\bar{J}}: \bar{J} \in P_0\}$. Moreover, for every $J \in P$ choose

$\bar{J} \in P_0$ such that $J \in U_{\bar{J}}$ and define $D_{n+1}^J = \bar{D}_{n+1}^{\bar{J}}$. It is easy to see that, by $(*)$, conditions (i) and (ii) are preserved. This completes the inductive construction.

Put $Y = \{s_n : n < \omega\}$. To see that it satisfies the lemma, pick an arbitrary $J \in P$. We will show that $F \cup Y^* \subset \bigcup_{n<\omega} \bigcap_{m>n} D_m^J$.

Clearly $F \subset \bigcup_{n<\omega} \bigcap_{m>n} D_m^J$ since, by (i), $f_n \in D_m^J$ for every $m > n$. So, fix an $A \in Y^*$. Then $A \setminus Y = F_i$ for some $i < \omega$. Let $n < \omega$ be such that $i < n$ and $s_n > \max F_i$. Then for every $m > n$ we have $i < m \le m+1$ and $(A \cap s_{m+1}) \setminus \{s_0, \ldots, s_m\} = F_i$. So, by (ii), we have $A \in D_m^J$ for every $m > n$. Thus, $A \in \bigcap_{m>n} D_m^J$. ∎

Lemma 2.3.9 If $F \subset \mathcal{P}(\omega)$ is countable, then the set

$$\mathcal{X}_F = \{J \in \mathcal{X} : J_n[\omega] \text{ covers } [F]^{k_n} \text{ for every } n < \omega\}$$

is Borel in $\mathcal{X}$.

PROOF. This follows from the fact that

$$\mathcal{X}_F = \bigcap_{n<\omega} \bigcap_{A \in [F]^{k_n}} \bigcup_{m<\omega} \bigcup_{A \subset B \in \mathcal{B}} \{J \in \mathcal{X} : J_n(m) = B\}$$

since each set $\{J \in \mathcal{X} : J_n(m) = B\}$ is clopen in $\mathcal{X}$. Thus, $\mathcal{X}_F$ is a G_δ-set. ∎

Theorem 2.3.10 CPA$_{\text{cube}}^{\text{game}}$ implies that there exists an uncountable strong γ-set in $\mathcal{P}(\omega)$.

PROOF. For $\alpha < \omega_1$ and an $\subseteq^*$-decreasing sequence $\mathcal{V} = \{V_\xi \in [\omega]^\omega : \xi < \alpha\}$ let $W(\mathcal{V}) \in [\omega]^\omega$ be such that $W(\mathcal{V}) \subsetneq^* V_\xi$ for all $\xi < \alpha$. Moreover, if $P \in \text{Perf}^*(\mathcal{X})$ is a cube, then we define a subcube $Q = Q(\mathcal{V}, P)$ of P and an infinite subset $Y = V(\mathcal{V}, P)$ of $X = W(\mathcal{V})$ as follows. Let $F = \mathcal{V} \cup [\omega]^{<\omega}$ and choose a subcube Q of P such that either $Q \cap \mathcal{X}_F = \emptyset$ or $Q \subset \mathcal{X}_F$. This can be done by Claim 1.1.5 since $\mathcal{X}_F$ is Borel. If $Q \cap \mathcal{X}_F = \emptyset$, we put $Y = X$; otherwise, we apply Lemma 2.3.8 to find Y.

Consider the following strategy S for Player II:

$$S(\langle\langle P_\eta, Q_\eta\rangle : \eta < \xi\rangle, P_\xi) = Q(\{V_\eta : \eta < \xi\}, P_\xi),$$

where sets V_η are defined inductively by $V_\eta = V(\{V_\zeta : \zeta < \eta\}, P_\eta)$. By CPA$_{\text{cube}}^{\text{game}}$, strategy S is not a winning strategy for Player II. So there exists a game $\langle\langle P_\xi, Q_\xi\rangle : \xi < \omega_1\rangle$ played according to S in which Player II loses, that is, $\mathcal{X} = \bigcup_{\xi<\omega_1} Q_\xi$. Let $\mathcal{V} = \{V_\xi : \xi < \omega_1\}$ be a sequence associated with this game, which is strictly $\subseteq^*$-decreasing, and let $T = \mathcal{V} \cup [\omega]^{<\omega}$. We claim that T is a strong γ-set.

Indeed, let $\langle \mathcal{U}_n \subset \mathcal{B} : n < \omega \rangle$ be such that $\mathcal{U}_n$ covers $[T]^{k_n}$ for every $n < \omega$. Then there is a $J \in \mathcal{X}$ such that $J_n[\omega] = \mathcal{U}_n$ for every $n < \omega$. Let $\alpha < \omega_1$ be such that $J \in Q_\alpha$. Then $J \in \mathcal{X}_{\{V_\eta : \eta < \alpha\} \cup [\omega]^{<\omega}}$, so we must have used Lemma 2.3.8 to get Q_α. In particular, there is a sequence $\langle D_n^J \in J_n[\omega] = \mathcal{U}_n : n < \omega \rangle$ such that

$$\left([\omega]^{<\omega} \cup \{V_\eta : \eta < \alpha\} \right) \cup (V_\alpha)^* \subset \bigcup_{n<\omega} \bigcap_{m>n} D_m^J.$$

So, $T \subset \bigcup_{n<\omega} \bigcap_{m>n} D_m^J$, as $\{V_\eta : \alpha \le \eta < \omega_1\} \subset (V_\alpha)^*$.   ∎

Since every homeomorphic image of a strong γ-set is evidently a strong γ-set, we immediately obtain the following conclusion.

Corollary 2.3.11 $\mathrm{CPA}_{\mathrm{cube}}^{\mathrm{game}}$ *implies that there exists an uncountable γ-set in $\mathbb{R}$ that is strongly meager.*

It is worth mentioning that a construction of an uncountable strong γ-set in $\mathcal{P}(\omega)$ under $\mathrm{CPA}_{\mathrm{cube}}^{\mathrm{game}}$ can also be obtained in a formalism similar to that used in the proofs of Theorems 2.3.4 and 2.3.6. In order to do this, we need the following definitions and facts. For a fixed sequence $\bar{k} = \langle k_n < \omega : n < \omega \rangle$ we say that $\mathcal{A} \subset (\mathcal{P}(\omega))^\omega$ is $\bar{k}$-*centered* provided for every $n < \omega$ any k_n many sets from $\{A(n) : A \in \mathcal{A}\}$ have a common point and $B \in \omega^\omega$ is a *quasi-intersection* of $\mathcal{A} \subset (\mathcal{P}(\omega))^\omega$ provided for every $A \in \mathcal{A}$ for all but finitely many $n < \omega$ we have $B(n) \in A(n)$. Now, if $\mathcal{K}^*$ is a family of all continuous functions from $\mathcal{P}(\omega)$ to $(\mathcal{P}(\omega))^\omega$, then the following is true:

A set $X \subset \mathcal{P}(\omega)$ is a strong γ-set if and only if there exists an increasing sequence $\bar{k} = \langle k_n < \omega : n < \omega \rangle$ such that for every $f \in \mathcal{K}^*$, if $f[X]$ is $\bar{k}$-centered, then $f[X]$ has a quasi-intersection.

With this characterization in hand we can construct an uncountable strong γ-set in $\mathcal{P}(\omega)$ by applying $\mathrm{CPA}_{\mathrm{cube}}^{\mathrm{game}}$ to the space $\mathcal{K}^*$.

2.4 Nowhere meager set $A \times A \subset \mathbb{R}^2$ intersecting continuous functions on a small set

The goal of this section is to prove the following theorem, in which id stands for the identity function.

Theorem 2.4.1 (K. Ciesielski and T. Natkaniec [33]) *Assume that* $\mathrm{CPA}^{\mathrm{game}}_{\mathrm{cube}}$ *holds. Then there exists a nowhere meager set* $A \subset \mathbb{R}$ *of cardinality less than* $\mathfrak{c}$ *such that for every* $f \in \mathcal{C}(\mathbb{R})$ *there exists a countable set* $Z \subset A$ *such that*

$$f \cap (A \times A) \subset (A \times Z) \cup \mathrm{id}.$$

In particular, for every nowhere constant continuous function $f \colon \mathbb{R} \to \mathbb{R}$

- *the set* $\{c \in \mathbb{R} \colon \pi[(f + c) \cap (A \times A)] \notin \mathcal{M}\}$ *is countable, and*
- *the set* $\{c \in \mathbb{R} \colon (f + c) \cap (A \times A) = \emptyset\}$ *is a complement of a set of cardinality* $\omega_1 < \mathfrak{c}$, *so it contains a Bernstein set.*

The study of the subsets $A \times A$ of the plain with similar properties (for the linear functions $f(x) = ax + b$) was initiated by J. Ceder and D.K. Ganguly in [22]. Similar examples have been also given by R.O. Davies [48] and T. Natkaniec [103, 104]. It is also known [33, thm. 4] that such a set cannot be constructed in ZFC even if we allow it to have cardinality $\mathfrak{c}$ and restrict our attention to functions f that are homeomorphisms.

We will apply the axiom to space $\mathcal{C} = \mathcal{C}(\mathbb{R})$ considered with the sup norm. Notice that for every $Q \subset \mathcal{C}$ the set $\bigcup Q \subset \mathbb{R}^2$ is a union of the graphs of all functions belonging to Q, since functions are identified with their graphs. For $A \in [\mathbb{R}]^{\leq \omega}$, let $\mathcal{E}(A)$ be the family of all $Q \in \mathrm{Perf}(\mathcal{C})$ such that

- $\bigcup Q$ is nowhere dense in $\mathbb{R}^2$, and
- for every $x \in A$ the vertical section $[\bigcup Q]_x$ of $\bigcup Q$ is nowhere dense in $\mathbb{R}$.

Lemma 2.4.2 *Family* $\mathcal{E}(A)$ *is cube-dense for every* $A \in [\mathbb{R}]^{\leq \omega}$.

PROOF. Without loss of generality we can assume that A is dense in $\mathbb{R}$. Let $P \in \mathrm{Perf}(\mathcal{C})$ and let μ be a countably additive Borel probability measure μ on P. By Claim 1.1.5 it is enough to show that there exists a $Q \in \mathrm{Perf}(P) \cap \mathcal{E}(A)$ such that $\mu(Q) > 0$. To see this, fix a countable base $\mathcal{B}$ for $\mathbb{R}$ and let $\langle \langle a_n, J_n \rangle \colon n < \omega \rangle$ be an enumeration of $A \times \mathcal{B}$. Notice that for every $n < \omega$ there exists a nonempty open set $U_n \subset J_n$ such that

$$\mu(\{f \in P \colon f(a_n) \in U_n\}) < 2^{-(n+2)}. \tag{2.2}$$

Indeed, if $\mathcal{U}_n$ is an infinite family of nonempty pairwise disjoint open subsets of J_n, then for each $U \in \mathcal{U}_n$ the set $\{f \in P \colon f(a_n) \in U\}$ is open in P (so μ-measurable) and so condition (2.2) must hold for some $U \in \mathcal{U}_n$. Let W be equal to $\bigcup_{n<\omega}\{f \in \mathcal{C} \colon f(a_n) \in U_n\}$. It is clear that W is open and dense in $\mathcal{C}$. So, $Q = P \setminus W = P \setminus \bigcup_{n<\omega}\{f \in P \colon f(a_n) \in U_n\}$ is nowhere dense (and therefore $\bigcup Q$ is nowhere dense in $\mathbb{R}^2$), and, by (2.2), it has μ-measure at least $1 - \sum_{n<\omega} 2^{-(n+2)} = 2^{-1} > 0$. It is also clear that for every $x \in A$ the set $\bigcup\{U_n : a_n = x\}$ is dense open in $\mathbb{R}$ and disjoint with $[\bigcup Q]_x$. Thus $Q \in \mathcal{E}(A)$. ∎

Proposition 2.4.3 *Assume that* $\mathrm{CPA}_{\mathrm{cube}}^{\mathrm{game}}$ *holds, let X be a Polish space, and let S be a mapping associating to every $\bar{P} \in \bigcup_{\alpha<\omega_1} (\mathrm{Perf}^*(X))^\alpha$ a cube-dense family $\mathcal{E}(\bar{P}) \subset \mathrm{Perf}^*(X)$. Then there is a sequence $\langle \langle P_\xi, Q_\xi \rangle : \xi < \omega_1 \rangle$ such that for every $\xi < \omega_1$ we have $Q_\xi \in \mathrm{Perf}^*(P_\xi) \cap \mathcal{E}(\langle P_\zeta : \zeta < \xi \rangle)$ and $X = \bigcup_{\xi<\omega_1} Q_\xi$.*

PROOF. This follows immediately from $\mathrm{CPA}_{\mathrm{cube}}^{\mathrm{game}}$. More precisely, it is enough to apply the axiom $\mathrm{CPA}_{\mathrm{cube}}^{\mathrm{game}}$ to the strategy S^* such that $S^*(\langle \langle P_\eta, Q_\eta \rangle : \eta < \xi \rangle, P_\xi)$ is a subcube of P_ξ from $S(\langle P_\eta : \eta < \xi \rangle)$. ∎

PROOF OF THEOREM 2.4.1. First recall that, under $\mathrm{CPA}_{\mathrm{cube}}$ (so also under $\mathrm{CPA}_{\mathrm{cube}}^{\mathrm{game}}$), we have $\mathrm{cof}(\mathcal{N}) = \omega_1$. Let $\mathcal{B}_0$ be a countable base for $\mathbb{R}$ and let $\{\langle M_\xi, J_\xi \rangle : \xi < \omega_1\}$ be an enumeration of $\mathcal{M}_0 \times \mathcal{B}_0$. By simultaneous induction on $\xi < \omega_1$, using Lemma 2.4.2, we will define the functions S, Q, and k on $(\mathrm{Perf}^*(\mathcal{C}))^\xi$ such that:

(i) $S(\langle P_\zeta : \zeta < \xi \rangle) = \mathcal{E}(\{a_\zeta : \zeta < \xi\})$, where $a_\zeta = k(\langle P_\eta : \eta \leq \zeta \rangle) \in \mathbb{R}$, and $Q_\xi = Q(\langle P_\zeta : \zeta < \xi \rangle) \in \mathcal{E}(\{a_\zeta : \zeta < \xi\})$.

(ii) $k(\langle P_\zeta : \zeta \leq \xi \rangle)$ belongs to J_ξ and the comeager set

$$\bigcap_{\zeta \leq \xi} \{z \in \mathbb{R} : (\bigcup Q_\zeta)^z \text{ and } (\bigcup Q_\zeta)_z \text{ are nowhere dense in } \mathbb{R}\}.$$

(iii) $k(\langle P_\zeta : \zeta \leq \xi \rangle)$ does not belong to the meager set

$$M_\xi \cup \bigcup_{\eta \leq \xi} (\{(\bigcup Q_\eta)_{a_\zeta} : \zeta < \xi\} \cup \{(\bigcup Q_\eta)^{a_\zeta} : \eta \leq \zeta < \xi\}).$$

The set as in (ii) is comeager by the Kuratowski-Ulam theorem, since each set $\bigcup Q_\zeta$ is nowhere dense, as Q_ζ belongs to some $\mathcal{E}(A)$. In (iii), for every $\eta \leq \zeta < \xi$ the set $(\bigcup Q_\eta)_{a_\zeta} \cup (\bigcup Q_\eta)^{a_\zeta}$ is nowhere dense by the choice of $a_\zeta = k(\langle P_\eta : \eta \leq \zeta \rangle)$ as in (ii). Finally, for $\zeta < \eta$, the set $(\bigcup Q_\eta)_{a_\zeta}$ is nowhere dense since, by (i), Q_η belongs to $S(\langle P_\zeta : \zeta < \eta \rangle) = \mathcal{E}(\{a_\zeta : \zeta < \eta\})$. Now, by the axiom $\mathrm{CPA}_{\mathrm{cube}}^{\mathrm{game}}$ and Proposition 2.4.3, there exists a sequence $\langle \langle P_\xi, Q_\xi, a_\xi \rangle : \xi < \omega_1 \rangle$ such that $\mathcal{C} = \bigcup_{\xi<\omega_1} Q_\xi$ and conditions (i)–(iii) are satisfied. We claim that $A = \{a_\xi : \xi < \omega_1\}$ satisfies Theorem 2.4.1. Clearly, A is nowhere meager since for every nonempty open set $U \subset \mathbb{R}$ and every meager set M there exists a $\xi < \omega_1$ such that $J_\xi \subset U$ and $M \subset M_\xi$.

But then $a_\xi \in (A \cap J_\xi) \setminus M_\xi \subset (A \cap U) \setminus M$, so $A \cap U \neq M$. To see the main part of Theorem 2.4.1, take an $f \in \mathcal{C}$. Then, there exists a $\eta < \omega_1$ such that $f \in Q_\eta$. We claim that for $Z = \{a_\beta : \beta < \eta\}$ we have $f \cap (A \times A) \subset (A \times Z) \cup \text{id}$. Indeed, let $\eta \leq \xi < \omega_1$ and $\zeta < \omega_1$ be such that $\zeta \neq \xi$. We need to show that $\langle a_\zeta, a_\xi \rangle \notin f$. But if $\zeta < \xi$, then, by (iii), a_ξ does not belong to $[\bigcup Q_\eta]_{a_\zeta} \ni f(a_\zeta)$, so $\langle a_\zeta, a_\xi \rangle \notin f$. Similarly, if $\xi < \zeta$, then, again by (iii), a_ζ does not belong to $[\bigcup Q_\eta]^{a_\xi} \supset f^{-1}(a_\xi)$ and once more $\langle a_\zeta, a_\xi \rangle \notin f$.   ∎

2.5 Remark on a form of CPA$_{\text{cube}}^{\text{game}}$

Notice also that, if CPA$_{\text{cube}}^{\text{game}}[X]$ stands for CPA$_{\text{cube}}^{\text{game}}$ for a fixed Polish space X, then, similarly as in Remark 1.8.3 we can also prove

Remark 2.5.1 For any Polish space X, axiom CPA$_{\text{cube}}^{\text{game}}[X]$ implies the full axiom CPA$_{\text{cube}}^{\text{game}}$.

Note also that if we remove from CPA$_{\text{cube}}^{\text{game}}$ the assumption that $\mathfrak{c} = \omega_2$, then the resulting statement becomes trivial:

Remark 2.5.2 If CH holds, then for any Polish space X Player I has a winning strategy in the game GAME$_{\text{cube}}(X)$.

To win Player I simply needs to play consecutively all constant cubes.

On the other hand, by Remark 1.8.2, the assumption that $\mathfrak{c} \geq \omega_2$ is all we really need to put in the axiom:

Remark 2.5.3 CPA$_{\text{cube}}^{\text{game}}$ follows from:

weak-CPA$_{\text{cube}}^{\text{game}}$: $\mathfrak{c} \geq \omega_2$, and for any Polish space X Player II has no winning strategy in the game GAME$_{\text{cube}}(X)$.

3

Prisms and axioms $\text{CPA}_{\text{prism}}^{\text{game}}$ and $\text{CPA}_{\text{prism}}$

The axioms $\text{CPA}_{\text{cube}}^{\text{game}}$ and CPA_{cube} deal with the notion of $\mathcal{F}_{\text{cube}}$-density, where $\mathcal{F}_{\text{cube}}$ is the family of all injections $f\colon C \to X$ with C being a perfect cube in $\mathfrak{C}^\omega$. In the applications of these axioms we were using the facts that different subfamilies of $\text{Perf}(X)$ are $\mathcal{F}_{\text{cube}}$-dense. Unfortunately, in many cases, the notion of $\mathcal{F}_{\text{cube}}$-density is too weak to do the job — in the applications that follow, the families $\mathcal{E} \subset \text{Perf}(X)$ will not be $\mathcal{F}_{\text{cube}}$-dense, but they will be dense in a weaker sense defined below. Luckily, this weaker notion of density still leads to consistent axioms.

To define this weaker notion of density, let us first take another look at the notion of "cube." Let A be a nonempty countable set of ordinal numbers. The notion of a perfect cube in $\mathfrak{C}^A$ can be defined the same way as it was done for $\mathfrak{C}^\omega$. However, it will be more convenient for us to define it as follows. Let $\Phi_{\text{cube}}(A)$ be the family of all continuous injections $f\colon \mathfrak{C}^A \to \mathfrak{C}^A$ such that

$$f(x)(\alpha) = f(y)(\alpha) \quad \text{for all } \alpha \in A \text{ and } x, y \in \mathfrak{C}^A \text{ with } x(\alpha) = y(\alpha).$$

In other words, $\Phi_{\text{cube}}(A)$ is the family of all functions of the form $f = \langle f_\alpha \rangle_{\alpha \in A}$, where each f_α is an injection from $\mathfrak{C}$ into $\mathfrak{C}$. Then the family of all perfect cubes in $\mathfrak{C}^A$ for an appropriate A is equal to

$$\text{CUBE}(A) = \{\text{range}(f)\colon f \in \Phi_{\text{cube}}(A)\},$$

while the family $\mathcal{F}_{\text{cube}}$ defined in the first chapter consists of all continuous injections $f\colon C \to X$ with $C \in \text{CUBE}(\omega)$.

In the definitions that follow, the notion of a "cube" will be replaced by that of a "prism." So, let $\Phi_{\text{prism}}(A)$ be the family of all continuous injections $f\colon \mathfrak{C}^A \to \mathfrak{C}^A$ with the property that

$$f(x) \restriction \alpha = f(y) \restriction \alpha \iff x \restriction \alpha = y \restriction \alpha \quad \text{for all } \alpha \in A \text{ and } x, y \in \mathfrak{C}^A \tag{3.1}$$

49

or, equivalently, such that, for every $\alpha \in A$,

$$f \upharpoonright\upharpoonright \alpha \overset{\text{def}}{=} \{\langle x \upharpoonright \alpha, y \upharpoonright \alpha \rangle : \langle x, y \rangle \in f\}$$

is a one to one function from $\mathfrak{C}^{A \cap \alpha}$ into $\mathfrak{C}^{A \cap \alpha}$. For example, if $A = \{0, 1, 2\}$, then function f belongs to $\Phi_{\text{prism}}(A)$ provided there exist continuous functions $f_0 \colon \mathfrak{C} \to \mathfrak{C}$, $f_1 \colon \mathfrak{C}^2 \to \mathfrak{C}$, and $f_2 \colon \mathfrak{C}^3 \to \mathfrak{C}$ such that $f(x_0, x_1, x_2) = \langle f_0(x_0), f_1(x_0, x_1), f_2(x_0, x_1, x_2) \rangle$ for all $x_0, x_1, x_2 \in \mathfrak{C}$ and maps f_0, $\langle f_0, f_1 \rangle$, and f are one to one. Functions f from $\Phi_{\text{prism}}(A)$ were first introduced, in a more general setting, in [73], where they are called *projection-keeping homeomorphisms*. Note that

$$\Phi_{\text{prism}}(A) \text{ is closed under compositions} \tag{3.2}$$

and tha,t for every ordinal number $\alpha > 0$,

$$\text{if } f \in \Phi_{\text{prism}}(A), \text{ then } f \upharpoonright\upharpoonright \alpha \in \Phi_{\text{prism}}(A \cap \alpha). \tag{3.3}$$

Let

$$\mathbb{P}_A = \{\text{range}(f) \colon f \in \Phi_{\text{prism}}(A)\}.$$

We will write Φ_{prism} for $\bigcup_{0 < \alpha < \omega_1} \Phi_{\text{prism}}(\alpha)$ and define

$$\mathbb{P}_{\omega_1} \overset{\text{def}}{=} \bigcup_{0 < \alpha < \omega_1} \mathbb{P}_\alpha = \{\text{range}(f) \colon f \in \Phi_{\text{prism}}\}.$$

Following [73], we will refer to elements of $\mathbb{P}_{\omega_1}$ as *iterated perfect sets*. (In [131], the elements of $\mathbb{P}_\alpha$ are called I-perfect, where I is the ideal of countable sets.)

Let $\mathcal{F}_{\text{prism}}(X)$ (or just $\mathcal{F}_{\text{prism}}$, if X is clear from the context) be the family of all continuous injections $f \colon E \to X$, where $E \in \mathbb{P}_{\omega_1}$ and X is a fixed Polish space. We adopt the shortcuts similar to those for cubes. Thus, we say that $P \in \text{Perf}(X)$ is a *prism* if we consider it with an (implicitly given) witness function $f \in \mathcal{F}_{\text{prism}}$ onto P. Then Q is a *subprism of a prism* P provided $Q = f[E]$, where $E \in \mathbb{P}_\alpha$ and $E \subset \text{dom}(f)$. Also, singletons $\{x\}$ in X will be identified with constant functions from $E \in \mathbb{P}_{\omega_1}$ to $\{x\}$, and these functions will be considered as elements of $\mathcal{C}_{\text{prism}} \subset \mathcal{F}_{\text{prism}}^*$, similarly as in (2.1).

Following the schema presented in (1.9), we say that a family $\mathcal{E} \subset \text{Perf}(X)$ is $\mathcal{F}_{\text{prism}}$-*dense* provided

$$\forall f \in \mathcal{F}_{\text{prism}} \ \exists g \in \mathcal{F}_{\text{prism}} \ (g \subset f \ \& \ \text{range}(g) \in \mathcal{E}).$$

Similarly as in Fact 1.0.2, using (3.2) we can also prove that:

Fact 3.0.1 $\mathcal{E} \subset \text{Perf}(X)$ *is* $\mathcal{F}_{\text{prism}}$*-dense if and only if*

$$\forall \alpha < \omega_1 \; \forall f \in \mathcal{F}_{\text{prism}}, \; \text{dom}(f) = \mathfrak{C}^\alpha \; \exists g \in \mathcal{F}_{\text{prism}} \; (g \subset f \; \& \; \text{range}(g) \in \mathcal{E}).$$

Notice also that $\Phi_{\text{cube}}(A) \subset \Phi_{\text{prism}}(A)$, so every cube is also a prism. From this and Fact 3.0.1 (see also Fact 1.8.5) it is also easy to see that

$$\text{if } \mathcal{E} \subset \text{Perf}(X) \text{ is } \mathcal{F}_{\text{cube}}\text{-dense, then } \mathcal{E} \text{ is also } \mathcal{F}_{\text{prism}}\text{-dense.} \qquad (3.4)$$

The converse of (3.4), however, is false. (See Remark 3.2.6.)

Now we are ready to state the next version of our axiom, in which the game GAME$_{\text{prism}}(X)$ is an obvious generalization of GAME$_{\text{cube}}(X)$.

CPA$_{\text{prism}}^{\text{game}}$: $\mathfrak{c} = \omega_2$, and for any Polish space X Player II has no winning strategy in the game GAME$_{\text{prism}}(X)$.

Remark 3.0.2 In order to apply CPA$_{\text{prism}}^{\text{game}}$, we will always construct some strategy S for Player II and then use the axiom to conclude that, since S is not winning, there exists a game $\langle\langle P_\xi, Q_\xi\rangle : \xi < \omega_1\rangle$ played according to S in which Player I wins. But in any such game, for every $\xi < \omega_1$, we have

$$Q_\xi = S(\langle\langle P_\eta, Q_\eta\rangle : \eta < \xi\rangle, P_\xi). \qquad (3.5)$$

Thus, in order to construct a meaningful Player II strategy S, for each sequence $\langle P_\xi \in \text{Perf}^*(X) : \xi < \omega_1\rangle$ we will be defining by induction a sequence $\langle Q_\xi : \xi < \omega_1\rangle$ such that each Q_ξ is a subprism of P_ξ and the definition of Q_ξ may depend only on $\langle P_\eta : \eta \leq \xi\rangle$, that is, it cannot depend on any P_η with $\xi < \eta < \omega_1$. If such a sequence $\langle Q_\xi : \xi < \omega_1\rangle$ is defined, then each Q_ξ can be expressed as in (3.5), where S is the function representing our inductive construction.

If we proceed as described above, then we say that the strategy S is associated with our inductive construction. (Such an S is not defined yet on all required sequences, but it is defined on all sequences relevant for us. So, we will be assuming that, on the other sequences, it is defined in some fixed, trivial way.)

Notice that if a prism $P \in \text{Perf}(X)$ is considered with a witness function $f \in \mathcal{F}_{\text{prism}}$ from $\mathfrak{C}^\alpha$ onto P, then P is also a cube and any subcube of P is also a subprism of P. Thus, any Player II strategy in a game GAME$_{\text{cube}}(X)$ can be translated to a strategy in a game GAME$_{\text{prism}}(X)$. (You need to identify appropriately $\mathfrak{C}^\alpha$ with $\mathfrak{C}^\omega$: First you identify $\mathfrak{C}^\alpha$ with the product $\mathfrak{C}^\omega \times \mathfrak{C}^{\alpha\setminus\{0\}}$, which is important for a finite α, and then this second space is identified with $\mathfrak{C}^\omega$ coordinatewise.) In particular, CPA$_{\text{prism}}^{\text{game}}$

implies $\text{CPA}^{\text{game}}_{\text{cube}}$. In addition, essentially the same argument as was used for Proposition 2.0.1 also gives the following.

Proposition 3.0.3 *Axiom* $\text{CPA}^{\text{game}}_{\text{prism}}$ *implies the following prism version of the axiom* CPA_{cube}:

$\text{CPA}_{\text{prism}}$: $\mathfrak{c} = \omega_2$, *and for every Polish space* X *and every* $\mathcal{F}_{\text{prism}}$-*dense family* $\mathcal{E} \subset \text{Perf}(X)$ *there is an* $\mathcal{E}_0 \subset \mathcal{E}$ *such that* $|\mathcal{E}_0| \le \omega_1$ *and* $|X \setminus \bigcup \mathcal{E}_0| \le \omega_1$.

By (3.4) it is also obvious that $\text{CPA}_{\text{prism}}$ implies CPA_{cube}. All these implications can be summarized by a graph.

We will prove the consistency of $\text{CPA}^{\text{game}}_{\text{prism}}$ in Chapter 7. For the remainder of this chapter we will concentrate on some basic consequences of $\text{CPA}_{\text{prism}}$. Most of the applications of the axioms $\text{CPA}_{\text{prism}}$ and $\text{CPA}^{\text{game}}_{\text{prism}}$ will be presented in the following two chapters. We finish this section with few simple but important general remarks.

Although we will not use this, it is illuminating to note that every iterated perfect set $E \in \mathbb{P}_\alpha$ comes with a *canonical* projection-keeping homeomorphism $f \in \Phi_{\text{prism}}(\alpha)$ for which range$(f) = E$. To see it, first note that for every $T \in \text{Perf}(\mathfrak{C})$ there is a canonical homeomorphism h_T from T onto $\mathfrak{C}$ defined by $h_T(t)(i) = t(m_T^i)$, where m_T^i is the i-th forking place of t in T, that is,

$$m_T^i = \min\left\{ k < \omega \colon k > m_T^{i-1} \ \& \ (\exists s \in T) \ s \upharpoonright k = t \upharpoonright k \ \& \ s(k) \neq t(k) \right\}.$$

Then for $x \in E$, $\beta < \alpha$, and $i < \omega$ we define

$$f^{-1}(x)(\beta)(i) = h_{\{y(\beta) : y \in E \ \& \ x \upharpoonright \beta = y \upharpoonright \beta\}}(x(\beta))(i).$$

In other wards, we obtain the value of $f^{-1}(x)$ by removing from each 0-1 sequence $x(\beta)$ its subsequences, where $x(\beta)$ does not branch. It is not difficult to see that such a defined f indeed belongs to $\Phi_{\text{prism}}(\alpha)$ and that range$(f) = E$.

Note also that for every $0 < \alpha < \omega_1$

$$\text{if } f \in \Phi_{\text{prism}}(\alpha) \text{ and } P \in \mathbb{P}_\alpha, \text{ then } f[P] \in \mathbb{P}_\alpha. \tag{3.6}$$

Indeed, if $P = g[\mathfrak{C}^\alpha]$ for some $g \in \Phi_{\text{prism}}(\alpha)$, then, by condition (3.2), we have $f[P] = f[g[\mathfrak{C}^\alpha]] = (f \circ g)[\mathfrak{C}^\alpha] \in \mathbb{P}_\alpha$.

In what follows, for a fixed $0 < \alpha < \omega_1$ and $0 < \beta \leq \alpha$, the symbol π_β will stand for the projection from $\mathfrak{C}^\alpha$ onto $\mathfrak{C}^\beta$, that is, $\mathfrak{C}^\alpha \ni x \overset{\pi_\beta}{\mapsto} x \restriction \beta \in \mathfrak{C}^\beta$. We will always consider $\mathfrak{C}^\alpha$ with the following metric ρ: Fix an enumeration $\{\langle \beta_k, n_k \rangle : k < \omega\}$ of $\alpha \times \omega$ and for distinct $x, y \in \mathfrak{C}^\alpha$ define

$$\rho(x, y) = 2^{-\min\{k < \omega : x(\beta_k)(n_k) \neq y(\beta_k)(n_k)\}}. \tag{3.7}$$

The open ball in $\mathfrak{C}^\alpha$ with a center at $z \in \mathfrak{C}^\alpha$ and radius $\varepsilon > 0$ will be denoted by $B_\alpha(z, \varepsilon)$. Notice that in this metric any two open balls are either disjoint or one is a subset of the other. Also, for every $\gamma < \alpha$ and $\varepsilon > 0$

$$\pi_\gamma[B_\alpha(x, \varepsilon)] = B_\gamma[(x \restriction \gamma, \varepsilon)] \quad \text{for every } x \in \mathfrak{C}^\alpha. \tag{3.8}$$

It is also easy to see that any $B_\alpha(z, \varepsilon)$ is a clopen set, and, in fact, it is a perfect cube in $\mathfrak{C}^\alpha$, so it belongs to $\mathbb{P}_\alpha$. In fact, more can be said:

$$\text{If } \mathcal{B}_\alpha \overset{\text{def}}{=} \{B \subset \mathfrak{C}^\alpha : B \text{ is clopen in } \mathfrak{C}^\alpha\}, \text{ then } \mathcal{B}_\alpha \subset \mathbb{P}_\alpha. \tag{3.9}$$

This is the case, since any clopen E in $\mathfrak{C}^\alpha$ is a finite union of disjoint open balls, each of which belongs to $\mathbb{P}_\alpha$, and it is easy to see that $\mathbb{P}_\alpha$ is closed under finite unions of open balls.

From this we conclude immediately that

$$\text{a clopen subset of } E \in \mathbb{P}_\alpha \text{ belongs to } \mathbb{P}_\alpha \tag{3.10}$$

and

$$\text{a clopen subset of a prism is its subprism,} \tag{3.11}$$

while (3.3) implies

$$\pi_\beta[E] \in \mathbb{P}_\beta \text{ for every } 0 < \beta < \alpha < \omega_1 \text{ and } E \in \mathbb{P}_\alpha. \tag{3.12}$$

Notice also that if $P \in \mathbb{P}_\alpha$ and $0 < \beta < \alpha$, then

$$P \cap \pi_\beta^{-1}(P') \in \mathbb{P}_\alpha \quad \text{for every } P' \in \mathbb{P}_\beta \text{ with } P' \subset \pi_\beta[P]. \tag{3.13}$$

Indeed, let $f \in \Phi_{\text{prism}}(\beta)$ and $g \in \Phi_{\text{prism}}(\alpha)$ be such that $f[\mathfrak{C}^\beta] = P'$ and $g[\mathfrak{C}^\alpha] = P$. Let $Q = (g \restriction\restriction \beta)^{-1}[P'] = (g \restriction\restriction \beta)^{-1} \circ f[\mathfrak{C}^\beta]$. Then, $Q \in \mathbb{P}_\beta$ since, by (3.3), $(g \restriction\restriction \beta)^{-1} \circ f \in \Phi_{\text{prism}}(\beta)$. Thus $\pi_\beta^{-1}(Q)$ belongs to $\mathbb{P}_\alpha$ and $P \cap \pi_\beta^{-1}(P') = g[\pi_\beta^{-1}(Q)] \in \mathbb{P}_\alpha$.

3.1 Fusion for prisms

One of the main technical tools used to prove that a family of perfect sets is dense is the so-called fusion lemma. It says that, for an appropriately chosen decreasing sequence $\{P_n : n < \omega\}$ of perfect sets, its intersection $P = \bigcap_{n<\omega} P_n$, called the *fusion*, is still a perfect set. The simple structure of perfect cubes makes it quite easy to formulate a "cube fusion lemma" in which the fusion set P is also a perfect cube. However, so far we have not had any need for such a lemma (at least in an explicit form), since its use was always hidden in the proofs of the results we quoted, like Claim 1.1.5 or Proposition 1.5.1. On the other hand, the new and more complicated structure of prisms does not leave us the option of avoiding fusion arguments any longer — we have to face it up front.

For a fixed $0 < \alpha < \omega_1$, let $\{\langle \beta_k, n_k \rangle : k < \omega\}$ be the enumeration of $\alpha \times \omega$ used in the definition (3.7) of the metric ρ and let

$$A_k = \{\langle \beta_i, n_i \rangle : i < k\} \quad \text{for every } k < \omega. \tag{3.14}$$

Lemma 3.1.1 (Fusion Sequence) *Let* $0 < \alpha < \omega_1$, *and for* $k < \omega$ *let* $\mathcal{E}_k = \{E_s : s \in 2^{A_k}\}$ *be a family of closed subsets of* $\mathfrak{C}^\alpha$. *Assume that for every* $k < \omega$, $s, t \in 2^{A_k}$, *and* $\beta < \alpha$ *we have:*

(i) *The diameter of* E_s *goes to 0 as the length of* s *goes to* ∞.

(ii) *If* $i < k$, *then* $E_s \subset E_{s \upharpoonright A_i}$.

(ag) *(agreement) If* $s \upharpoonright (\beta \times \omega) = t \upharpoonright (\beta \times \omega)$, *then* $\pi_\beta[E_s] = \pi_\beta[E_t]$.

(sp) *(split) If* $s \upharpoonright (\beta \times \omega) \neq t \upharpoonright (\beta \times \omega)$, *then* $\pi_\beta[E_s] \cap \pi_\beta[E_t] = \emptyset$.

Then $Q = \bigcap_{k<\omega} \bigcup \mathcal{E}_k$ *belongs to* $\mathbb{P}_\alpha$.

PROOF. For $x \in \mathfrak{C}^\alpha$, let $\bar{x} \in 2^{\alpha \times \omega}$ be defined by $\bar{x}(\beta, n) = x(\beta)(n)$.

First note that, by conditions (i) and (sp), for every $k < \omega$ the sets in $\mathcal{E}_k$ are pairwise disjoint. Thus, taking into account (ii), the function $h : \mathfrak{C}^\alpha \to \mathfrak{C}^\alpha$ defined by

$$h(x) = r \quad \Longleftrightarrow \quad \{r\} = \bigcap_{k<\omega} E_{\bar{x} \upharpoonright A_k}$$

is well defined and is one to one. It is also easy to see that h is continuous and that $Q = h[\mathfrak{C}^\alpha]$. Thus, we need to prove only that h is projection-keeping.

To show this, fix $\beta < \alpha$, put $S = \bigcup_{i<\omega} 2^{A_i}$, and notice that, by (i) and (ag), for every $x \in \mathfrak{C}^\alpha$ we have

$$\{h(x) \upharpoonright \beta\} = \pi_\beta \left[\bigcap \{E_{\bar{x} \upharpoonright A_k} : k < \omega\} \right]$$

$$= \bigcap \{\pi_\beta [E_{\bar{x} \upharpoonright A_k}] : k < \omega\}$$

$$= \bigcap \{\pi_\beta [E_s] : s \in S \,\&\, s \subset \bar{x}\}$$

$$= \bigcap \{\pi_\beta [E_s] : s \in S \,\&\, s \upharpoonright (\beta \times \omega) \subset \bar{x}\}.$$

Now, if $x \upharpoonright \beta = y \upharpoonright \beta$, then for every $s \in S$

$$s \upharpoonright (\beta \times \omega) \subset \bar{x} \iff s \upharpoonright (\beta \times \omega) \subset \bar{y},$$

so $h(x) \upharpoonright \beta = h(y) \upharpoonright \beta$.

On the other hand, if $x \upharpoonright \beta \neq y \upharpoonright \beta$, then there is a $k < \omega$ large enough such that for $s = \bar{x} \upharpoonright A_k$ and $t = \bar{y} \upharpoonright A_k$ we have $s \upharpoonright (\beta \times \omega) \neq t \upharpoonright (\beta \times \omega)$. But then $\{h(x) \upharpoonright \beta\}$ and $\{h(y) \upharpoonright \beta\}$ are subsets of $\pi_\beta [E_s]$ and $\pi_\beta [E_t]$, respectively, which, by (sp), are disjoint. So, $h(x) \upharpoonright \beta \neq h(y) \upharpoonright \beta$. ∎

In most of our applications the task of constructing sequences $\langle \mathcal{E}_k : k < \omega \rangle$ satisfying the specific conditions (ag) and (sp) can be reduced to checking some simple density properties listed in our next lemma. In its statement we consider $\mathbb{P}_\alpha$ as ordered by inclusion and use the standard terminology from the theory of partially ordered sets: $D \subset \mathbb{P}_\alpha$ is *dense* provided for every $E \in \mathbb{P}_\alpha$ there is an $E' \in D$ with $E' \subset E$; it is *open* provided for every $E \in D$ if $E' \in \mathbb{P}_\alpha$ and $E' \subset E$, then $E' \in D$. Moreover, for a family $\mathcal{E}$ of pairwise disjoint subsets of $\mathbb{P}_\alpha$, we say that $\mathcal{E}' \subset \mathbb{P}_\alpha$ is a *refinement of* $\mathcal{E}$ provided $\mathcal{E}' = \{P_E : E \in \mathcal{E}\}$, where $P_E \subset E$ for all $E \in \mathcal{E}$.

Lemma 3.1.2 *Let* $0 < \alpha < \omega_1$ *and* $k < \omega$. *If* $\mathcal{E}_k = \{E_s \in \mathbb{P}_\alpha : s \in 2^{A_k}\}$ *satisfies (ag) and (sp), then:*

(A) *There exists an* $\mathcal{E}_{k+1} = \{E_s \in \mathbb{P}_\alpha : s \in 2^{A_{k+1}}\}$ *of sets of diameter less than* $2^{-(k+1)}$ *such that (ii), (ag), and (sp) hold for all* $s, t \in 2^{A_{k+1}}$ *and* $r \in 2^{A_k}$.

Moreover, if $\mathcal{D} \subset [\mathbb{P}_\alpha]^{<\omega}$ *is a family of pairwise disjoint sets such that* $\emptyset \in \mathcal{D}$, $\mathcal{D}$ *is closed under refinements, and*

(†) *for every* $\mathcal{E} \in \mathcal{D}$ *and* $E \in \mathbb{P}_\alpha$ *which is disjoint with* $\bigcup \mathcal{E}$ *there exists an* $E' \in \mathbb{P}_\alpha \cap \mathcal{P}(E)$ *such that* $\{E'\} \cup \mathcal{E} \in \mathcal{D}$,

then

(B) *there exists a refinement* $\mathcal{E}'_k \in \mathcal{D}$ *of* $\mathcal{E}_k$ *satisfying (ag) and (sp);*

(C) *there exists an* $\mathcal{E}_{k+1}$ *as in (A) such that* $\mathcal{E}_{k+1} \in \mathcal{D}$.

PROOF. For $s \in 2^{A_k}$ and $j < 2$, let $s\hat{\,}j$ stand for $s \cup \{\langle\langle\beta_k, n_k\rangle, j\rangle\} \in 2^{A_{k+1}}$.

Let $\{s_i : i < 2^{k+1}\}$ be an enumeration of $2^{A_{k+1}}$. By induction on $i < 2^{k+1}$, we will construct a sequence $\langle x_{s_i} \in \mathfrak{C}^\alpha : i < 2^{k+1}\rangle$ such that for every $i < 2^{k+1}$

(a) $x_{s_i} \in E_{s_i \restriction A_k}$,
(b) for every $m < i$, if $\beta = \max\{\bar{\beta}: s_i \restriction (\bar{\beta} \times \omega) = s_m \restriction (\bar{\beta} \times \omega)\}$, then

$$x_{s_i} \restriction \beta = x_{s_m} \restriction \beta \text{ and } x_{s_i}(\beta) \neq x_{s_m}(\beta).$$

The point x_{s_0} is chosen arbitrarily from $E_{s_0 \restriction A_k}$. To make an inductive step, if for some $0 < i \leq 2^{k+1}$ points $\{x_{s_m} : m < i\}$ are already constructed, choose an $\bar{m} < i$ for which β as in (b) is maximal. Notice that by the inductive assumption and the condition (ag) we have $x_{s_{\bar{m}}} \restriction \beta \in \pi_\beta[E_{s_{\bar{m}} \restriction A_k}] = \pi_\beta[E_{s_i \restriction A_k}]$. So we can choose an $x_{s_i} \in E_{s_i \restriction A_k}$ extending $x_{s_{\bar{m}}} \restriction \beta$ and such that $x_{s_i}(\beta) \neq x_{s_m}(\beta)$ for all $m < i$. It is easy to see that such an x_{s_i} satisfies (a) and condition (b) for $m = \bar{m}$. For other $m < i$, condition (b) follows from the maximality of β and the assumption that $\mathcal{E}_k$ satisfies (ag) and (sp).

Conditions (a) and (b) imply that $\mathcal{E}'_{k+1} = \{\{x_s\}: s \in 2^{A_{k+1}}\}$ satisfy condition (A) except for being a subset of $\mathbb{P}_\alpha$. Let $\varepsilon \in \left(0, 2^{-(k+1)}\right)$ be small enough that for every $m < i < 2^{k+1}$ and β as in (b) we have $\pi_{\beta+1}[B_\alpha(x_{s_i}, \varepsilon)] \cap \pi_{\beta+1}[B_\alpha(x_{s_m}, \varepsilon)] = \emptyset$. For $s \in 2^{A_k}$ and $j < 2$, define

$$E_{s\hat{\,}j} = E_s \cap B_\alpha(x_{s\hat{\,}j}, \varepsilon).$$

Then $\mathcal{E}_{k+1} = \{E_s: s \in 2^{A_{k+1}}\}$ is a subset of $\mathbb{P}_\alpha$ by (3.10). Condition (ii) is clear from the construction, while (ag) for $\mathcal{E}_{k+1}$ follows from (b) and (3.8). Property (sp) holds by (b) and the choice of ε, since (sp) was true for $\mathcal{E}'_{k+1}$. We have completed the proof of (A).

To prove condition (B), fix an enumeration $\{s_i : i < 2^k\}$ of 2^{A_k} and define $\gamma = \max\{\beta_0, \ldots, \beta_k\} < \alpha$. Also, for $i, m < 2^k$, put $E_{s_i}^{-1} = E_{s_i}$ and

$$\beta_i^m = \max\{\beta \leq \gamma: s_i \restriction (\beta \times \omega) = s_m \restriction (\beta \times \omega)\}.$$

By induction we will construct the sequences $\langle\{E_{s_i}^m \in \mathbb{P}_\alpha : i < 2^k\}: m < 2^k\rangle$ and $\langle P_m \in \mathbb{P}_\alpha : m < 2^k\rangle$ such that, for every $j, m < 2^k$,

(a) $\mathcal{E}^m = \{E_{s_i}^m : i < 2^k\}$ satisfies (ag);
(b) $E_{s_j}^m \subset E_{s_j}^{m-1}$ and if $x \in E_{s_j}^{m-1}$ and $\pi_\gamma(x) \in \pi_\gamma[E_{s_j}^m]$, then $x \in E_{s_j}^m$;
(c) $\pi_\gamma[P_m] = \pi_\gamma[E_{s_m}^m]$;
(d) $P_m \subset E_{s_m}^{m-1}$ and $\{P_i : i \leq m\} \in \mathcal{D}$.

So, assume that for some $m < 2^k$ the sequence $\langle P_i : i < m \rangle$ and the family $\mathcal{E}^{m-1}$ satisfying (ag) are already constructed. Notice that, by (b), sets in $\mathcal{E}^{m-1}$ are pairwise disjoint, since this was the case for $\mathcal{E}^{-1} = \mathcal{E}_k$. Thus, by condition (†) applied to the family $\mathcal{E} = \{P_i : i < m\}$, we can choose a $P_m \in \mathbb{P}_\alpha \cap \mathcal{P}(E_{s_m}^{m-1})$ such that $\{P_m\} \cup \{P_i : i < m\} \in \mathcal{D}$. This guarantees (d).

Next, for $i < 2^k$ define

$$E_{s_i}^m = E_{s_i}^{m-1} \cap \pi_{\beta_i^m}^{-1}(\pi_{\beta_i^m}[P_m]) = \{x \in E_{s_i}^{m-1} : x \upharpoonright \beta_i^m \in \pi_{\beta_i^m}[P_m]\}$$

and notice that $\pi_{\beta_i^m}[P_m] \subset \pi_{\beta_i^m}[E_{s_m}^{m-1}] = \pi_{\beta_i^m}[E_{s_i}^{m-1}]$. So, by (3.13), $E_{s_i}^m \in \mathbb{P}_\alpha$. Also, the definition ensures (b) since $\beta_i^m \leq \gamma$.

Note that, by the inductive assumption (a), for all $i < 2^k$ we have

$$\pi_{\beta_i^m}[E_{s_i}^m] = \pi_{\beta_i^m}[E_{s_i}^{m-1}] \cap \pi_{\beta_i^m}[P_m] = \pi_{\beta_i^m}[E_{s_m}^{m-1}] \cap \pi_{\beta_i^m}[P_m] = \pi_{\beta_i^m}[P_m].$$

Since $\beta_m^m = \gamma$, this implies (c). To prove (a), pick $\beta < \alpha$ and different $i, j < 2^k$ such that $s_i \upharpoonright (\beta \times \omega) = s_j \upharpoonright (\beta \times \omega)$. If $\beta \leq \beta_i^m$, then also $\beta \leq \beta_j^m$ and $\pi_\beta[E_{s_i}^m] = \pi_\beta[P_m] = \pi_\beta[E_{s_j}^m]$. So, assume that $\beta > \beta_i^m$ and $\beta > \beta_j^m$. Then $\beta_i^m = \beta_j^m$ and

$$
\begin{aligned}
\pi_\beta[E_{s_i}^m] &= \{\pi_\beta(x) : x \in E_{s_i}^{m-1} \ \& \ x \upharpoonright \beta_i^m \in \pi_{\beta_i^m}[P_m]\} \\
&= \{\pi_\beta(x) : x \in E_{s_j}^{m-1} \ \& \ x \upharpoonright \beta_j^m \in \pi_{\beta_j^m}[P_m]\} \\
&= \pi_\beta[E_{s_j}^m].
\end{aligned}
$$

So $\mathcal{E}^m$ satisfies (a). This finishes the construction.

Notice that by the maximality of γ and properties (a) and (c), the family $\mathcal{E}'_k = \{P_m : m < 2^k\}$ satisfies (ag). Since it is a refinement of $\mathcal{E}_k$, it also satisfies (sp). So (B) is proved.

To find $\mathcal{E}_{k+1}$ as in (C), first take an $\mathcal{E}'_{k+1}$ satisfying (A) and then use (B) to find its refinement $\mathcal{E}_{k+1} \in \mathcal{D}$ satisfying (ag) and (sp). ∎

One of the most important consequences of Lemma 3.1.2 is the following.

Corollary 3.1.3 *Let $0 < \alpha < \omega_1$ and let $\{D_k : k < \omega\}$ be a collection of dense open subsets of $\mathbb{P}_\alpha$. If for every $k < \omega$*

$$D_k^* = \left\{ \bigcup \mathcal{D} : \mathcal{D} \in [D_k]^{<\omega} \text{ and the sets in } \mathcal{D} \text{ are pairwise disjoint} \right\},$$

then $\bar{D} = \bigcap_{k<\omega} D_k^*$ is open and dense in $\mathbb{P}_\alpha$.

PROOF. It is clear that $\bar{D}$ is open. To see its density, notice that the families

$$\mathcal{D}_k = \left\{ \mathcal{D} \in [D_k]^{<\omega} \text{ and sets in } \mathcal{D} \text{ are pairwise disjoint} \right\}$$

satisfy condition (†). Let $E \in \mathbb{P}_\alpha$, choose an $E_\emptyset \in D_0 \subset D_0^*$ below E, and put $\mathcal{E}_0 = \{E_\emptyset\}$. Applying (C) from Lemma 3.1.2 by induction we can define families $\mathcal{E}_k \in \mathcal{D}_k$, $k < \omega$, such that conditions (i), (ii), (ag), and (sp) from Lemma 3.1.1 are satisfied. But then $Q = \bigcap_{k<\omega} \bigcup \mathcal{E}_k \subset E$ belongs to $\bar{D}$. ■

3.2 On $\mathcal{F}$-independent prisms

The following variant of the Kuratowski-Ulam theorem will be useful in what follows.

Lemma 3.2.1 *Let X be a Polish space and consider X^T with the product topology, where $T \neq \emptyset$ is an arbitrary set. Fix at most countable family $\mathcal{K}$ of sets $K \subsetneq T$. Then for every comeager set $H \subset X^T$ there exists a comeager set $G \subset H$ such that for every $x \in G$ and $K \in \mathcal{K}$ the set*

$$G_{x \restriction K} = \left\{ y \in X^{T \setminus K} : (x \restriction K) \cup y \in G \right\}$$

is comeager in $X^{T \setminus K}$.

PROOF. Let $\{K_i : i < \omega\}$ be an enumeration of $\mathcal{K}$ with infinite repetitions. We construct, by induction on $i < \omega$, a decreasing sequence $\langle G_i : i < \omega \rangle$ of comeager subsets of H such that for every $i < \omega$:

(i) The set $(G_i)_{x \restriction K_i}$ is comeager in $X^{T \setminus K_i}$ for every $x \in G_i$.

Put $G_{-1} = H$ and assume that for some $i < \omega$ the comeager set G_{i-1} is already constructed. To define G_i identify X^T with $X^{K_i} \times X^{T \setminus K_i}$. Then, by the Kuratowski-Ulam theorem, the set

$$A = \left\{ y \in X^{K_i} : (G_{i-1})_y \text{ is comeager in } X^{T \setminus K_i} \right\}$$

is comeager in X^{K_i}. Put $G_i = G_{i-1} \cap (A \times X^{T \setminus K_i})$.

Clearly $G_i \subset G_{i-1}$ is comeager in X^T. If $x \in G_i$, then $x \restriction K_i \in A$ and so $(G_i)_{x \restriction K_i} = (G_{i-1})_{x \restriction K_i}$ is comeager in $X^{T \setminus K_i}$. So, (i) holds. This completes the definition of the sequence $\langle G_i : i < \omega \rangle$.

Let $G = \bigcap_{i<\omega} G_i$. Clearly $G \subset H$ is comeager in X^T. To see the additional part, take a $K \in \mathcal{K}$. Since $G = \bigcap \{G_i : i < \omega \ \& \ K_i = K\}$, for every $x \in G$ the set

$$G_{x \restriction K} = \bigcap \{(G_i)_{x \restriction K_i} : i < \omega \ \& \ K_i = K\}$$

is comeager in $X^{T \setminus K}$. ■

Applying Lemma 3.2.1 to $X = \mathfrak{C}$, $T = \alpha$, and $\mathcal{K} = \alpha$ we immediately obtain the following corollary.

Corollary 3.2.2 *Let $0 < \alpha < \omega_1$. For every comeager set $H \subset \mathfrak{C}^\alpha$ there exists a comeager set $G \subset H$ such that for every $x \in G$ and $\beta < \alpha$ the set*

$$G_{x \restriction \beta} = \left\{ y \in \mathfrak{C}^{\alpha \setminus \beta} \colon (x \restriction \beta) \cup y \in G \right\}$$

is comeager in $\mathfrak{C}^{\alpha \setminus \beta}$.

Let X be a Polish space, $0 < n < \omega$, and $F \subset X^n$ be an n-ary relation. We say that a set $S \subset X$ is F-*independent* provided $F(x(0), \dots, x(n-1))$ does not hold for any one to one $x \colon n \to S$. For a family $\mathcal{F}$ of finitary relations on X (i.e., relations $F \subset X^n$, where $0 < n < \omega$) we say that $S \subset X$ is $\mathcal{F}$-independent provided S is F-independent for every $F \in \mathcal{F}$. We will use the term *unary relation* for any 1-ary relation.

Proposition 3.2.3 *Let $0 < \alpha < \omega_1$ and $\mathcal{F}$ be a countable family of closed finitary relations on $\mathfrak{C}^\alpha$. Assume that every unary relation in $\mathcal{F}$ is nowhere dense in $\mathfrak{C}^\alpha$ and that for every $F \in \mathcal{F}$ there exists a comeager subset G_F of $\mathfrak{C}^\alpha$ such that:*

(ex) *For every F-independent finite set $S \subset G_F$, $x \in S$, and $\beta < \alpha$ the set*

$$\left\{ z \in \mathfrak{C}^{\alpha \setminus \beta} \colon S \cup \{z \cup x \restriction \beta\} \subset G_F \text{ is } F\text{-independent} \right\}$$

is dense in $\mathfrak{C}^{\alpha \setminus \beta}$.

Then there is an $E \in \mathbb{P}_\alpha$ that is $\mathcal{F}$-independent.

Note that, without the assumption that the unary relations in $\mathcal{F}$ are nowhere dense the proposition is false: The unary relation $F = \mathfrak{C}^\alpha$ satisfies the condition (ex) (with $G_F = \mathfrak{C}^\alpha$) and no nonempty set is F-independent. On the other hand, for any n-ary relation $F \in \mathcal{F}$ with $n > 1$, condition (ex) implies that F is nowhere dense in $(\mathfrak{C}^\alpha)^n$. However, not every nowhere dense binary relation satisfies (ex). For example, $F = \{\langle x, y \rangle \colon x(0) = y(0)\}$ is nowhere dense and it does not satisfy (ex) if $\alpha > 1$.

PROOF. First notice that, applying Corollary 3.2.2, if necessary, we can assume that for every $F \in \mathcal{F}$, $x \in G_F$, and $\beta < \alpha$ the set $(G_F)_{x \restriction \beta}$ is comeager in $\mathfrak{C}^{\alpha \setminus \beta}$. But this implies that each set from the condition (ex) is comeager in $\mathfrak{C}^{\alpha \setminus \beta}$ since it is an intersection of $(G_F)_{x \restriction \beta}$ and an open set $\{z \in \mathfrak{C}^{\alpha \setminus \beta} \colon S \cup \{z \cup x \restriction \beta\}$ is F-independent$\}$. In particular, if we put $G = \bigcap_{F \in \mathcal{F}} G_F$, then G is comeager in $\mathfrak{C}^\alpha$ and it is easy to see that it satisfies the following condition.

(EX) For every $\mathcal{F}$-independent finite set $S \subset G$, $x \in S$, and $\beta < \alpha$ the set

$$\left\{ z \in \mathfrak{C}^{\alpha \backslash \beta} : \ S \cup \{z \cup x \upharpoonright \beta\} \subset G \text{ is } \mathcal{F}\text{-independent} \right\}$$

is dense in $\mathfrak{C}^{\alpha \backslash \beta}$.

Let $\{F_k : k < \omega\}$ be an enumeration of $\mathcal{F}$ with infinite repetitions. Also, for $k < \omega$, let $A_k = \{\langle \beta_i, n_i \rangle : i < k\}$ be as in condition (3.14). By induction on $k < \omega$ we will construct two sequences: $\langle \varepsilon_k > 0 : k < \omega \rangle$ converging to 0 and $\langle \{x_s \in G : s \in 2^{A_k}\} : k < \omega \rangle$ of $\mathcal{F}$-independent sets such that for every $\beta < \alpha$, $k < \omega$, and $s, t \in 2^{A_k}$:

(a) $x_s \upharpoonright \beta = x_t \upharpoonright \beta$ if and only if $s \upharpoonright \beta \times \omega = t \upharpoonright \beta \times \omega$.
(b) If $E_s = B_\alpha(x_s, \varepsilon_k)$ and $\mathcal{E}_k = \{E_s : s \in 2^{A_k}\}$, then the $\mathcal{E}_k$'s satisfy (ii), (ag), and (sp) from Lemma 3.1.1.
(c) If F_k is an n-ary relation, then $F_k(z_0, \ldots, z_{n-1})$ does not hold provided each z_i is chosen from a different ball from $\mathcal{E}_k$.

Before we construct such sequences, let us first note that $E = \bigcap_{k<\omega} \bigcup \mathcal{E}_k$ is as desired. Indeed, $E \in \mathbb{P}_\alpha$ by Lemma 3.1.1. To see that E is $\mathcal{F}$-independent, pick an n-ary relation $F \in \mathcal{F}$, $\{z_0, \ldots, z_{n-1}\} \in [E]^n$, and find a $k < \omega$ with $F_k = F$ that is big enough so that ε_k is smaller than the distance between z_i and z_j for all $i < j < n$. Then the z_i's must belong to distinct elements of $\mathcal{E}_k$; so, by (c), $F(z_0, \ldots, z_{n-1})$ does not hold.

For $k = 0$ we pick an arbitrary $\mathcal{F}$-independent $x_\emptyset \in G$ by choosing an arbitrary element of G that does not belong to any nowhere dense unary relation from $\mathcal{F}$. Also, we choose an $\varepsilon_0 \in (0, 1]$ ensuring (c), which can be done since F_0 is closed. (This is a nontrivial requirement only when F_0 is a unary relation.) Clearly (a)–(c) are satisfied.

Assume that for some $k < \omega$ the construction is done up to the level k. For $s \in 2^{A_k}$ and $j < 2$, let $s\hat{\ }j = s \cup \{\langle\langle \beta_k, n_k \rangle, j \rangle\} \in 2^{A_{k+1}}$ and define $x_{s\hat{\ }0} = x_s$. Let $\{s_i : i < 2^k\}$ be an enumeration of 2^{A_k} and put $S = \{x_{s\hat{\ }0} : s \in 2^{A_k}\}$. Points $x_{s_i\hat{\ }1} \in G \cap E_{s_i}$ will be chosen by induction on $i \le 2^k$ such that the set $S_i = S \cup \{x_{s_j\hat{\ }1} : j < i\}$ is $\mathcal{F}$-independent and condition (a) is satisfied for the elements of S_i. Clearly, by the inductive assumption, (a) is satisfied for the elements of $S_0 = S$. So, assume that for some $i \le 2^k$ the set S_i is already constructed. We need to find an appropriate $x_{s_i\hat{\ }1} \in G \cap E_{s_i}$. Let $\beta < \alpha$ be maximal such that there is an $s \in \{s\hat{\ }0 : s \in 2^{A_k}\} \cup \{s_j\hat{\ }1 : j < i\}$ with $s \upharpoonright \beta \times \omega = (s_i\hat{\ }1) \upharpoonright \beta \times \omega$ and let $x = x_s \upharpoonright \beta$. We will choose $x_{s_i\hat{\ }1}$ extending x and such that $x_{s_i\hat{\ }1}(\beta) \neq x_t(\beta)$ for all $x_t \in S_i$. Notice that this will ensure that condition (a) is satisfied for the elements of S_{i+1}. Surprisingly, a more difficult condition to ensure will be that $x_{s_i{}^\frown 1} \in E_{s_i} = B_\alpha(x_{s_i{}^\frown 0}, \varepsilon_k)$, since at the first glance it is not even obvious that

$$B_\alpha(x_{s_i{}^\frown 0}, \varepsilon_k) \text{ contains an extension of } x. \tag{3.15}$$

To argue for this, first notice that maximality of β ensures that $\beta \geq \beta_k$, since $s_i{}^\frown 0 \in S_i$ and $(s_i{}^\frown 0) \upharpoonright \beta_k \times \omega = (s_i{}^\frown 1) \upharpoonright \beta_k \times \omega$. If $\beta = \beta_k$, we have $x = x_{s_i{}^\frown 0} \upharpoonright \beta$ and (3.15) is obvious. So, assume that $\beta > \beta_k$. Then there is a $j < i$ such that $s = s_j{}^\frown 1$. We also have $s_j \upharpoonright \beta \times \omega = s_i \upharpoonright \beta \times \omega$; so, by the inductive assumption, $x_{s_j} \upharpoonright \beta = x_{s_i} \upharpoonright \beta$.

Now, let $n < \omega$ be the smallest such that $2^{-n} < \varepsilon_k$. Then, by the definition of the metric on $\mathfrak{C}^\alpha$, the fact that $x_s = x_{s_j{}^\frown 1} \in E_{s_j} = B_\alpha(x_{s_j}, \varepsilon_k)$ means that $x_s(\gamma)(m) = x_{s_j}(\gamma)(m)$ for every $\langle \gamma, m \rangle \in A_n$. Therefore, we have $x(\gamma)(m) = x_s(\gamma)(m) = x_{s_j}(\gamma)(m) = x_{s_i}(\gamma)(m)$ for every $\langle \gamma, m \rangle \in A_n$ with $\gamma < \beta$. Thus, we can extend x to an element $y \in \mathfrak{C}^\alpha$ for which $y(\gamma)(m) = x_{s_i}(\gamma)(m)$ for every $\langle \gamma, m \rangle \in A_n$. But this y witnesses (3.15).

To finish the construction of $x_{s_i{}^\frown 1}$, notice that by (3.15) we can find an open ball B in $\mathfrak{C}^{\alpha \setminus \beta}$ such that $\{x\} \times B \subset B_\alpha(x_{s_i{}^\frown 0}, \varepsilon_k)$. Decreasing B, if necessary, we can also insure that $y(\beta) \neq x_t(\beta)$ for every $t \in S_i$ and $y \in \{x\} \times B$. By condition (EX) we can find a $z \in B$ such that $S_i \cup \{x \cup z\} \subset G$ is $\mathcal{F}$-independent. We put $x_{s_i{}^\frown 1} = x \cup z$.

Thus, we constructed an $\mathcal{F}$-independent set $\{x_{s{}^\frown j} : s \in 2^{A_k} \& j < 2\} \subset G$ satisfying (a) and such that $x_{s{}^\frown 0}, x_{s{}^\frown 1} \in E_s$ for every $s \in 2^{A_k}$. To finish the construction insuring (a)–(c) we need to choose an $\varepsilon_{k+1} \leq 2^{-(k+1)}$ small enough to guarantee the following properties.

- $E_{s{}^\frown j} = B_\alpha(x_{s{}^\frown 0}, \varepsilon_{k_1}) \subset E_s$ for every $s \in 2^{A_k}$ and $j < 2$. This will ensure condition (ii).

- Condition (sp) holds. This can be done, since (a) is satisfied.

- Condition (c) is satisfied. This can be done since $\{x_s : s \in 2^{A_{k+1}}\}$ is $\mathcal{F}$-independent and F_{k+1} is a closed relation.

Note that (ag) is guaranteed by (a) and our definition of the E_s's. This finishes the proof of Proposition 3.2.3. $\blacksquare$

It worth mentioning that Proposition 3.2.3 can be viewed as a generalization of J. Mycielski's theorem [101]. (Compare also [102].) Also, in the case when $\mathcal{F}$ consists of one binary relation R, a slight modification of the argument for Proposition 3.2.3 gives us the following.

Proposition 3.2.4 *Let $0 < \alpha < \omega_1$ and let R be a closed binary relation on $\mathfrak{C}^\alpha$. Assume that there exists a comeager subset G of $\mathfrak{C}^\alpha$ such that:*

(bin) *For every $x \in G$, $\varepsilon > 0$, and $\beta < \alpha$ there is a $y \in G \cap B_\alpha(x, \varepsilon) \setminus \{x\}$ such that $y \restriction \beta = x \restriction \beta$ and $\{x, y\}$ is R-independent.*

Then there is an $E \in \mathbb{P}_\alpha$ that is R-independent.

PROOF. Let $\mathcal{F} = \{R\}$. We will just indicate the modifications needed in the proof of Proposition 3.2.3 to obtain the current result. Thus, we construct the sequences $\langle \varepsilon_k > 0 \colon k < \omega \rangle$ and $\langle \{x_s \in G \colon s \in 2^{A_k}\} \colon k < \omega \rangle$ subject to the same requirements. As before, this will give us the desired $E \in \mathbb{P}_\alpha$.

First notice that, by Corollary 3.2.2, decreasing G if necessary, we can assume that

$$G_{x \restriction \beta} = \left\{ y \in \mathfrak{C}^{\alpha \setminus \beta} \colon (x \restriction \beta) \cup y \in G \right\}$$

is comeager for every $x \in G$ and $\beta < \alpha$.

The choice of $x_\emptyset$ is trivial, since any point from G together with any ε_0 satisfy the requirements. To make an inductive step assume that for some $k < \omega$ the construction is done up to the level k. As before, for $s \in 2^{A_k}$ we put $x_{s\char`\^0} = x_s$, fix an enumeration $\{s_i \colon i < 2^k\}$ of $2^{A_k}$, and put $S = \{x_{s\char`\^0} \colon s \in 2^{A_k}\}$. Before we choose points $x_{s_i\char`\^1}$ we need to make some preparations.

Let $\bar{\beta} = \max\{\beta_i \colon i \leq k\} < \alpha$ and notice that, by (bin), for every $i < 2^k$ we can find a $y_i \in G \cap E_{s_i} \setminus \{x_{s_i}\}$ such that $y_i \restriction \bar{\beta} = x_{s_i} \restriction \bar{\beta}$ and $\{x_{s_i}, y_i\}$ is R-independent. Note that this, together with condition (c) for the step k, implies that the set $T = S \cup \{y_i \colon i < 2^k\}$ is R-independent. Now, by the closure assumption about R, we can find an $\varepsilon \in (0, \varepsilon_k]$ such that the balls $\{B_\alpha(t, \varepsilon) \colon t \in T\}$ are pairwise disjoint and any selector from this family is R-independent. Note also that $B_\alpha(x_{s_i}, \varepsilon) \cup B_\alpha(y_i, \varepsilon) \subset E_{s_i}$ for every $i < 2^k$ as $\varepsilon \leq \varepsilon_k$. Since we will choose $x_{s_i\char`\^1} \in B_\alpha(y_i, \varepsilon) \cap G$, the resulting set $\{x_s \colon s \in 2^{A_{k+1}}\} \subset G$ will be R-independent. We just need to insure that it satisfies (a).

Points $x_{s_i\char`\^1}$ will be chosen by induction on $i < 2^k$ such that the elements of the set $S_i = S \cup \{x_{s_j\char`\^1} \colon j < i\}$ satisfy condition (a). For this, we proceed precisely as in the proof of Proposition 3.2.3. We take a maximal $\beta < \alpha$ for which there exists an $s \in \{s\char`\^0 \colon s \in 2^{A_k}\} \cup \{s_j\char`\^1 \colon j < i\}$ with $s \restriction \beta \times \omega = (s_i\char`\^1) \restriction \beta \times \omega$ and let $x = x_s \restriction \beta$. We will choose $x_{s_i\char`\^1}$ extending x and such that $x_{s_i\char`\^1}(\beta) \neq x_t(\beta)$ for all $x_t \in S_i$, ensuring that condition (a) is satisfied for the elements of S_{i+1}. For this we need to notice that, similarly as for (3.15), we can prove that

$$B_\alpha(y_i, \varepsilon) \text{ contains an extension of } x.$$

The same argument works here, since we have shown there that $x(\gamma)(m) = x_s(\gamma)(m) = x_{s_j}(\gamma)(m) = x_{s_i}(\gamma)(m)$ for all appropriate pairs $\langle \gamma, m \rangle \in A_n$ with $\gamma < \beta$, while we have $\beta \le \bar{\beta}$ and $y_i \restriction \bar{\beta} = x_{s_i} \restriction \bar{\beta}$. In other words, $x(\gamma)(m) = x_{t_i}(\gamma)(m)$ for all appropriate pairs $\langle \gamma, m \rangle \in A_n$ with $\gamma < \beta$, which gives us the above condition.

Now, since $G_x = G_{x_s \restriction \beta}$ is comeager in $\mathfrak{C}^{\alpha \setminus \beta}$, there exists an

$$x_{s_i \hat{\ } 1} \in B_\alpha(y_i, \varepsilon) \cap (\{x\} \times \{z \in G_x : z(\beta) \ne x_t(\beta) \text{ for all } x_t \in S_i\}).$$

It is easy to see that such an $x_{s_i \hat{\ } 1}$ satisfies all the requirements.

The choice of an appropriate ε_{k+1} is done as in Proposition 3.2.3. ∎

In what follows we will also need the following fact, which is one of the most important properties of prisms (or, more precisely, iterated perfect sets) and distinguishes them from cubes. (Compare also [73, thm. 20].)

Lemma 3.2.5 *For every $0 < \alpha < \omega_1$, $E \in \mathbb{P}_\alpha$, a Polish space X, and a continuous function $f : E \to X$ there exists a $P \in \mathbb{P}_\alpha$ such that $P \subset E$ and either f is constant on P or else there exists a $0 < \beta \le \alpha$ such that $f \circ \pi_\beta^{-1}$ is a one to one function on $\pi_\beta[P] \in \mathbb{P}_\beta$.*

Notice that the property that "$f \circ \pi_\beta^{-1}$ *is a function on* $\pi_\beta[P]$" means simply that the value of $f \restriction P$ at $x \in P$ depends only on $x \restriction \beta$, the first β coordinates of x. Also, we could eliminate from the lemma the case "f *is constant on* P" if we allow $\beta = 0$, but then we could not claim that $\pi_\beta[P] = \{\emptyset\}$ belongs to $\mathbb{P}_\beta$ because $\mathbb{P}_0$ is not defined.

PROOF. Since $E \in \mathbb{P}_\alpha$, there is a $g \in \Phi_{\text{prism}}(\alpha)$ mapping $\mathfrak{C}^\alpha$ onto E. Notice that it is enough to prove the lemma for $\bar{f} = f \circ g$ and $\bar{E} = \mathfrak{C}^\alpha$: If $\bar{P}$ satisfies the lemma for this pair, then $P = g[\bar{P}]$ satisfies it for the original pair. Thus, without loss of generality we can assume that $E = \mathfrak{C}^\alpha$.

If there is a $P \in \mathbb{P}_\alpha$ on which f is constant, then we are done. So, assume that this is not the case, that is, that

$$\text{there is no } P \in \mathbb{P}_\alpha \text{ such that } f \text{ is constant on } P. \tag{3.16}$$

First we prove that under the additional assumption that

($\bullet$) $f \circ \pi_\gamma^{-1}$ is a function on $\pi_\beta[\hat{E}]$ for no $\gamma < \alpha$ and $\hat{E} \in \mathbb{P}_\alpha$

we can find a $P \in \mathbb{P}_\alpha$ on which f is one to one.

To see this, for every $0 < \gamma < \alpha$ consider the closed set

$$F_\gamma = \left\{ z \in \mathfrak{C}^\gamma : f \text{ is constant on } \{z\} \times \mathfrak{C}^{\alpha \setminus \gamma} \right\}$$

and notice that it must be nowhere dense: If it had contained a nonempty open ball B, then $P = B \times \mathfrak{C}^{\alpha \setminus \gamma} \in \mathbb{P}_\alpha$ would have contradicted ($\bullet$). Thus, the set $F = \bigcup_{0 < \gamma < \alpha} F_\gamma \times \mathfrak{C}^{\alpha \setminus \gamma}$ is meager, and so, by Corollary 3.2.2, we can find a comeager set $G \subset \mathfrak{C}^\alpha \setminus F$ such that for every $x \in G$ and $\gamma < \alpha$ the set $G_{x \restriction \gamma}$ is comeager in $\mathfrak{C}^{\alpha \setminus \gamma}$. But this implies that

$$f \text{ is not constant on } \{x \restriction \gamma\} \times G_{x \restriction \gamma} \text{ for any } \gamma < \alpha \text{ and } x \in G,$$

since otherwise f would be constant on the closure of $\{x \restriction \gamma\} \times G_{x \restriction \gamma}$, which is equal to $\{x \restriction \gamma\} \times \mathfrak{C}^{\alpha \setminus \gamma}$, and x would belong to F, contradicting $x \in G$. So the relation $R = \{\langle x_0, x_1 \rangle : f(x_0) = f(x_1)\}$ and G satisfy the condition (bin) from Proposition 3.2.4. Therefore, there exists an R-independent $P \in \mathbb{P}_\alpha$ and it is easy to see that f is one to one on such a P. Thus ($\bullet$) implies what we promised.

To prove the lemma in a general case, let $\beta \leq \alpha$ be the smallest ordinal such that $\hat{f} = f \circ \pi_\beta^{-1}$ is a function on $\pi_\beta[\hat{E}] \in \mathbb{P}_\beta$ for some $\hat{E} \in \mathbb{P}_\alpha$. Note that, by (3.16), $\beta > 0$. Using the argument from the first paragraph of the proof we can assume that $\hat{E} = \mathfrak{C}^\alpha$. (Recall that the function g is in this argument is projection-keeping.) Then, the minimality of β implies that $\hat{f}$ satisfies ($\bullet$). Thus, from what we have already proved, we can conclude that there is a $\hat{P} \in \mathbb{P}_\beta$ such that $\hat{f}$ is one to one on $\hat{P}$. Then $P = \hat{P} \times \mathfrak{C}^{\alpha \setminus \beta} \in \mathbb{P}_\alpha$ and β are as desired. ∎

Remark 3.2.6 Notice that Lemma 3.2.5 is false if we replace $\mathbb{P}_\alpha$ with CUBE(α). Indeed, this is obviously the case if we take $f : \mathfrak{C}^2 \to \mathfrak{C}$ given by $f(x_0, x_1) = x_1$.

The result presented in the reminder of this section will be used only in Section 5.1. We say that an n-ary relation F on a Polish space X is *symmetric* provided for any sequence $\langle x_i \in X : i < n \rangle$ and any permutation π of n

$$F(x_0, \ldots, x_{n-1}) \text{ holds if and only if } F\left(x_{\pi(0)}, \ldots, x_{\pi(n-1)}\right) \text{ holds.}$$

For such an F and $A \subset X$ we put

$$F * A = A \cup \{x \in X : (\exists a_1, \ldots, a_{n-1} \in A) \ F(x, a_1, \ldots, a_{n-1})\}.$$

If F is unary relation, we interpret the above as $F * A = A \cup F$. If $\mathcal{F}$ is a family of symmetric finitary relations on X, then we put $\mathcal{F} * A = \bigcup_{F \in \mathcal{F}} F * A$. Also, an $\mathcal{F}$-*closure of* A, denoted by $\mathrm{cl}_\mathcal{F}(A)$, is the least $B \subset X$ containing A such that $\mathcal{F} * B = B$. Note that $\mathrm{cl}_\mathcal{F}(A) = \bigcup_{n < \omega} \mathcal{F}^n * A$, where $\mathcal{F}^0 * A = A$ and $\mathcal{F}^{n+1} * A = \mathcal{F} * (\mathcal{F}^n * A)$. Thus, if $\mathcal{F}$ is a countable family of closed symmetric finitary relations, then $\mathrm{cl}_\mathcal{F}(A)$ is F_σ in X for a σ-compact $A \subset X$ since $F * K$ is closed for every $F \in \mathcal{F}$ and compact $K \subset X$.

We are most interested in these notions when we are concerned with either linear independence (over $\mathbb{Q}$) or algebraic independence in $\mathbb{R}$. In the first case, $\mathcal{F}$ is defined as the family of all relations F_w of all $\langle x_0, \ldots, x_{n-1} \rangle$ for which

$$w(x_{\pi(0)}, \ldots, x_{\pi(n-1)}) = 0 \text{ for some permutation } \pi \text{ of } n,$$

where w is a nonzero linear function with rational coefficients. In this case $\mathcal{F}$-independence stands for linear independence (over $\mathbb{Q}$) and $\mathrm{cl}_\mathcal{F}(A)$ is the linear span of A. When $\mathcal{F}$ is the family of all relations F_w, where w spans over all nonzero polynomials with rational coefficients, then $\mathcal{F}$-independence stands for algebraic independence while $\mathrm{cl}_\mathcal{F}(A)$ is the algebraic closure of $\mathbb{Q}(A)$.

We will also need one more notion. For a family $\mathcal{F}$ of closed symmetric finitary relations on X and an $M \subset X$ we define $\mathcal{F}_M$ as the collection of all possible projections of the relations from $\mathcal{F}$ along M. In other words, $\mathcal{F}_M$ is the collection of all (symmetric) relations

$$\{\langle x_0, \ldots, x_{k-1} \rangle : (\exists a_k, \ldots, a_{n-1} \in M)\ F(x_0, \ldots, x_{k-1}, a_k, \ldots, a_{n-1})\},$$

where $F \in \mathcal{F}$ is an n-ary relation and $0 < k \leq n$. Note that if M is compact, then each relation in $\mathcal{F}_M$ is still closed and for every $A \subset X$ we have

$$\mathrm{cl}_\mathcal{F}(M \cup A) = \mathrm{cl}_{\mathcal{F}_M}(A). \tag{3.17}$$

Also, if M is $\mathcal{F}$-independent, then

$$A \cup M \text{ is } \mathcal{F}\text{-independent provided } A \text{ is } \mathcal{F}_M\text{-independent.} \tag{3.18}$$

The following lemma will be the crucial for our applications presented in Section 5.1. We will also present there, in Remark 5.1.6, an example showing that in the lemma we cannot require $R = Q$.

Lemma 3.2.7 *Let $\mathcal{F}$ be an arbitrary family of closed symmetric finitary relations in a Polish space X. Then for every prism P in X there exists a subprism Q of P and a compact $\mathcal{F}$-independent set $R \subset P$ such that $Q \subset \mathrm{cl}_{\mathcal{F}}(R)$.*

PROOF. For $0 < \alpha < \omega_1$, let I_α be the statement:

I_α: The lemma holds for any prism P with witness function $f: \mathfrak{C}^\alpha \to P$.

We will prove I_α by induction on α.

First notice that I_α implies the following:

I_α^*: For every $k < \omega$ and continuous functions $g_0, \dots, g_k: \mathfrak{C}^\alpha \to X$, there exist an $E \in \mathbb{P}_\alpha$ and a compact $\mathcal{F}$-independent set $R \subset \bigcup_{i \le k} g_i[\mathfrak{C}^\alpha]$ such that $\bigcup_{i \le k} g_i[E] \subset \mathrm{cl}_{\mathcal{F}}(R)$.

To see that I_α^* holds true for $k = 0$, for every n-ary relation $F \in \mathcal{F}$ define $F^0 = \{\langle x_0, \dots, x_{n-1}\rangle \in (\mathfrak{C}^\alpha)^n : F(g_0(x_0), \dots, g_0(x_{n-1}))\}$. By I_α applied to $\mathcal{F}_0 = \{F^0 : F \in \mathcal{F}\}$ we can find an $\mathcal{F}_0$-independent set $R_0 \subset \mathfrak{C}^\alpha$ and an $E \in \mathbb{P}_\alpha$ such that $E \subset \mathrm{cl}_{\mathcal{F}_0}(R)$. But then $R = g_0[R_0]$ is compact $\mathcal{F}$-independent and $g_0[E] \subset \mathrm{cl}_{\mathcal{F}}(g_0[R_0]) = \mathrm{cl}_{\mathcal{F}}(R)$.

To make an inductive step assume that I_α^* holds for some $k < \omega$ and take continuous functions $g_0, \dots, g_{k+1}: \mathfrak{C}^\alpha \to X$. By the inductive assumption we can find an $E_0 \in \mathbb{P}_\alpha$ and a compact $\mathcal{F}$-independent set $R_0 \subset \bigcup_{i \le k} g_i[\mathfrak{C}^\alpha]$ such that $\bigcup_{i \le k} g_i[E_0] \subset \mathrm{cl}_{\mathcal{F}}(R_0)$. Let $h \in \Phi_{\mathrm{prism}}(\alpha)$ be a mapping from $\mathfrak{C}^\alpha$ onto E_0. Using the case $k = 0$ to the function $g_{k+1} \circ h$ and the family $\mathcal{F}_{R_0}$ we can find an $E_1 \in \mathbb{P}_\alpha$ and a compact $\mathcal{F}_{R_0}$-independent set $R_1 \subset (g_{k+1} \circ h)[\mathfrak{C}^\alpha]$ such that $(g_{k+1} \circ h)[E_1] \subset \mathrm{cl}_{\mathcal{F}_{R_0}}(R_1)$. Then, by (3.18), we conclude that $R = R_0 \cup R_1$ is $\mathcal{F}$-independent. Put $E = h[E_1] \in \mathbb{P}_\alpha$. Then, by (3.17), we have $g_{k+1}[E] \subset \mathrm{cl}_{\mathcal{F}_{R_0}}(R_1) = \mathrm{cl}_{\mathcal{F}}(R_0 \cup R_1) = \mathrm{cl}_{\mathcal{F}}(R)$, while clearly $\bigcup_{i \le k} g_i[E] \subset \bigcup_{i \le k} g_i[E_0] \subset \mathrm{cl}_{\mathcal{F}}(R_0) \subset \mathrm{cl}_{\mathcal{F}}(R)$. Thus, E and R satisfy I_α^*.

Now we are ready to prove I_α. So, fix $0 < \alpha < \omega_1$ and assume that I_γ is true for all $0 < \gamma < \alpha$. Let P be a prism in X with witness function $f: \mathfrak{C}^\alpha \to P$. We need to find an appropriate Q and R.

Let W be the set of all $\beta \le \alpha$ for which there exists an $E \in \mathbb{P}_\alpha$ and an $F \in \mathcal{F}$ such that for every $z \in \pi_\beta[E]$ there is a finite set $R_z \subset P$ for which

$$f[\{x \in E : z \subset x\}] \subset F * R_z. \tag{3.19}$$

Notice that W is nonempty since $\alpha \in W$. So $\beta = \min W$ is well defined. Let $E \in \mathbb{P}_\alpha$ be such that (3.19) holds for β. As usual, replacing f with its composition with an appropriate function from $\Phi_{\mathrm{prism}}(\alpha)$, if necessary, we can assume that $E = \mathfrak{C}^\alpha$.

If $\beta = 0$, then $f[\mathfrak{C}^\alpha] \subset \mathrm{cl}_{\mathcal{F}}(R_0)$ for some finite set $R_0 \subset P$, and we can find an $\mathcal{F}$-independent finite $R \subset R_0$ with $f[\mathfrak{C}^\alpha] \subset \mathrm{cl}_{\mathcal{F}}(R)$. (Note that if T is $\mathcal{F}$-independent and $x \in X \setminus \mathrm{cl}_{\mathcal{F}}(T)$, then $T \cup \{x\}$ is also $\mathcal{F}$-independent.) Thus, $Q = f[\mathfrak{C}^\alpha]$ and R satisfy I_α. So, for the rest of the proof we will assume that $\beta > 0$.

Next, assume that $0 < \beta < \alpha$. Let $\mathcal{B}_\beta$ be a countable basis of $\mathfrak{C}^{\alpha \setminus \beta}$ consisting of nonempty clopen sets and assume that F satisfying (3.19) is $(n+1)$-ary. For every $B \in \mathcal{B}_\beta$ consider the set

$$K_B = \{z \in \mathfrak{C}^\beta \colon (\exists \langle x_1, \ldots, x_n \rangle \in P^n)\, (\forall y \in B)\ F(f(z \cup y), x_1, \ldots, x_n)\}.$$

It is easy to see that each set K_B is closed. Notice also that

$$\mathfrak{C}^\beta = \bigcup_{B \in \mathcal{B}_\beta} K_B. \tag{3.20}$$

To see this, fix a $z \in \mathfrak{C}^\beta$. By (3.19), there exists a finite set $S_z \subset \mathfrak{C}^\alpha$ such that $\mathfrak{C}^{\alpha \setminus \beta} = \bigcup_{x_1, \ldots, x_n \in f[S_z]} \{y \in \mathfrak{C}^{\alpha \setminus \beta} \colon F(f(z \cup y), x_1, \ldots, x_n)\}$. Since each set $\{y \in \mathfrak{C}^{\alpha \setminus \beta} \colon F(f(z \cup y), x_1, \ldots, x_n)\}$ is closed, one of them must contain a $B \in \mathcal{B}_\beta$, and so $z \in K_B$.

Thus, by (3.20), there exists a $B \in \mathcal{B}_\beta$ such that K_B has a nonempty interior. In particular, there is a nonempty clopen set $U \subset K_B$. But then for every $z \in U$ there is a $g(z) = \langle g_1(z), \ldots, g_n(z) \rangle \in P^n$ such that $F(f(z \cup y), g_1(z), \ldots, g_n(z))$ holds for every $y \in B$. Now

$$T = \{\langle z, \bar{p} \rangle \in U \times P^n \colon (\forall y \in B)\ F(f(z \cup y), \bar{p})\}$$

is a compact subset of $U \times P^n$ and g constitutes a selector of T. Thus, we can choose g to be Borel. In particular, there is a dense G_δ subset W of U such that $g \upharpoonright W$ is continuous. So, by Claim 1.1.5, we can find a perfect cube $C \subset W \subset \mathfrak{C}^\beta$. Now, identifying C with $\mathfrak{C}^\beta$, we conclude that the functions $g_1, \ldots, g_n \colon \mathfrak{C}^\beta \to P$ are continuous and that the relation $F(f(z \cup y), g_1(z), \ldots, g_n(z))$ holds for every $z \in \mathfrak{C}^\beta$ and $y \in B$.

Since, by the inductive hypothesis, I_β is true, condition I_β^* holds as well. Thus, there exist an $E \in \mathbb{P}_\beta$ and a compact $\mathcal{F}$-independent set $R \subset P$ such that $\bigcup_{i=1}^n g_i[E] \subset \mathrm{cl}_{\mathcal{F}}(R)$. Since $Q = f[E \times B]$ is a subprism of P, we just need to show that $Q \subset \mathrm{cl}_{\mathcal{F}}(R)$. To see this just note that for every $z \in E$ we have $f[\{z\} \times B] \subset F * \{g_1(z), \ldots, g_n(z)\} \subset \mathrm{cl}_{\mathcal{F}}\left(\bigcup_{i=1}^n g_i[E]\right) \subset \mathrm{cl}_{\mathcal{F}}(R)$. This finishes the proof of the case $0 < \beta < \alpha$.

For the reminder of the proof we will assume that $\beta = \alpha$. This means that there is no $E \in \mathbb{P}_\alpha$ such that for some $F \in \mathcal{F}$ and $\beta < \alpha$

$$(\forall z \in \pi_\beta[E]) \, (\exists R_z \in [P]^{<\omega}) \; f[\{x \in E: z \subset x\}] \subset F * R_z. \qquad (3.21)$$

For every n-ary $F \in \mathcal{F}$, let $F^* = \{\langle x_0, \ldots, x_{n-1}\rangle: F(f(x_0), \ldots, f(x_{n-1}))\}$ and let $\mathcal{F}^* = \{F^*: F \in \mathcal{F}\}$. We will apply Proposition 3.2.3 to find an $\mathcal{F}^*$-independent $E \in \mathbb{P}_\alpha$. Then $Q = f[E]$ is an $\mathcal{F}$-independent subprism of P and together with $R = Q$ they satisfy the lemma.

To see that the assumptions of Proposition 3.2.3 are satisfied, first notice that unary relations in $\mathcal{F}^*$ are nowhere dense. Indeed, otherwise there is a unary relation $F^* \in \mathcal{F}^*$ and a nonempty clopen set $E \subset F^*$. But then E contradicts (3.21), as $f[E] \subset F * \emptyset$. Thus, we just need to show that the condition (ex) is satisfied.

So, fix an $F \in \mathcal{F}$. For $0 < \beta < \alpha$ and $B \in \mathcal{B}_\beta$ let

$$K(B) = \{z \in \mathfrak{C}^\beta: (\exists R_z \in [P]^{<\omega}) \; f[\{z\} \times B] \subset F * R_z\}.$$

Clearly $K(B)$ is F_σ. Notice also that it is meager, since otherwise there would exist a nonempty clopen $U \subset K(B)$ and $E = U \times B$ would contradict (3.21). Thus, each set $K_\beta = \bigcup_{B \in \mathcal{B}_\alpha} K(B)$ is meager. Also, for every $z \in \mathfrak{C}^\beta \setminus K_\beta$ and for every finite $R \subset P$ the set $\{y \in \mathfrak{C}^{\alpha \setminus \beta}: f(z \cup y) \notin F * R\}$ is dense and open. In particular, if R is a finite F-independent subset of P, then

$$W_R = \{y \in \mathfrak{C}^{\alpha \setminus \beta}: R \cup \{f(z \cup y)\} \text{ is } F\text{-independent}\} \qquad (3.22)$$

is dense and open. Let

$$H = \bigcap_{0 < \beta < \alpha} \left((\mathfrak{C}^\beta \setminus K_\beta) \times \mathfrak{C}^{\alpha \setminus \beta} \right)$$

and notice that H is comeager since each K_β is meager in $\mathfrak{C}^\beta$. By Corollary 3.2.2 we can find a comeager set $G \subset H$ such that

$$G_{x \restriction \beta} = \{y \in \mathfrak{C}^{\alpha \setminus \beta}: (x \restriction \beta) \cup y \in G\}$$

is comeager for every $x \in G$ and $\beta < \alpha$. To finish the proof it is enough to show that G satisfies (ex) for F^*. So, take an F^*-independent finite set $S \subset G$, an $x \in S$, and a $\beta < \alpha$.

First let us assume that $\beta > 0$. Then $x \in S \subset G \subset H$ implies that $z = x \restriction \beta \in \mathfrak{C}^\beta \setminus K_\beta$. In particular, the set $W_{f[S]}$ from (3.22) is comeager, and so is $W_{f[S]} \cap G_{x \restriction \beta}$. To get (ex) it is enough to notice that $W_{f[S]} \cap G_{x \restriction \beta}$ is a subset of $\{y \in \mathfrak{C}^{\alpha \setminus \beta}: S \cup \{y \cup z\} \subset G \text{ is } F^*\text{-independent}\}$.

Finally, assume that $\beta = 0$. We need to show that the set

$$\{y \in G: S \cup \{y\} \text{ is } F^*\text{-independent}\}$$

is dense. But this set must be comeager, since otherwise its complement would contain a nonempty clopen set E, which wold contradict (3.21) with $\beta = 0$.   ∎

3.3 CPA$_{\text{prism}}$, additivity of s_0, and more on (A)

The results presented in this section come from K. Ciesielski and J. Pawlikowski [43].

We will start with noticing that the axiom CPA$_{\text{prism}}$ leads in a natural way to the following generalization of the ideal s_0^{cube}:

$$s_0^{\text{prism}} = \left\{ X \setminus \bigcup \mathcal{E} \colon \mathcal{E} \text{ is } \mathcal{F}_{\text{prism}}\text{-dense in } \text{Perf}(X) \right\}.$$

Similarly as for Proposition 1.0.3, it can be shown that

Proposition 3.3.1 *If* CPA$_{\text{prism}}$ *holds, then* $s_0^{\text{prism}} = [X]^{\leq \omega_1}$.

It can be also shown, refining the argument for Fact 1.0.4, that

Fact 3.3.2 *For a Polish space X we have* $[X]^{<\mathfrak{c}} \subset s_0^{\text{cube}} \subset s_0^{\text{prism}} \subset s_0$.

However, we will not use these facts in the rest of this text.

The next lemma and its corollaries represent a very useful application of Lemma 3.2.5. For a fixed Polish space X and $0 < \alpha < \omega_1$, let $\mathcal{F}^\alpha$ denote the family of all continuous injections from $\mathfrak{C}^\alpha$ into X. Note that if we consider $\mathcal{F}^\alpha$ with the topology of uniform convergence, then

$$\mathcal{F}^\alpha \text{ is a Polish space.} \tag{3.23}$$

To prove (3.23) it is enough to show that $\mathcal{F}^\alpha$ is a G_δ subset of the space $\mathcal{C} = \mathcal{C}(\mathfrak{C}^\alpha, X)$ of all continuous functions from $\mathfrak{C}^\alpha$ into X. But $\mathcal{F}^\alpha$ is the intersection of the open sets G_n, $n < \omega$, where the sets G_n are constructed as follows. Fix a finite partition $\mathcal{P}_n$ of $\mathfrak{C}^\alpha$ into clopen sets each of the diameter less than 2^{-n}, and let $\mathcal{H}_n$ be the family of all mappings h from $\mathcal{P}_n$ into the topology of X such that $h(P) \cap h(P') = \emptyset$ for distinct $P, P' \in \mathcal{P}_n$. We put

$$G_n = \bigcup_{h \in \mathcal{H}_n} \{f \in \mathcal{C} : (\forall P \in \mathcal{P}_n)(\forall x \in P)\ f(x) \in h(P)\}.$$

This completes the argument for (3.23).

Lemma 3.3.3 *Let X be a Polish space and $0 < \alpha < \omega_1$. Then every function $f: \mathfrak{C}^\beta \to \mathcal{F}^\alpha$ from $\mathcal{F}_{\mathrm{prism}}(\mathcal{F}^\alpha)$ has a restriction $f^* \in \mathcal{F}_{\mathrm{prism}}(\mathcal{F}^\alpha)$ with the property that there exists an $\hat{f} \in \mathcal{F}_{\mathrm{prism}}(X)$ defined on a subset of $\mathfrak{C}^{\beta+\alpha}$ such that*

(a) *$\hat{f}(s,t) = f^*(s)(t)$ for all $\langle s, t \rangle \in \left(\mathfrak{C}^\beta \times \mathfrak{C}^\alpha \right) \cap \mathrm{dom}(\hat{f})$, and*

(b) *for each $s \in \mathrm{dom}(f^*)$ function $\hat{f}(s, \cdot): \{t \in \mathfrak{C}^\alpha: \langle s, t \rangle \in \mathrm{dom}(\hat{f})\} \to X$ is a restriction of $f^*(s)$ and belongs to $\mathcal{F}_{\mathrm{prism}}(X)$.*

PROOF. Let $f: \mathfrak{C}^\beta \to \mathcal{F}^\alpha$, $f \in \mathcal{F}_{\mathrm{prism}}(\mathcal{F}^\alpha)$, and define a function g from a set $\mathfrak{C}^\beta \times \mathfrak{C}^\alpha = \mathfrak{C}^{\beta+\alpha}$ into X by $g(s,t) = f(s)(t)$ for $\langle s, t \rangle \in \mathfrak{C}^\beta \times \mathfrak{C}^\alpha$. It is easy to see that g is continuous.

Apply Lemma 3.2.5 to $E = \mathfrak{C}^{\beta+\alpha} \in \mathbb{P}_{\beta+\alpha}$ and to the function g to find a $\gamma \le \beta + \alpha$ and a subset $P \in \mathbb{P}_{\beta+\alpha}$ of E such that $g \circ \pi_\gamma^{-1}$ is a function on $\pi_\gamma[P] \in \mathbb{P}_\gamma$ that is either one to one or constant. Let $f^* = f \restriction \pi_\beta[P]$. We will show that it is as desired.

First note that

$$\gamma = \beta + \alpha \text{ and } g \text{ is one to one on } P.$$

Indeed, if $z \in \mathrm{range}(f^*) \cap \mathcal{F}_{\mathrm{prism}}(X)$ and $z = f^*(s)$, then for every different $t_0, t_1 \in \mathfrak{C}^\alpha$ with $\langle s, t_0 \rangle, \langle s, t_1 \rangle \in P$ we have $g(s, t_0) = f(s)(t_0) = z(t_0) \ne z(t_1) = g(s, t_1)$. So, g cannot be constant and if $\gamma < \beta + \alpha$, then we can find t_0 and t_1 such that $\pi_\gamma(\langle s, t_0 \rangle) = \pi_\gamma(\langle s, t_1 \rangle)$, contradicting the above calculation.

It is easy to see that $\hat{f} = g \restriction P$ is as desired. ∎

Lemma 3.3.3 implies the following useful fact.

Proposition 3.3.4 $\mathrm{CPA}_{\mathrm{prism}}$ *implies that for every Polish space X there exists a family $\mathcal{H}$ of continuous functions from compact subsets of X onto $\mathfrak{C} \times \mathfrak{C}$ such that $|\mathcal{H}| \le \omega_1$ and*

- *for every prism P in X there are $h \in \mathcal{H}$ and $c \in \mathfrak{C}$ such that $h^{-1}(\{c\} \times \mathfrak{C})$ and $h^{-1}(\langle c, d \rangle)$ are subprisms of P for every $d \in \mathfrak{C}$.*

In particular, $\mathcal{F} = \{h^{-1}(\{c\} \times \mathfrak{C}): h \in \mathcal{H} \ \& \ c \in \mathfrak{C}\}$ is $\mathcal{F}_{\mathrm{prism}}$-dense in X.

PROOF. Let $0 < \alpha < \omega_1$. We will use the notation as in Lemma 3.3.3.

Since the family of all sets $\mathrm{range}(f^*)$ is $\mathcal{F}_{\mathrm{prism}}$-dense in $\mathcal{F}^\alpha$ by $\mathrm{CPA}_{\mathrm{prism}}$, we can find a $\mathcal{G}_\alpha = \{f_\xi^*: \xi < \omega_1\}$ such that $R_\alpha = \mathcal{F}^\alpha \setminus \bigcup_{\xi < \omega_1} \mathrm{range}(f_\xi^*)$ has cardinality less than or equal to ω_1.

If $f^* \in \mathcal{G}_\alpha$, then $\hat{f}$ maps injectively a $P = P_f \in \mathbb{P}_{\beta+\alpha}$ onto $Q = Q_f \subset X$.

Moreover, for every $z \in \mathcal{F}^\alpha \setminus R_\alpha$ there are $f^* \in \mathcal{G}_\alpha$ and $s \in \text{dom}(f^*)$ such that $z = f^*(s)$ and $\hat{f}(s, \cdot) \in \mathcal{F}_{prism}(X)$ is a restriction of z.

Now, let $H_f \in \Phi_{prism}(\beta + \alpha)$ be from $\mathfrak{C}^{\beta+\alpha}$ onto P and consider the composition $\hat{f} \circ H_f : \mathfrak{C}^{\beta+\alpha} \to Q$. Then the functions $(\hat{f} \circ H_f)^{-1} : Q_f \to \mathfrak{C}^{\beta+\alpha}$ are our desired functions modulo some projections. More precisely, let $k_0 : \mathfrak{C}^\beta \to \mathfrak{C}$ be a homeomorphism and choose a mapping $k_1 : \mathfrak{C} \to \mathfrak{C}$ be such that $k_1^{-1}(c) \in \text{Perf}(\mathfrak{C})$ for every $c \in \mathfrak{C}$. Define $h_f^\alpha : Q_f \to \mathfrak{C} \times \mathfrak{C}$ by

$$h_f^\alpha(x) = \langle (k_0 \circ \pi_\beta)((\hat{f} \circ H_f)^{-1}(x)), k_1([(\hat{f} \circ H_f)^{-1}(x)](\beta)) \rangle.$$

Then the family $\mathcal{H}_0 = \{h_f^\alpha : \alpha < \omega_1 \ \& \ f^* \in \mathcal{G}_\alpha\}$ works for all functions not in $R = \bigcup_{0 < \alpha < \omega_1} R_\alpha$. Also, for every function $g \in R$ it is easy to find a continuous function h_g from $\text{range}(g)$ onto $\mathfrak{C} \times \mathfrak{C}$ such that $h_g^{-1}(\{c\} \times \mathfrak{C})$ and $h_g^{-1}(\langle c, d \rangle)$ are subprisms of $\text{range}(g)$ for every $c, d \in \mathfrak{C}$. Then the family $\mathcal{H} = \mathcal{H}_0 \cup \{h_g : g \in R\}$ is as desired.   ∎

Proposition 3.3.4 implies the following stronger version of property (A). This can be considered as a version of a remark due to A. Miller [95, p. 581], who noticed that in the iterated perfect set model functions coded in the ground model can be taken as a family $\mathcal{G}$.

Corollary 3.3.5 *Assume that* CPA$_{prism}$ *holds. Then:*

(A*) *There exists a family $\mathcal{G}$ of uniformly continuous functions from $\mathbb{R}$ to $[0,1]$ such that $|\mathcal{G}| = \omega_1$ and for every $S \in [\mathbb{R}]^{\mathfrak{c}}$ there exists a $g \in \mathcal{G}$ with $g[S] = [0,1]$.*

PROOF. Let $\mathcal{H}$ be as in Proposition 3.3.4 for $X = \mathbb{R}$, $k : \mathfrak{C} \to [0,1]$ be continuous surjection, and for every $h = \langle h_0, h_1 \rangle \in \mathcal{H}$ let $g_h : \mathbb{R} \to [0,1]$ be a continuous extension of a function $h^* : \text{dom}(h) \to [0,1]$ defined by $h^*(x) = k(h_1(x))$. We claim that $\mathcal{G} = \{g_h : h \in \mathcal{H}\}$ is as desired.

Since, by Proposition 3.3.1, $s_0^{prism} = [\mathbb{R}]^{\leq \omega_1}$, there exists a prism P in $\mathbb{R}$ such that S intersects every subprism of P. Let $h \in \mathcal{H}$ and $c \in \mathfrak{C}$ be such that $h^{-1}(\{c\} \times \mathfrak{C})$ and $h^{-1}(\langle c, d \rangle)$ are subprisms of P for every $d \in \mathfrak{C}$. Then S intersects each $h^{-1}(\langle c, d \rangle)$ and so $h[S]$ contains $\{c\} \times \mathfrak{C}$. Thus $g_h[S] = [0,1]$.   ∎

Next we will show that CPA$_{prism}$ implies[1] that the additivity of the ideal s_0

$$\text{add}(s_0) = \min\left\{|F| : F \subset s_0 \ \& \ \bigcup F \notin s_0\right\},$$

[1] In fact, this also follows from CPA$_{cube}$.

is equal to ω_1. This stays in contrast with Proposition 6.1.1, in which we will show that CPA implies $\mathrm{cov}(s_0) = \omega_2$.

Notice that the numbers $\mathrm{add}(s_0)$, $\mathrm{cov}(s_0)$, $\mathrm{non}(s_0)$, and $\mathrm{cof}(s_0)$ have been intensively studied. (See, e.g., [71].) It is known that $\mathrm{cof}(s_0) > \mathfrak{c}$ (see [71, thm. 1.3]) and that $\mathrm{non}(s_0) = \mathfrak{c}$ since there are s_0-sets of cardinality $\mathfrak{c}$. There are models of ZFC+MA with $\mathfrak{c} = \omega_2$ and $\mathrm{cov}(s_0) = \omega_1$, while the proper forcing axiom, PFA, implies that $\mathrm{add}(s_0) = \mathfrak{c}$.

In what follows we need the following useful fact, which is essentially [71, lem. 1.1]. (Compare also [111, thm 2.4(1)].)

Fact 3.3.6 *For any open dense subset $\mathcal{D}$ of $\mathrm{Perf}(\mathfrak{C})$ (considered as ordered by inclusion) there exists a maximal antichain $\mathcal{A} \subset \mathcal{D}$ consisting of pairwise disjoint sets such that every $P \in \mathrm{Perf}(\bigcup \mathcal{A})$ is covered by less than continuum many sets from $\mathcal{A}$.*

PROOF. Let $\mathrm{Perf}(\mathfrak{C}) = \{P_\alpha : \alpha < \mathfrak{c}\}$. We will build inductively a sequence $\langle \langle A_\alpha, x_\alpha \rangle \in \mathcal{D} \times \mathfrak{C} : \alpha < \mathfrak{c} \rangle$ aiming for $\mathcal{A} = \{A_\alpha : \alpha < \mathfrak{c}\}$. At step $\alpha < \mathfrak{c}$, given already $\langle \langle A_\beta, x_\beta \rangle : \beta < \alpha \rangle$, we look at P_α.

Choice of x_α: If $P_\alpha \subset \bigcup_{\beta < \alpha} A_\beta$, we take x_α as an arbitrary element of $\mathfrak{C}$; otherwise we pick $x_\alpha \in P_\alpha \setminus \bigcup_{\beta < \alpha} A_\beta$.

Choice of A_α: If there is a $\beta < \alpha$ such that $P_\alpha \cap A_\beta$ is uncountable, we let $A_\alpha = A_\beta$; otherwise pick $A_\alpha \in \mathcal{D}$ below P_α and notice that we can refine it, if necessary, to be disjoint with $\bigcup_{\beta < \alpha} A_\beta \cup \{x_\beta : \beta \leq \alpha\}$.

It is easy to see that $\mathcal{A} = \{A_\alpha : \alpha < \mathfrak{c}\}$ is as required. ∎

Corollary 3.3.7 $\mathrm{CPA}_{\mathrm{prism}}$ *implies that* $\mathrm{add}(s_0) = \omega_1$.

PROOF. Let $\mathcal{H} = \{h_\xi : \xi < \omega_1\}$ be as in Proposition 3.3.4 with $X = \mathfrak{C}$. For every $\xi < \omega_1$ put $\mathcal{A}^0_\xi = \{h^{-1}_\xi(\{c\} \times \mathfrak{C}) : c \in \mathfrak{C}\}$. Then each $\mathcal{A}^0_\xi$ is a family of pairwise disjoint sets and $\mathcal{A}^0 = \bigcup_{\xi < \omega_1} \mathcal{A}^0_\xi$ is dense in $\mathrm{Perf}(\mathfrak{C})$.

For each $\xi < \omega_1$ let $\mathcal{A}^*_\xi$ be a maximal antichain extending $\mathcal{A}^0_\xi$, define $\mathcal{D}_\xi = \{P \in \mathrm{Perf}(\mathfrak{C}) : P \subset A \text{ for some } A \in \mathcal{A}^*_\xi\}$, and let $\mathcal{A}_\xi \subset \mathcal{D}_\xi$ be as in Fact 3.3.6. Then $\mathcal{A} = \bigcup_{\xi < \omega_1} \mathcal{A}_\xi$ is still dense in $\mathrm{Perf}(\mathfrak{C})$.

For each $\xi < \omega_1$ let $\{P^\alpha_\xi : \alpha < \mathfrak{c}\}$ be an enumeration of $\mathcal{A}_\xi$. (Note that each $\mathcal{A}_\xi$ has cardinality $\mathfrak{c}$, since this was the case for the sets $\mathcal{A}^0_\xi$.) Pick x^α_ξ from each P^α_ξ and put $A_\xi = \{x^\alpha_\xi : \alpha < \mathfrak{c}\}$. Then $A_\xi \in s_0$ for every $\xi < \omega_1$. However, $A = \bigcup_{\xi < \omega_1} A_\xi \notin s_0$ since it intersects every element of a dense set $\mathcal{A}$. ∎

It can be also shown that $\mathrm{CPA}_{\mathrm{prism}}$, with the help of Proposition 3.3.4, implies that the Sacks forcing $\mathbb{P} = \langle \text{Perf}(\mathfrak{C}), \subset \rangle$ collapses $\mathfrak{c}$ to ω_1. However, this also follows immediately from a theorem of P. Simon [122], that $\mathbb{P}$ collapses $\mathfrak{c}$ to $\mathfrak{b}$ while CPA_{cube} already implies that $\mathfrak{b} \leq \text{cof}(\mathcal{N}) = \omega_1$.

Note also that although the family $\mathcal{F}$ from Proposition 3.3.4 is $\mathcal{F}_{\text{prism}}$-dense, we certainly cannot repeat the proof of Corollary 3.3.7 to show that, under $\text{CPA}_{\text{prism}}$, $\text{add}(s_0^{\text{prism}}) = \omega_1$ – this clearly contradicts Proposition 3.3.1. The place where the proof breaks is Fact 3.3.6, which cannot be proved for a simple density being replaced by an $\mathcal{F}_{\text{prism}}$-density.

3.4 Intersections of ω_1 many open sets

The results presented in this section come from K. Ciesielski and J. Pawlikowski [41].

For a Polish space X let G_{ω_1} be the collection of the intersections of ω_1 many open subsets of X. Next we are going to prove the following theorem.

Theorem 3.4.1 $\text{CPA}_{\text{prism}}$ *implies that the following property holds for every Polish space X.*

(N*) *If G is a G_{ω_1} subset of X and $|G| = \mathfrak{c}$, then G contains a perfect set.*

Theorem 3.4.1 provides an affirmative answer to a question of J. Brendle, who asked us, in [11], whether (N*) can be deduced from our axiom CPA. The fact that (N*) holds in the iterated perfect set model is proved by J. Brendle, P. Larson, and S. Todorcevic [12]. The argument presented below is considerably simpler.

Before we prove Theorem 3.4.1, we would like to note that in property (N*) we can replace the class of open sets with a considerably larger class Π_2^1.

Corollary 3.4.2 *Assume that $\text{CPA}_{\text{prism}}$ holds and X is a Polish space.*

- *If G is an intersection of ω_1 many Π_2^1 sets from X and $|G| = \mathfrak{c}$, then G contains a perfect set.*

PROOF. Let $G = \bigcap_{\xi < \omega_1} T_\xi$, where each $T_\xi \subset X$ is a Π_2^1 set. Then we have $X \setminus G = \bigcup_{\xi < \omega_1} (X \setminus T_\xi)$ and each set $X \setminus T_\xi$ is in the class Σ_2^1; so, by Fact 1.1.7, it is a union of ω_1 many compact sets. Thus, each T_ξ is an intersection of ω_1 open sets. ∎

Theorem 3.4.1 follows easily from the following combinatorial fact concerning iterated perfect sets.

Proposition 3.4.3 *Let* $0 < \alpha < \omega_1 < \mathfrak{c}$ *and* $H = \mathfrak{C}^\alpha \setminus \bigcup_{\xi < \omega_1} F_\xi$, *where each set* F_ξ *is compact. Then either* H *contains a perfect set* P, *or else there exists an* $E \in \mathbb{P}_\alpha$ *disjoint with* H.

PROOF OF THEOREM 3.4.1. Let $G = X \setminus \bigcup_{\xi < \omega_1} T_\xi$, where each set T_ξ is closed in X, and assume that G does not contain a perfect set.

Let $\mathcal{E} = \{P \in \text{Perf}(X) : P \cap G = \emptyset\}$. We will show that $\mathcal{E}$ is $\mathcal{F}_{\text{prism}}$-dense. This will finish the proof since then, by CPA$_{\text{prism}}$, $X \setminus \bigcup \mathcal{E} \supset G$ has cardinality $\leq \omega_1 < \mathfrak{c}$.

So, let $f : \mathfrak{C}^\alpha \to X$ be a continuous injection. We need to find an $E \in \mathbb{P}_\alpha$ for which $f[E] \in \mathcal{E}$, that is, $f[E] \cap G = \emptyset$. Let $F_\xi = f^{-1}(T_\xi)$ for $\xi < \omega_1$. Then $H = \mathfrak{C}^\alpha \setminus \bigcup_{\xi < \omega_1} F_\xi$ is equal to $f^{-1}(G)$. If H contains a perfect set P, then so does $G \supset f[P]$, contradicting our assumption. Thus, by Proposition 3.4.3, there exists an $E \in \mathbb{P}_\alpha$ disjoint with $H = f^{-1}(G)$. So, $f[E]$ is disjoint with G. ∎

To prove Proposition 3.4.3 we need some auxiliary terminology. For a proper ideal I of subsets of $\mathfrak{C}$ and an ordinal $0 < \alpha < \omega_1$ we say that a tree $T \subset \mathfrak{C}^{\leq \alpha}$ (ordered by inclusion) is a *co-I tree*, $T \in \mathcal{T}_I^\alpha$, provided:

- For every $\beta < \alpha$ and $t \in T \cap \mathfrak{C}^\beta$ we have $\mathfrak{C} \setminus \text{succ}_T(t) \in I$, where $\text{succ}_T(t) = \{s(\beta) : t \subset s \in T \cap \mathfrak{C}^{\beta+1}\}$ is the set of all immediate successors of t in T.
- For every limit ordinal $\lambda \leq \alpha$ the λ-th level $T \cap \mathfrak{C}^\lambda$ of T consists all the branches of $T \cap \mathfrak{C}^{<\lambda}$, that is, $t \in T \cap \mathfrak{C}^\lambda$ if and only if $t \restriction \gamma \in T$ for every $\gamma < \lambda$.

Also let $\mathcal{K}_I^\alpha = \{T \cap \mathfrak{C}^\alpha : T \in \mathcal{T}_I^\alpha\}$ and let $I^\alpha = \bigcup_{K \in \mathcal{K}_I^\alpha} \mathcal{P}(\mathfrak{C}^\alpha \setminus K)$. It is easy to see that I^α is an ideal on $\mathfrak{C}^\alpha$. We will call I^α the α-th *Fubini power* of the ideal I. This notion, with a slightly different emphasis, also appears in J. Zapletal [131].

We will be interested in these notions only for I of the form $I_\kappa = [\mathfrak{C}]^{\leq \kappa}$, where $\kappa < \mathfrak{c}$ is equal to either ω or ω_1. It is easy to see that the families $\mathcal{T}_{I_\kappa}^\alpha$ and $\mathcal{K}_{I_\kappa}^\alpha$ are closed under the intersection of κ many of their elements. So $(I_\kappa)^\alpha$ is a κ^+-additive ideal on $\mathfrak{C}^\alpha$.

A big part of the difficulty behind our proof of Theorem 3.4.1 lies in the following lemma generalizing the Cantor-Bendixon theorem, that is, that an uncountable closed set contains a perfect set. In fact, this result is true not only for closed sets, but also for analytic sets. This version of the lemma, which can be viewed as a generalization of the Suslin theorem that an uncountable analytic set contains a perfect set, can be found in [131, lem. 4.1.29(2)]. (See also [131, lem. 3.3.1].)

Lemma 3.4.4 *Let* $0 < \alpha < \omega_1$. *If F is a closed subset of $\mathfrak{C}^\alpha$ and $F \notin (I_\omega)^\alpha$, then F contains an iterated perfect set, that is, there exists an $E \in \mathbb{P}_\alpha$ such that $E \subset F$.*

PROOF OF PROPOSITION 3.4.3. If there is a $\xi < \omega_1$ such that $F_\xi \notin (I_\omega)^\alpha$, then, by Lemma 3.4.4, there is an $E \in \mathbb{P}_\alpha$ such that $E \subset F_\xi$. Clearly this E is disjoint with $H = \mathfrak{C}^\alpha \setminus \bigcup_{\xi < \omega_1} F_\xi$. So, assume that $F_\xi \in (I_\omega)^\alpha$ for every $\xi < \omega$. We will show that H contains a perfect set.

For every $\xi < \omega_1$, since $F_\xi \in (I_\omega)^\alpha$, there is a $T_\xi \in T^\alpha_{I_\omega}$ disjoint with F_ξ. But then $T = \bigcap_{\xi < \omega_1} T_\xi \in T^\alpha_{I_{\omega_1}}$ is disjoint with $\bigcup_{\xi < \omega_1} F_\xi$, so T is a subset of H. Thus, it is enough to show that T contains a perfect set. But this follows immediately from the assumption that $\omega_1 < \mathfrak{c}$.

If α is a successor ordinal, say $\alpha = \beta + 1$, then take $t \in \pi_\beta[T]$ and a perfect set $C \subset \mathrm{succ}_T(t)$. Then $P = \{t\} \times C$ is a perfect subset of T.

If α is a limit ordinal, take an increasing sequence $\langle \alpha_n < \alpha : n < \omega \rangle$ cofinal with α. Choose a sequence $\langle t_\sigma \in \pi_{\alpha_n}[T] : n < \omega \ \& \ \sigma \in 2^n \rangle$, by induction on $n < \omega$, such that $t_{\sigma \smallfrown 0}$ and $t_{\sigma \smallfrown 1}$ are distinct extensions of t_σ. Then

$$P = \{t \in \mathfrak{C}^\alpha : t \restriction \alpha_n \in \{t_\sigma : \sigma \in 2^n\} \text{ for every } n < \omega\}$$

is a perfect subset of T.   ∎

J. Brendle, P. Larson, and S. Todorcevic conjectured in [12, conj. 5.11] that, in the iterated perfect set model, the complement of the G_{ω_1} set is also G_{ω_1}, that is, that any G_{ω_1} set is a union of ω_1 many closed sets. Below we will show that the conjecture is false, by proving that the existence of a counterexample follows from $\mathrm{CPA}_{\mathrm{prism}}$. More precisely, this follows from the following theorem, which is interesting on its own.

Let $\mathcal{X}$ be the family of all continuous functions f from an uncountable G_δ subset of $\mathfrak{C}$ into $\mathfrak{C}$.

Theorem 3.4.5 *Assume that $\mathrm{CPA}_{\mathrm{prism}}$ holds. Then there is a sequence $\langle g_\alpha \in \mathcal{X} : \alpha < \omega_1 \rangle$ such that for every $g \in \mathcal{X}$ there exists an $\alpha < \omega_1$ such that $|g \cap g_\alpha| = \mathfrak{c}$.*

Before we prove the theorem we notice that it easily implies the following corollary, which refutes [12, conj. 5.11]. It also gives a negative answer to [12, question 5.13]. Note that in the proof of the corollary we use from Theorem 3.4.5 only the fact that $g \cap g_\alpha \neq \emptyset$.

Corollary 3.4.6 *If* $\omega_1 < \mathfrak{c}$ *and* $G = \mathfrak{C}^2 \setminus \bigcup_{\alpha < \omega_1} g_\alpha$, *where the* g_α*'s are from Theorem 3.4.5, then* G *is not a union of* ω_1 *many closed sets.*

In particular, $\mathrm{CPA}_{\mathrm{prism}}$ *implies that* G *is* G_{ω_1} *but* $\mathfrak{C}^2 \setminus G$ *is not.*

PROOF. To see that G is not a union of ω_1 many closed sets, notice that

$$\text{if } P \subset G \text{ is compact, then } \pi_0[P] \text{ is countable.} \tag{3.24}$$

To see this, take a compact subset P of $\mathfrak{C}^2$ for which $\pi_0[P]$ is uncountable. We need to show that $P \cap \bigcup_{\alpha < \omega_1} g_\alpha \neq \emptyset$. So, take a Borel selection $g \colon \pi_0[P] \to \mathfrak{C}$ for P. (For example, put $g(x) = \min\{y \in \mathfrak{C} \colon \langle x, y \rangle \in P\}$.) Then, there exists a dense G_δ subset D of $\pi_0[P]$ such that $g \upharpoonright D$ is continuous. Thus, $g \upharpoonright D \in \mathcal{X}$ and $(g \upharpoonright D) \cap g_\alpha \neq \emptyset$ for some $\alpha < \omega_1$. In particular, $P \cap \bigcup_{\alpha < \omega_1} g_\alpha \supset (g \upharpoonright D) \cap g_\alpha \neq \emptyset$.

Next, let $\mathcal{P} = \{P_\xi \colon \xi < \omega_1\}$ be a family of compact subsets of G. We need to show that $G \neq \bigcup \mathcal{P}$. But, by (3.24), there exists an $x \in \mathfrak{C} \setminus \bigcup_{\xi < \omega_1} \pi_0(P_\xi)$. Let $p \in (\{x\} \times \mathfrak{C}) \setminus \bigcup_{\alpha < \omega_1} g_\alpha$. Then $p \in G \setminus \bigcup \mathcal{P}$.

To see the additional part, first note that $\mathfrak{C}^2 \setminus G$ is not G_{ω_1} by the above argument. To see that G is G_{ω_1}, it is enough to show that its complement $\bigcup_{\alpha < \omega_1} g_\alpha$ is a union of ω_1 many compact sets. But every g_α is a G_δ subset of $\mathfrak{C}^2$, so it is a Polish space. Thus g_α is a union of ω_1 compact sets — this follows directly from the formulation of $\mathrm{CPA}_{\mathrm{prism}}$ (as well as from $\mathrm{CPA}_{\mathrm{cube}}$). ∎

The proof of Theorem 3.4.5 is based on the following simple observation. Note also that if we want to prove the theorem only with the conclusion that $g \cap g_\alpha \neq \emptyset$ (i.e., a version needed to prove Corollary 3.4.6), then in Lemma 3.4.7 we need only require that $F \cap f \neq \emptyset$ for every continuous $f \colon \mathfrak{C} \to \mathfrak{C}$. The argument for this version is a bit simpler, and in this form Lemma 3.4.7 is more likely to be known.

Lemma 3.4.7 *There exists a continuous function* F *from a* G_δ *subset* T *of* $\mathfrak{C}$ *into* $\mathfrak{C}$ *such that* $|F \cap f| = \mathfrak{c}$ *for every continuous* $f \colon \mathfrak{C} \to \mathfrak{C}$.

PROOF. Let $\mathcal{C}$ be the family of all continuous functions $f \colon \mathfrak{C} \to \mathfrak{C}$, considered with the sup norm. Then $\mathcal{C}$ is homeomorphic to the Baire space ω^ω, and so is $\mathcal{C} \times \mathfrak{C}$. (To see this use, for example, [74, thm. 7.7].) Let T be a G_δ subset of $\mathfrak{C}$ homeomorphic to ω^ω and let $h = \langle h_1, h_2 \rangle \colon T \to \mathcal{C} \times \mathfrak{C}$ be a homeomorphism. Define $F \colon T \to \mathfrak{C}$ by

$$F(t) = [h_1(t)](t).$$

Clearly F is continuous. To see the other part, take an $f \in \mathcal{C}$ and notice that $P = (h_1)^{-1}(f)$ is uncountable. It is enough to show that $F \restriction P = f \restriction P$.

Indeed, if $t \in P$, then $h_1(t) = f$ and $F(t) = [h_1(t)](t) = f(t)$.   ∎

We will also need the following version of Proposition 3.3.4.

Proposition 3.4.8 *Assume that* $\text{CPA}_{\text{prism}}$ *holds. Then there exists a family* $\{f_\alpha : \alpha < \omega_1\}$ *of continuous injections from* $\mathfrak{C}^2$ *into* $\mathfrak{C}$ *such that* $\{f_\alpha[\{x\} \times \mathfrak{C}] : x \in \mathfrak{C} \ \& \ \alpha < \omega_1\}$ *is dense in* $\text{Perf}(\mathfrak{C})$.

PROOF. We will use the notation as in the proof of Proposition 3.3.4, where we put $X = \mathfrak{C}$. For $0 < \alpha < \omega_1$ let $\mathcal{G}_\alpha = \{f_\xi^* : \xi < \omega_1\}$ be such that $R_\alpha = \mathcal{F}^\alpha \setminus \bigcup_{\xi < \omega_1} \text{range}(f_\xi^*)$ has cardinality less than or equal to ω_1.

Fix an $f^* \in \mathcal{G}_\alpha$. Then $\hat{f}$ maps injectively a $P = P_f \in \mathbb{P}_{\beta + \alpha}$ onto $Q = Q_f \subset \mathfrak{C}$ and for every $z \in \mathcal{F}^\alpha \setminus R_\alpha$ there are $f^* \in \mathcal{G}_\alpha$ and $s \in \text{dom}(f^*)$ such that $z = f^*(s)$ and $\hat{f}(s, \cdot) \in \mathcal{F}_{\text{prism}}(X)$ is a restriction of z. Let $H_f \in \Phi_{\text{prism}}(\beta + \alpha)$ be from $\mathfrak{C}^{\beta + \alpha}$ onto P. Then $\hat{f} \circ H_f$ is an injection from $\mathfrak{C}^{\beta + \alpha} = \mathfrak{C}^\beta \times \mathfrak{C}^\alpha$ onto Q. Let k_γ be a homeomorphism from $\mathfrak{C}$ onto $\mathfrak{C}^\gamma$ and let $\bar{f} : \mathfrak{C}^2 \to \mathfrak{C}$ be defined by

$$\bar{f}(x, y) = (\hat{f} \circ H_f)(k_\beta(x), k_\alpha(y)).$$

Then the family $\mathcal{F} = \{\bar{f} : f \in \bigcup_{0 < \alpha < \omega_1} \mathcal{G}_\alpha\}$ is almost as desired: Every $P \in \text{Perf}(\mathfrak{C}) \setminus \{\text{range}(z) : z \in \bigcup_{0 < \alpha < \omega_1} R_\alpha\}$ contains a set $f[\{x\} \times \mathfrak{C}]$ for some $x \in \mathfrak{C}$ and $f \in \mathcal{F}$. Since we are missing at most ω_1 sets, we can easily extend $\mathcal{F}$ to be as required.   ∎

PROOF OF THEOREM 3.4.5. Let $\{f_\alpha : \alpha < \omega_1\}$ be as in Proposition 3.4.8 and let $F : T \to \mathfrak{C}$ be as in Lemma 3.4.7. For $\alpha < \omega_1$ let $K_\alpha = f_\alpha[\mathfrak{C} \times T]$. Define $g_\alpha : K_\alpha \to \mathfrak{C}$ by

$$g_\alpha(f_\alpha(x, t)) = F(t) \quad \text{for every } \langle x, t \rangle \in \mathfrak{C} \times T.$$

Clearly, functions g_α are continuous and defined on G_δ sets.

Fix a $g \in \mathcal{X}$. We need to find an $\alpha < \omega_1$ such that $|g \cap g_\alpha| = \mathfrak{c}$. Since the domain of g is uncountable, it contains a perfect set P. So, there are $\alpha < \omega_1$ and $x_0 \in \mathfrak{C}$ such that $f_\alpha[\{x_0\} \times \mathfrak{C}] \subset P$. Let $f \in \mathcal{C}$ be defined by $f(y) = g(f_\alpha(x_0, y))$. Then, by Lemma 3.4.7, there is a $Q \in \text{Perf}(T)$ such that $F \restriction Q = f \restriction Q$. Thus, for every $t \in Q \subset T$ we have

$$g_\alpha(f_\alpha(x_0, t)) = F(t) = f(t) = g(f_\alpha(x_0, t)).$$

Thus, $g_\alpha \restriction f_\alpha[\{x_0\} \times Q] = g \restriction f_\alpha[\{x_0\} \times Q]$.   ∎

3.5 α-prisms and separately nowhere constant functions

The results presented in this section come from a paper [30] by K. Ciesielski and A. Millán. Similar to the notion of an n-cube density defined in Section 1.8, we can consider α-prism density for every $0 < \alpha < \omega_1$: A family $\mathcal{F} \subset \mathrm{Perf}(X)$ is α-*prism dense* provided that for every injection $f \colon \mathfrak{C}^\alpha \to X$ there is an iterated perfect set $C \subset \mathfrak{C}^\alpha$ such that $f[C] \in \mathcal{F}$. Essentially the same proof as for Fact 1.8.5 shows that for arbitrary $1 \le \beta + 1 < \alpha \le \omega_1$

$$\text{every } \alpha\text{-prism dense family is also } (\beta + 1)\text{-prism dense.} \qquad (3.25)$$

We do not know if for a limit ordinal $\lambda < \omega_1$ the $(\lambda + 1)$-prism density implies λ-prism density.

The main goal of this section is to prove the following two theorems. The first of them shows that essentially all notions of α-prism and n-cube densities are different. The second theorem shows that any strengthening of the axiom $\mathrm{CPA}_{\mathrm{prism}}$ obtained by replacing the prism density with a proper subclass of these densities leads to a statement that is false in ZFC.

Theorem 3.5.1 *For a Polish space X, a family $\mathcal{F} \subset \mathrm{Perf}(X)$, and an ordinal $1 < \alpha < \omega_1$ consider the following sentences:*

C_α: *family $\mathcal{F}$ is β-cube dense for every $0 < \beta < \alpha$;*
P_α: *family $\mathcal{F}$ is β-prism dense for every $0 < \beta < \alpha$.*

Then, for $2 < m < n < \omega$ and $\omega + 1 < \alpha < \gamma < \omega_1$, they are related by the following implications.

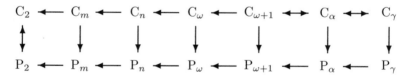

Moreover, none of these implications can be reversed.

For $\alpha < \omega_1$ we say that a family $\mathcal{F} \subset \mathrm{Perf}(X)$ is α-*prism* dense* provided $\mathcal{F}$ is β-prism dense for every $0 < \beta < \alpha$ and it is n-cube dense for every $0 < n < \omega$.

Theorem 3.5.2 *For every $\alpha < \omega_1$ and for every Polish space X there is an α-prism* dense family $\mathcal{F} \subset \mathrm{Perf}(X)$ for which $|X \setminus \bigcup \mathcal{F}| = \mathfrak{c}$.*

A big part of the proof of Theorem 3.5.1 will be to distinguish between different n-cube densities. For this we will need a notion of separately nowhere constant function. For a topological space X, a function $f \colon X \to Y$

is *nowhere constant* if f is not constant on any nonempty open subset of X. For a subset G of a product space $X = \prod_{i \in I} X_i$ we say that a function $f \colon G \to Y$ is *separately nowhere constant* if for every $t \in G$ and $k \in I$ the function f restricted to the section $G_k^t = \{x \in G \colon x \restriction I \setminus \{k\} = t \restriction I \setminus \{k\}\}$ is nowhere constant. This notion is the most natural when $G = X$. In this case it is related in a natural way to the notion of a *separately continuous* function $f \colon X \to Y$, that is, such that f restricted to every section X_k^t is continuous.

Note that every separately nowhere constant function is nowhere constant. However, the converse implication is false, as shown by the polynomial functions from $\mathbb{R}^2$ into $\mathbb{R}$ defined by $w_0(x, y) = xy$ and $w_1(x, y) = x$. This implication pattern stays in contrast with the implications for separate continuity: Continuity implies separate continuity, but the converse implication is false.

The main theorem on separately nowhere constant functions used here is the following result.

Theorem 3.5.3 *Let G be a dense G_δ subset of a product $\prod_{i \in I} X_i$ of Polish spaces and let f be a continuous function from G into a Polish space Y. If f is separately nowhere constant, then there is a perfect cube P in $\prod_{i \in I} X_i$ such that $P \subset G$ and $f \restriction P$ is one to one.*

It is not difficult to see that the conclusion of the theorem remains true for the function $w_0(x, y) = xy$, despite the fact that w_0 is not separately nowhere constant. On the other hand, the theorem's conclusion is false for the nowhere constant function $w_1(x, y) = x$.

We will start here with the proofs of the first two theorems, leaving the proof of Theorem 3.5.3 to the end of this section. First, note the following properties of the iterated perfect sets. Property (3.27) follows easily from (3.3):

$$\text{If } P \in \mathbb{P}_{\alpha+1} \text{ and } x \in \pi_\alpha[P], \text{ then } |(\{x\} \times \mathfrak{C}) \cap P| = \mathfrak{c}. \tag{3.26}$$

$$\text{If } 0 < \beta < \alpha \text{ and } P \in \mathbb{P}_\alpha, \text{ then } \pi_\beta[P] \in \mathbb{P}_\beta. \tag{3.27}$$

Lemma 3.5.4 *For $0 < n < \omega$ and any continuous $f \colon \mathfrak{C}^n \to Y$ there exist a basic clopen subset $U = \prod_{i < n} U_i$ of $\mathfrak{C}^n$, an $A \subset n$, and, if $A \neq n$, a dense G_δ subset G of $W = \prod_{i \in n \setminus A} U_i$ such that*

- *$f \restriction U$ does not depend on the variables x_j for $j \in A$;*
- *if $A \neq n$, then $f \restriction U$, considered as a function of the variables x_i with $i \in n \setminus A$, is separately nowhere constant on G.*

PROOF. We proceed by induction on n. If $n = 1$, the lemma is true by the definition of a nowhere constant function. Suppose the lemma is true for n and let $f: \mathfrak{C}^{n+1} \to Y$ be continuous. Denote by $\mathcal{B}_0$ a countable basis for the topology on $\mathfrak{C}$ consisting of nonempty clopen sets. For $i \leq n$ and $V \in \mathcal{B}_0$ consider the closed set $S_i(V) = \{\vec{x} \in \mathfrak{C}^{n+1 \setminus \{i\}} : f \upharpoonright \{\vec{x}\} \times V \text{ is constant}\}$.

First assume that for every $i \leq n$ and $V \in \mathcal{B}_0$ the set $S_i(V)$ has an empty interior. Then each set $H_i = \mathfrak{C}^{\{i\}} \times \left(\mathfrak{C}^{n+1 \setminus \{i\}} \setminus \bigcup\{S_i(V) : B \in \mathcal{B}_0\} \right)$ is comeager in $\mathfrak{C}^{n+1}$. Therefore, we can apply Lemma 3.2.1 to $X = \mathfrak{C}$, $T = n + 1$, $\mathcal{K} = \{n + 1 \setminus \{i\} : i \leq n\}$, and $H = \bigcap_{i \leq n} H_i$ to find a comeager set $G \subset H$ with the property that for every $x \in G$ and $i \leq n$ the set $G_{x \upharpoonright n+1 \setminus \{i\}} = \{y \in \mathfrak{C}^{\{i\}} : (x \upharpoonright n + 1 \setminus \{i\}) \cup y \in G\}$ is comeager in $\mathfrak{C}^{\{i\}}$. Note also that this last property implies that $f \upharpoonright G$ is separately nowhere constant, since for every $x \in G$ its restriction $x \upharpoonright n + 1 \setminus \{i\}$ does not belong to $\bigcup\{S_i(V) : B \in \mathcal{B}_0\}$. Thus, in this case the lemma is satisfied with $U = \mathfrak{C}^{n+1}$, $A = \emptyset$, and the above chosen G.

So, assume that there exist $i \leq n$ and $V_i \in \mathcal{B}_0$ such that the set $S_i(V_i)$ has a nonempty interior. Let $V^* \subset S_i(V_i)$ be a nonempty basic clopen subset of $\mathfrak{C}^{n+1 \setminus \{i\}}$. Then $V^* = \prod_{j \neq i} V_j$, where $V_j \subset \mathfrak{C}$ is a basic clopen set for every $j \neq i$. If $V = \prod_{j \leq n} V_j$, then V is homeomorphic to $\mathfrak{C}^{n+1}$, $f \upharpoonright V$ does not depend on the variable x_i, and we can consider $f \upharpoonright V$ as a function g from V^* to Y. By applying our inductive hypothesis to g we can find a basic clopen subset $U^* = \prod_{j \neq i} U_j$ of V^*, a set $A^* \subset n + 1 \setminus \{i\}$, and, if $A^* \neq n + 1 \setminus \{i\}$, a dense G_δ subset G of $W = \prod_{n+1 \setminus A} U_j$, where $A = A^* \cup \{i\}$, satisfying the lemma for g. But then $U = \prod_{j \leq n} U_j$, where $U_i = V_i$, and the sets A and G are as desired. $\blacksquare$

Here is the main example of the section.

Example 3.5.5 *For every $0 < \alpha < \omega_1$ there is a family $\mathcal{G}_\alpha \subset \text{Perf}(\mathfrak{C}^\alpha)$ such that:*

(a) $\mathcal{G}_\alpha$ *does not contain any iterated perfect set, that is, $\mathcal{G}_\alpha \cap \mathbb{P}_\alpha = \emptyset$.*
(b) $\mathcal{G}_\alpha$ *is γ-prism dense for every $0 < \gamma < \alpha$.*
(c) $\mathcal{G}_\alpha$ *is n-cube dense for every $0 < n < \min\{\alpha, \omega\}$.*
(d) *If $\mathcal{G}^* \in [\mathcal{G}_\alpha]^{<\mathfrak{c}}$, then $|\mathfrak{C}^\alpha \setminus \bigcup \mathcal{G}^*| = \mathfrak{c}$.*

PROOF. For $\xi < \alpha$ let $\mathcal{K}_\xi = \{P \in \text{Perf}(\mathfrak{C}^\alpha) : \pi_\xi \upharpoonright \pi_{\xi+1}[P] \text{ is one to one}\}$, where for $\xi = 0$ we understand $\mathcal{K}_0$ as $\{P : \pi_1[P] \text{ is a singleton}\}$. It is worth noticing that $\{P \in \text{Perf}(\mathfrak{C}^\alpha) : \pi_\xi \upharpoonright P \text{ is one to one}\} \subset \mathcal{K}_\xi$. We define $\mathcal{G}_\alpha = \bigcup_{\xi < \alpha} \mathcal{K}_\xi$.

To see (a) take $P \in \mathbb{P}_\alpha$ and $\xi < \alpha$. We need to show that $P \notin \mathcal{K}_\xi$. But by (3.27) we have $\pi_{\xi+1}[P] \in \mathbb{P}_{\xi+1}$, and then (3.26) shows that $P \notin \mathcal{K}_\xi$.

We prove (b) by induction on α. Clearly it holds for $\alpha = 1$. So, assume that for some $1 < \alpha < \omega_1$ condition (b) holds for every nonzero $\alpha' < \alpha$. To see that (b) holds for α, fix $0 < \gamma < \alpha$ and a continuous injection $f: \mathfrak{C}^\gamma \to \mathfrak{C}^\alpha$. We need to find a $Q \in \mathbb{P}_\gamma$ for which $f[Q] \in \mathcal{G}_\alpha$.

Let $g = \pi_\gamma \circ f$. By Lemma 3.2.5 there exist $P \in \mathbb{P}_\gamma$ and $0 < \beta \leq \gamma$ such that $h = g \circ \pi_\beta^{-1}$ is a function on $\pi_\beta[P]$ (i.e., $g \restriction P$ does not depend on coordinates $\delta \geq \beta$) and this function is either one to one or constant. If h is constant, then $\pi_1[f[P]]$ is a singleton and $f[P] \in \mathcal{K}_0 \subset \mathcal{G}_\alpha$. So, assume that h is one to one.

If $\beta = \gamma$, then $g = \pi_\gamma \circ f$ is one to one on P and so π_γ is one to one on $f[P]$. Then $f[P] \in \mathcal{K}_\gamma \subset \mathcal{G}_\alpha$. So, assume that $\beta < \gamma$. Then h is an injection from $\pi_\beta[P] \in \mathbb{P}_\beta$ into $\mathfrak{C}^\gamma$. Let $\varphi \in \Phi_\beta$ witness $\pi_\beta[P] \in \mathbb{P}_\beta$. Then φ maps $\mathfrak{C}^\beta$ onto $\pi_\beta[P]$. Since $h \circ \varphi: \mathfrak{C}^\beta \to \mathfrak{C}^\gamma$ is a continuous injection, by the inductive hypothesis used for $\alpha' = \gamma$ there exists an $E \in \mathbb{P}_\beta$ such that $Z = h \circ \varphi[E] \in \mathcal{G}_\gamma$, that is, there exists a $\xi < \gamma$ for which $\pi_\xi \restriction \pi_{\xi+1}[Z]$ is one to one.

Next notice that $R = \varphi[E] \in \mathbb{P}_\beta$, since Φ_β is closed under the composition, and $R \subset \pi_\beta[P]$. So, by (3.13), $Q = \{x \in P: \pi_\beta(x) \in R\} \in \mathbb{P}_\alpha$. Moreover,

$$Z = h \circ \varphi[E] = h[R] = (g \circ \pi_\beta^{-1})[\pi_\beta[Q]] = g[P] = \pi_\gamma[f[Q]]$$

and so $\pi_{\xi+1}[Z] = \pi_{\xi+1}[\pi_\gamma[f[Q]]] = \pi_{\xi+1}[f[Q]]$. Thus, $\pi_\xi \restriction \pi_{\xi+1}[f[Q]]$ is one to one and so $f[Q] \in \mathcal{K}_\xi \subset \mathcal{G}_\alpha$.

To show (c) we will prove by induction on $0 < n < \omega$ the statement:

For every $0 < \alpha < \omega_1$, if $n < \alpha$, then $\mathcal{G}_\alpha$ is n-cube dense.

So, take $0 < n < \omega$ and assume that the statement holds for all nonzero $k < n$. Take an $\alpha > n$. To prove that $\mathcal{G}_\alpha$ is n-cube dense, fix a continuous injection $f: \mathfrak{C}^n \to \mathfrak{C}^\alpha$. Then $\pi_n \circ f: \mathfrak{C}^n \to \mathfrak{C}^n$ is continuous. Apply Lemma 3.5.4 to $\pi_n \circ f$ to find $U = \prod_{i<n} U_i \subset \mathfrak{C}^n$, $A \subset n$, and G, satisfying the lemma.

If $A = n$, then $\pi_n[f[U]] = (\pi_n \circ f)[U]$ is a singleton and $f[U] \in \mathcal{K}_0 \subset \mathcal{G}_\alpha$.

If $A = \emptyset$, then $\pi_n \circ f \restriction G$ is continuous and separately nowhere constant. So, by Theorem 3.5.3, there exist perfect sets $\{P_i \subset U_i: i < n\}$ such that $(\pi_n \circ f) \restriction \prod_{i<n} P_i$ is one to one. Then, $\pi_n \restriction f\left[\prod_{i<n} P_i\right]$ is one to one and so $f\left[\prod_{i<n} P_i\right] \in \mathcal{K}_n \subset \mathcal{G}_\alpha$.

Therefore, assume that $\emptyset \neq A \neq n$ and let $k = |n \setminus A|$. Then $0 < k < n$. Since $(\pi_n \circ f) \restriction U$ does not depend on the variables x_j for $j \in A$, it can be considered as a function g on $W = \prod_{i \in n \setminus A} U_i$. Moreover, $g \restriction G$ is separately nowhere constant. Thus, by Theorem 3.5.3, we can find a perfect cube $P = \prod_{i \in n \setminus A} P_i \subset G \subset \prod_{i \in n \setminus A} U_i$ on which g is one to one. Thus, g is a continuous injection from P, which can be identified with $\mathfrak{C}^k$, into $\mathfrak{C}^n$. Since, by the inductive assumption, $\mathcal{G}_n$ is k-cube dense, there exists a perfect cube $C = \prod_{i \in n \setminus A} C_i \subset \prod_{i \in n \setminus A} P_i$ such that $g[C] \in \mathcal{G}_n$. Let $C_i = U_i$ for $i \in A$. Then $Q = \prod_{i < n} C_i \subset \prod_{i < n} U_i$ is a perfect cube and $\pi_n[f[Q]] = (\pi_n \circ f)[Q] = g[C] \in \mathcal{G}_n$. So, there exists a $\xi < n$ such that π_ξ is one to one on $\pi_{\xi+1}[\pi_n[f[Q]]] = \pi_{\xi+1}[f[Q]]$, and so $f[Q] \in \mathcal{K}_\xi \subset \mathcal{G}_\alpha$.

Now, to argue for (d), fix a $\mathcal{G}^* \in [\mathcal{G}_\alpha]^{<\mathfrak{c}}$. We need to show that $|\mathfrak{C}^\alpha \setminus \bigcup \mathcal{G}^*| = \mathfrak{c}$. For $\xi < \alpha$, let $\mathcal{G}_\xi^* = \mathcal{G}^* \cap \mathcal{K}_\xi$. By induction on $\xi < \alpha$. choose

$$x(\xi) \in \mathfrak{C} \setminus \{z(\xi) \colon z \in \mathcal{G}_\xi^* \ \& \ z(\eta) = x(\eta) \text{ for every } \eta < \xi\}.$$

Note that at each step we have less than continuum many restricted points since for every $z \in \mathcal{K}_\xi$ the set $\{z(\xi) \colon z(\eta) = x(\eta) \text{ for every } \eta < \xi\}$ may have at most one element. It is easy to see that $x = \langle x(\xi) \colon \xi < \alpha \rangle$ belongs to $\mathfrak{C}^\alpha \setminus \bigcup_{\xi < \alpha} \mathcal{G}_\xi^* = \mathfrak{C}^\alpha \setminus \mathcal{G}^*$. To finish the proof it is enough to notice that the value of $x(0)$ can be chosen in continuum many ways, so indeed $|\mathfrak{C}^\alpha \setminus \bigcup \mathcal{G}^*| = \mathfrak{c}$. ∎

To transport the above example into an arbitrary Polish space we will use the following simple fact.

Fact 3.5.6 *Let h be a homeomorphic embedding of a Polish space Y into a Polish space X, let $\mathcal{F} \subset \mathrm{Perf}(Y)$, and define the family $\mathcal{F}^*$ by $\mathcal{F}^* = \{h[F] \colon F \in \mathcal{F}\} \cup \mathrm{Perf}(X \setminus h[Y[)$. Then, for every $1 \le \alpha \le \omega_1$, the following conditions are equivalent:*

(a) *$\mathcal{F}$ is α-cube (α-prism) dense in Y.*

(b) *$\mathcal{F}^*$ is α-cube (α-prism) dense in X.*

PROOF. (a)$\Longrightarrow$(b) Let $f \colon \mathfrak{C}^\alpha \to X$ be injective and continuous. Since $h[Y]$ is a G_δ-set in X, we can apply Claim 1.1.5 to find a perfect cube $C \subset \mathfrak{C}^\alpha$ such that either $f[C] \subset h[Y]$ or $f[C] \cap h[Y] = \emptyset$.

If $f[C] \cap h[Y] = \emptyset$, then $f[C] \in \mathcal{F}^*$ and we are done. If $f[C] \subset h[Y]$, then $h^{-1} \circ f \colon C \to Y$ is a continuous injection. Identifying C with $\mathfrak{C}^\alpha$ and using to $h^{-1} \circ f$ the α-cube (α-prism) density of $\mathcal{F}$ in Y we can find a $C' \subset C$ such that C' is a perfect cube (belongs to $\mathbb{P}_\alpha$) and $F = (h^{-1} \circ f)[C'] \in \mathcal{F}$. So $f[C'] = h[F] \in \mathcal{F}^*$. Thus, the family $\mathcal{F}^*$ is as desired.

The other implication is easy. ∎

Corollary 3.5.7 *For every* $1 < \alpha < \omega_1$ *and every Polish space* X *there exists a family* $\mathcal{F}_\alpha \subset \mathrm{Perf}(X)$ *such that:* $\mathcal{F}_\alpha$ *is not* α-*prism dense;* $\mathcal{F}_\alpha$ *is* β-*prism dense for every* $0 < \beta < \alpha$; $\mathcal{F}_\alpha$ *is* n-*cube dense for every number* $0 < n < \min\{\alpha, \omega\}$; *and* $|X \setminus \bigcup \mathcal{F}_\alpha| = \mathfrak{c}$.

PROOF. First note that it is enough to find such an $\mathcal{F}_\alpha$ for $X = \mathfrak{C}^\alpha$. Indeed, if $\mathcal{F}$ is such a family, X is an arbitrary Polish space, and h is an embedding from $\mathfrak{C}^\alpha$ into X, then the family $\mathcal{F}^*$ from Fact 3.5.6 is as desired.

Thus, it is enough to notice that the family $\mathcal{G}_\alpha$ from Example 3.5.5 is not α-prism dense. But this is the case since for the identity function f on $\mathfrak{C}^\alpha$ there is no $P \in \mathbb{P}$ for which $f[P] = P \in \mathcal{G}_\alpha$.   ∎

PROOF OF THEOREM 3.5.2. Use Corollary 3.5.7 with $\alpha \geq \omega$.   ∎

PROOF OF THEOREM 3.5.1. The vertical implications, that α-cube density implies α-prism density, follows from the fact that every perfect cube in $\mathfrak{C}^\alpha$ is also in $\mathbb{P}_\alpha$. For $0 < \beta < \alpha < \omega_1$, the implications $\mathrm{C}_\alpha \Longrightarrow \mathrm{C}_\beta$ and $\mathrm{P}_\alpha \Longrightarrow \mathrm{P}_\beta$ are obvious. P_2 implies C_2 since 1-prism density is just perfect set density ($\mathbb{P}_1 = \mathrm{Perf}(\mathfrak{C}^1)$, as Φ_1 consists just of autohomeomorphisms of $\mathfrak{C}^1$) and so it implies 1-cube density.

To see that for $\omega < \alpha < \omega_1$ we have $\mathrm{C}_{\omega+1} \Longrightarrow \mathrm{C}_\alpha$ it is enough to notice that any ω-cube dense family is also β-cube dense for any $\omega \leq \beta < \omega_1$. This is the case since the coordinatewise homeomorphism between $\mathfrak{C}^\omega$ and $\mathfrak{C}^\beta$ preserves perfect cubes.

The fact that no other horizontal implication can be reversed is justified by the family $\mathcal{F}_\alpha$ from Corollary 3.5.7 for different values of α. Indeed, $\mathcal{F}_\alpha$ clearly justifies $\mathrm{P}_\alpha \nRightarrow \mathrm{P}_\gamma$ for any $1 < \alpha < \gamma$ since it satisfies P_α but not P_γ as it is not α-prism dense. If $1 < m < n \leq \omega$, then $\mathcal{F}_m$ also witnesses $\mathrm{C}_m \nRightarrow \mathrm{C}_n$ since it satisfies C_m but not C_n, since it cannot be m-cube dense without being m-prism dense.

The fact that none of the vertical implications $\mathrm{C}_\alpha \Longrightarrow \mathrm{P}_\alpha$, for $2 < \alpha < \omega_1$, can be reversed is justified by any family that is α-prism dense for every α but is not 2-cube dense. There are many such families. For example, this is the case for the family $\mathcal{F}$ of all linearly independent (over $\mathbb{Q}$) subsets of $\mathbb{R}$. It is shown in Corollary 5.1.2 that this $\mathcal{F}$ is α-prism dense for every $0 < \alpha < \omega_1$. On the other hand, it is not 2-cube dense, as shown by the function f from Remark 5.1.4.

Another such example is a family $\mathcal{F}$ of all $P \in \mathrm{Perf}(\mathfrak{C}^2)$ such that the projection on one of the coordinates is one to one. It follows quite easily from Lemma 3.2.5 that $\mathcal{F}$ is α-prism dense for every α. (See Proposition 4.1.3.)

It is not 2-cube dense since for the identity function $f: \mathfrak{C}^2 \to \mathfrak{C}^2$ there is no perfect cube C for which $f[P] \in \mathcal{F}$. ∎

Now we are ready for the proof of Theorem 3.5.3. For the rest of this section we will consider $\mathbb{R}^\omega$ as a vector space over $\mathbb{R}$ with the operations defined pointwise from the usual operations in $\mathbb{R}$. In this context we consider for every $k < \omega$ the canonical unit vectors $\vec{e}_k \in \mathbb{R}^\omega$: $\vec{e}_k(k) = 1$ and $\vec{e}_k(i) = 0$ for all other $i < \omega$. If $S \subset \mathbb{R}^\omega$ and $\delta \in \mathbb{R}$, then $\delta \cdot \vec{e}_k + S = \{\delta \cdot \vec{e}_k + s : s \in S\}$. If $\varepsilon > 0$ and $x \in \mathbb{R}^\omega$, then $B(x, \varepsilon)$ denotes the open ball with center x and radius ε and $\overline{B}(x, \varepsilon)$ is the corresponding closed ball. If $m < \omega$, then we identify $\mathbb{R}^\omega$ with $\mathbb{R} \times \mathbb{R}^{\omega \setminus \{m\}}$; if $y \in \mathbb{R}^{\omega \setminus \{m\}}$ and $G \subset \mathbb{R}^\omega$, then the section of G along y is the set $(G)^y = \{x \in \mathbb{R} : \langle x, y \rangle \in G\}$.

The next lemma is an immediate consequence of Lemma 3.2.1 applied to $X = \mathbb{R}$, $T = \omega$, and $\mathcal{K} = \{\omega \setminus \{n\} : n < \omega\}$.

Lemma 3.5.8 *For every comeager set $G \subset \mathbb{R}^\omega$ there exists a comeager set $H \subset G$ such that for every $x \in H$ and $n < \omega$ the set $H \cap (x + \mathbb{R} \cdot \vec{e}_n)$ is comeager in $x + \mathbb{R} \cdot \vec{e}_n$.*

The following lemma will facilitate the inductive step in the next theorem.

Lemma 3.5.9 *Let G be a comeager subset of $\mathbb{R}^\omega$ such that:*

($\bullet$) *$G \cap (x + \mathbb{R} \cdot \vec{e}_k)$ is comeager in $x + \mathbb{R} \cdot \vec{e}_k$ for every $x \in G$ and $k < \omega$.*

Let f be a continuous separately nowhere constant function from G into a Polish space Y. If $S \in [G]^{<\omega}$ is such that f is one to one on S, then for every $k < \omega$ and $\varepsilon > 0$ there exists a $\delta \in (0, \varepsilon)$ such that $(S + \delta \cdot \vec{e}_k) \subset G$ and f is one to one on $S \cup (S + \delta \cdot \vec{e}_k)$.

PROOF. Let $S = \{x_i : i < n\} \subset G$ be such that $f \upharpoonright S$ is one to one. Since f is continuous, decreasing ε if necessary, we can assume that:

(∗) *If $S^* = \{x_i^* : i < n\}$ is such that $x_i^* \in G \cap B(x_i, \varepsilon)$ for every $i < n$, then f is also one to one on S^*.*

For each $x \in S$ consider the sets $M_x = \{\delta \in \mathbb{R} : x + \delta \cdot \vec{e}_k \in G\}$, which by ($\bullet$) are comeager, and $N_x = \{\delta \in M_x : f(x + \delta \cdot \vec{e}_k) \in f[S]\}$. Since $f \upharpoonright G \cap (x + \mathbb{R} \cdot \vec{e}_k)$ is nowhere constant, the set N_x is meager in $\mathbb{R}$. So, $B = \bigcap_{x \in S} M_x \setminus \bigcup_{x \in S} N_x$ is comeager in $\mathbb{R}$.

Pick a $\delta \in (0, \varepsilon) \cap B$. Then $S + \delta \cdot \vec{e}_k \subset G$ as $\delta \in \bigcap_{x \in S} M_x$. To see that f is one to one on $S \cup (S + \delta \cdot \vec{e}_k)$ take $x \neq y$ in this set. We need to show that $f(x) \neq f(y)$. This follows from the assumption when $x, y \in S$, from (∗) when $x, y \in S + \delta \cdot \vec{e}_k$, and from $\delta \notin \bigcup_{x \in S} N_x$ otherwise. ∎

Theorem 3.5.10 *Let G be a comeager subset of $\mathbb{R}^\omega$ and let f be a continuous separately nowhere constant function from G into a Polish space Y. Then there is a perfect cube P in $\mathbb{R}^\omega$ such that $P \subset G$ and f is one to one on P.*

PROOF. Let $\{m_k : k < \omega\}$ be an enumeration of ω where every natural number appears infinitely often. By Lemma 3.5.8, shrinking G if necessary, we can assume that G satisfies the condition $(\bullet)$ from Lemma 3.5.9. Since G is a dense G_δ subset of $\mathbb{R}^\omega$, we have $G = \bigcap_{n<\omega} G_n$, where each G_n is open and dense subset of $\mathbb{R}^\omega$. We will construct by induction on $k < \omega$ the sequences $\langle S_k \in [G]^{2^k} : k < \omega \rangle$, $\langle \varepsilon_k : k < \omega \rangle$, and $\langle \delta_k : k < \omega \rangle$ such that for every $k < \omega$:

(1) $0 < \delta_k < \varepsilon_k \leq 2^{-k}$,
(2) $S_{k+1} = S_k \cup (\delta_k \cdot \vec{e}_{m_k} + S_k) \subset \bigcup \{B(x, \varepsilon_k) : x \in S_k\}$,
(3) $\overline{B}(x, \varepsilon_k) \subset G_k$ for every $x \in S_k$,
(4) $f[\overline{B}(x, \varepsilon_k)] \cap f[\overline{B}(x^*, \varepsilon_k)] = \emptyset$ for every distinct $x, x^* \in S_k$.

We start the construction with an arbitrary $S_0 = \{s\} \subset G$ and $\varepsilon_0 \leq 1$ ensuring (3). If for some $k < \omega$ the set S_k and ε_k are already constructed, we choose δ_k using Lemma 3.5.9 with $k = m_k$ and $\varepsilon \leq \varepsilon_k$ small enough that it ensures (2) and $|S_{k+1}| = 2^{k+1}$. Then f is one to one on $S_{k+1} \subset G$, and, using continuity of f, we can choose ε_{k+1} satisfying (1), (3), and (4). This finishes the construction.

If for $n, k < \omega$ we put $A_{k,n} = \{x(n) : x \in S_k\}$, then it is easy to see that:

(a) $S_k = \prod_{n<\omega} A_{k,n}$,
(b) $A_{k+1,n} = A_{k,n}$ for every $n \neq m_{k+1}$,
(c) $A_{k+1,m_{k+1}} = A_{k,m_{k+1}} \cup (\delta_k + A_{k,m_{k+1}})$.

We define $P_n = \mathrm{cl}\left(\bigcup_{k<\omega} A_{k,n}\right)$ and put $P = \prod_{n<\omega} P_n$. We will show that each P_n is a perfect subset of $\mathbb{R}$, $P \subset G$, and f is one to one on P. Notice that this will finish the proof, because as a final adjustment we can shrink each P_n to a subset from $\mathrm{Perf}(\mathbb{R})$.

Clearly each P_n is closed and, by (1) and (2), it has no isolated points. We need to show that

$$\mathrm{cl}\left(\bigcup_{k<\omega} S_k\right) = \prod_{n<\omega} P_n.$$

The inclusion $\mathrm{cl}\left(\bigcup_{k<\omega} S_k\right) \subset \prod_{n<\omega} P_n$ follows from (a). In order to prove the other inclusion, pick an $x \in \prod_{n<\omega} P_n$. Then for every $n < \omega$ there is a sequence $\{a_i^n : i < \omega\} \subset \bigcup_{k<\omega} A_{k,n}$ with distinct terms such that

$\lim_{i \to \infty} a_i^n = x(n)$. For every $m < \omega$ let $x_m \in \mathbb{R}^\omega$ be defined as $x_m(i) = a_i^m$ if $i \leq m$ and $x_m(i) = s(i)$ if $i > m$. Then $\{x_m : m < \omega\} \subset \bigcup_{k < \omega} S_k$ and $\lim_{m \to \infty} x_m = x$. This proves that $\prod_{n < \omega} P_n \subset \text{cl}\left(\bigcup_{k < \omega} S_k\right)$.

Next notice that if for $k < \omega$ we put $T_k = \bigcup\{\overline{B}(x, \varepsilon_k) : x \in S_k\}$, then condition (2) gives us $\text{cl}\left(\bigcup_{k < \omega} S_k\right) \subset \bigcap_{k < \omega} T_k$ while the other inclusion is obvious. In particular, we have

$$\prod_{n < \omega} P_n = \text{cl}\left(\bigcup_{k < \omega} S_k\right) = \bigcap_{k < \omega} T_k.$$

In order to prove that f is one to one on $\prod_{n < \omega} P_n$, pick distinct x and y from $\prod_{n < \omega} P_n$. Then there are sequences $\{x_m\}$ and $\{y_m\}$ such that for every $m < \omega$ we have $x_m, y_m \in S_m$, $x \in \overline{B}(x_m, \varepsilon_{m+1})$, and $y \in \overline{B}(y_m, \varepsilon_{m+1})$. Since $x \neq y$, there exists an $m < \omega$ such that $x_m \neq y_m$. So, by condition (5), we have $f[\overline{B}(x_m, \varepsilon_{m+1})] \cap f[\overline{B}(y_m, \varepsilon_{m+1})] = \emptyset$. Hence $f(x) \neq f(y)$. This shows that f is one to one on $\prod_{n < \omega} P_n$.

Finally note that by (3) we have $\prod_{n < \omega} P_n = \bigcap_{k < \omega} T_k \subset G$. ∎

Corollary 3.5.11 *Let $\{X_n : n < \omega\}$ be a family of Polish spaces, let G be a dense G_δ subset of $\prod_{n < \omega} X_n$, and let f be a continuous separately nowhere constant function from G into a Polish space Y. Then there exist perfect sets $P_n \in \text{Perf}(X_n)$, $n < \omega$, such that f is one to one on $\prod_{n < \omega} P_n$.*

PROOF. For every $n < \omega$ let G_n be a dense G_δ subset of X_n homeomorphic to the Baire space ω^ω.[1] Since ω^ω is homeomorphic to $\mathbb{R} \setminus \mathbb{Q}$, there is a homeomorphism $h_n : G_n \to \mathbb{R} \setminus \mathbb{Q}$. Then, $h : \prod_{n < \omega} G_n \to (\mathbb{R} \setminus \mathbb{Q})^\omega$ defined by $h = \langle h_n : n < \omega \rangle$ is a cube-preserving homeomorphism. We can apply Theorem 2.1.1 to the function $f \circ h^{-1}$ on a dense G_δ subset $h\left[G \cap \prod_{n < \omega} G_n\right]$ of $\mathbb{R}^\omega$ to obtain a perfect cube $\prod_{n < \omega} Q_n$ on which $f \circ h^{-1}$ is one to one. Then, $h^{-1}\left[\prod_{n < \omega} Q_n\right]$ is a perfect cube in $\prod_{n < \omega} X_n$ on which f is one to one. ∎

PROOF OF THEOREM 3.5.3. We can assume that the index set I is a cardinal number κ. Let $X = \prod_{i \in \kappa} X_i$.

The case $\kappa = \omega$ is true by Corollary 3.5.11.

If $\kappa = n < \omega$ and $f : G \to Y$ is a continuous and separately nowhere

---

[1] Every Polish space X has a dense subspace G homeomorphic to ω^ω constructed as follows. Let $\{B_n : n < \omega\}$ be a basis for X. Then $Y = X \setminus \bigcup_{n < \omega} \text{bd}(B_n)$ is a zero-dimensional dense G_δ subspace of X, where $\text{bd}(B_n)$ denotes the boundary of B_n. Take a countable dense subset D of Y and put $G = Y \setminus D$. Then G is a dense G_δ subspace of X (as X has no isolated points). Also, G is Polish, zero-dimensional, and every compact subset of G has an empty interior. So, by the Alexandrov-Urysohn theorem [74, thm. 7.7], it is homeomorphic to ω^ω.

constant function, consider the mapping $F: G \times \mathfrak{C}^{\omega \setminus n} \to Y \times \mathfrak{C}^{\omega \setminus n}$ defined by $F(x) = (f(x \restriction n), x \restriction \omega \setminus n)$. Then, F is a continuous and separately nowhere constant function defined on a dense G_δ subset of $\prod_{i \in \omega} X_i$, where $X_i = \mathfrak{C}$ for every $i \in \omega \setminus n$. Therefore, by the case $\kappa = \omega$, there are $\{P_i \in \mathrm{Perf}(X_i): i < \omega\}$ such that F is one to one on $\prod_{i < \omega} P_i \subset G \times \mathfrak{C}^{\omega \setminus n}$. This implies that f is one to one on $\prod_{i < n} P_i \subset G$.

If $\kappa > \omega$, then the result is trivial because in this case f cannot be simultaneously continuous and separately nowhere constant on a dense G_δ subset G of X. To see this, first notice that G contains a subset of the form $H \times \prod_{i \in \kappa \setminus A} X_i$, where A is a countable subset of κ and H is a dense G_δ subset of $\prod_{i \in A} X_i$. This is the case, since every dense open subset U of X contains a dense open subset in a similar form: a union of a maximal pairwise disjoint family of basic open subsets of U. So, we can assume that G is in this form. Pick an $x_0 \in G$. By the continuity of f at x_0, for every $n < \omega$ there exists an open subset U_n in X containing x_0 such that the diameter of $f[G \cap U_n]$ is less than 2^{-n}. By the definition of the product topology, each U_n contains a set of the form $\prod_{i \in F_n} \{x_0(i)\} \times \prod_{i \in \kappa \setminus F_n} X_i$, where each $F_n \subset \kappa$ is finite. Put $F = A \cup \bigcup_{n < \omega} F_n$ and notice that $Z = \prod_{i \in F} \{x_0(i)\} \times \prod_{i \in \kappa \setminus F} X_i \subset G \cap \bigcap_{n < \omega} U_n$. So, $f[Z]$ has a diameter equal to 0, that is, f is constant on Z. But this contradicts the fact that f is separately nowhere constant on G, since, for $\xi \in \kappa \setminus F$, set Z contains the section $\{x \in X: x \restriction \kappa \setminus \{\xi\} = x_0 \restriction \kappa \setminus \{\xi\}\}$. ∎

Finally, the next example shows that Lemma 3.5.4 fails, in a strong way, for functions defined on an infinite product.

Example 3.5.12 *There exists a continuous function $f: \mathfrak{C}^\omega \to \mathfrak{C}^\omega$ such that for every perfect cube $P = \prod_{n < \omega} P_n$ there is an $n < \omega$ such that $f \restriction P$ is one to one on some section of an n-th variable and is constant on some other sections of the same variable.*

PROOF. For $n < \omega$ let $f_n: \mathfrak{C}^2 \to \mathfrak{C}$ be defined by $f_n(x, y)(i) = y(n) \cdot x(i)$. Clearly f_n is continuous. Moreover, if $y(n) = 1$, then $f_n(\cdot, y)$ is the identity function, so it is one to one; if $y(n) = 0$, then $f_n(\cdot, y)$ is constant and equal to 0.

For $\langle x_n: n < \omega \rangle \in \mathfrak{C}^\omega$ define

$$f(\langle x_n: n < \omega \rangle) = \langle f_n(x_{n+1}, x_0): n < \omega \rangle.$$

Then f is clearly continuous. Consider f restricted to a perfect cube $P = \prod_{n < \omega} P_n$. Let $a, b \in P_0$ be distinct and let $n < \omega$ be such that

$a(n) \neq b(n)$. Assume that $a(n) = 0$ and let $z \in \mathfrak{C}^{\omega \setminus \{n+1\}}$. Look at $f \restriction P$ on a section given by z and note that

- if $z(0) = a$, then $f \restriction P$ is constant on this section;
- if $z(0) = b$, then $f \restriction P$ is one to one on this section.   ■

3.6 Multi-games and other remarks on $\mathrm{CPA}^{\mathrm{game}}_{\mathrm{prism}}$ and $\mathrm{CPA}_{\mathrm{prism}}$

First note that if $\mathrm{CPA}^{\mathrm{game}}_{\mathrm{prism}}[X]$ and $\mathrm{CPA}_{\mathrm{prism}}[X]$ stand, respectively, for the axioms $\mathrm{CPA}^{\mathrm{game}}_{\mathrm{prism}}$ and $\mathrm{CPA}_{\mathrm{prism}}$ for a fixed Polish space X, then, similarly as in Remark 1.8.3 we can also prove

Remark 3.6.1 For any Polish space X:

- $\mathrm{CPA}^{\mathrm{game}}_{\mathrm{prism}}[X]$ implies $\mathrm{CPA}^{\mathrm{game}}_{\mathrm{prism}}$, and
- $\mathrm{CPA}_{\mathrm{prism}}[X]$ implies $\mathrm{CPA}_{\mathrm{prism}}$.

For a nonempty collection $\mathcal{X}$ of pairwise disjoint Polish spaces consider the following "simultaneous" two-player game $\mathrm{GAME}_{\mathrm{prism}}(\mathcal{X})$ of length ω_1. At each stage $\xi < \omega_1$ of the game Player I can play a prism $P_\xi \in \mathrm{Perf}^*(X)$ from an arbitrarily chosen $X \in \mathcal{X}$. Player II responds with a subprism Q_ξ of P_ξ. The game $\langle\langle P_\xi, Q_\xi \rangle : \xi < \omega_1 \rangle$ is won by Player I provided

$$\bigcup_{\xi < \omega_1} Q_\xi = \bigcup \mathcal{X};$$

otherwise the game is won by Player II. Thus, for any Polish space X the games $\mathrm{GAME}_{\mathrm{prism}}(X)$ and $\mathrm{GAME}_{\mathrm{prism}}(\{X\})$ are identical.

Theorem 3.6.2 *Let $\mathcal{X}$ of size $\leq \omega_1$ be a nonempty collection of pairwise disjoint Polish spaces. Then $\mathrm{CPA}^{\mathrm{game}}_{\mathrm{prism}}$ is equivalent to*

$\mathrm{CPA}^{\mathrm{game}}_{\mathrm{prism}}(\mathcal{X})$: *Player II has no winning strategy in $\mathrm{GAME}_{\mathrm{prism}}(\mathcal{X})$.*

PROOF. To see that $\mathrm{CPA}^{\mathrm{game}}_{\mathrm{prism}}(\mathcal{X})$ implies $\mathrm{CPA}^{\mathrm{game}}_{\mathrm{prism}}$, take an $X \in \mathcal{X}$. It is easy to see that $\mathrm{CPA}^{\mathrm{game}}_{\mathrm{prism}}(\mathcal{X})$ implies $\mathrm{CPA}^{\mathrm{game}}_{\mathrm{prism}}[X]$, while $\mathrm{CPA}^{\mathrm{game}}_{\mathrm{prism}}[X]$ implies $\mathrm{CPA}^{\mathrm{game}}_{\mathrm{prism}}$ by Remark 3.6.1.

To see the converse implication assume that $\mathrm{CPA}^{\mathrm{game}}_{\mathrm{prism}}$ holds and let $I = [0,1]$. Let $L = \{x_\xi : \xi < \omega_1\}$ be a Luzin set in I, that is, such that $|L \cap N| \leq \omega$ for every nowhere dense subset N of I. The existence of such a set under $\mathrm{CPA}^{\mathrm{game}}_{\mathrm{prism}}$ follows from Corollary 1.3.3.

Let $\kappa = |\mathcal{X}| \leq \omega_1$ and let $\{X_\eta : \eta < \kappa\}$ be an enumeration of $\mathcal{X}$. We will identify each X_η, $\eta < \kappa$, with a G_δ subset of $\{x_\eta\} \times I^\omega$ homeomorphic to it.

Now, let S_0 be a Player II strategy in the game GAME$_{\text{prism}}(\mathcal{X})$. We will modify it to a Player II strategy S in the game GAME$_{\text{prism}}(I \times I^\omega)$ in the following way. First, for every prism P in $I \times I^\omega$, let $R(P)$ be its subprism such that

$$\text{either } R(P) \subset X_\eta \text{ for some } \eta < \kappa \text{ or } R(P) \cap \bigcup \mathcal{X} = \emptyset.$$

To choose such an $R(P)$, first choose a subprism R of P such that its first coordinate projection $\pi[R]$ is nowhere dense in I. (This can be done, for example, by applying Lemma 3.2.5.) So, $\pi[R]$ contains at most countably many points x_η. Thus, by Claim 1.1.5, there is a subprism R_1 of R such that either $\pi[R_1]$ is disjoint with L or there is an $\eta < \kappa$ such that $\pi[R_1] = \{x_\eta\}$. In the first case we put $R(P) = R_1$. In the second case we use Claim 1.1.5 to find a subprism $R(P)$ of R_1 such that either $R(P) \subset X_\eta$ or $R(P) \cap X_\eta = \emptyset$.

Now strategy S is defined by induction on ξ, the step level of the game. Thus, if a sequences $\bar{T} = \langle \langle P_\eta, Q_\eta \rangle : \eta < \xi \rangle$ represents a "legal" sequence (a sequence that could have been produced by S defined so far), we define $S(\bar{T}, P_\xi)$ as follows. If $R(P_\xi) \cap \bigcup \mathcal{X} = \emptyset$, we just put $S(\bar{T}, P_\xi) = R(P_\xi)$. For the other case, let $J = \{\eta < \xi : R(P_\eta) \subset X \text{ for some } X \in \mathcal{X}\}$ and define

$$S(\bar{T}, P_\xi) = S_0(\langle \langle R(P_\eta), Q_\eta \rangle : \eta \in J \rangle, R(P_\xi)),$$

where $\langle \langle R(P_\eta), Q_\eta \rangle : \eta \in J \rangle$ is identified with $\langle \langle R(P_{i(\eta)}), Q_{i(\eta)} \rangle : \eta < \alpha \rangle$, while i is an order isomorphism between an ordinal α and J.

Since, by CPA$_{\text{prism}}^{\text{game}}$, S is not winning in GAME$_{\text{prism}}(I \times I^\omega)$ for Player II, there is a game $\langle \langle P_\xi, Q_\xi \rangle : \xi < \omega_1 \rangle$ played according to S in which Player I wins. To finish the proof put $K = \{\xi < \omega_1 : R(P_\xi) \subset \bigcup \mathcal{X}\}$ and notice that $\langle \langle R(P_\xi), Q_\xi \rangle : \xi \in K \rangle$ is a game in GAME$_{\text{prism}}(\mathcal{X})$ played according to S_0, in which Player I wins. Thus, S_0 is not winning for Player II. ∎

Corollary 3.6.3 CPA$_{\text{prism}}^{\text{game}}$ *implies that*

strong-CPA$_{\text{prism}}^{\text{game}}$: *For every Polish space X and every Player II strategy S for* GAME$_{\text{prism}}(X)$ *there is a game $\langle \langle P_\xi, Q_\xi \rangle : \xi < \omega_1 \rangle$ played according to S for which $\bigcup_{\eta < \xi < \omega_1} Q_\xi = X$ for every $\eta < \omega_1$.*

PROOF. Let $\mathcal{X} = \{\{\alpha\} \times X : \alpha < \omega_1\}$. Define a strategy S_0 in the game GAME$_{\text{prism}}(\mathcal{X})$ as follows. If $\bar{T} = \langle \langle \{\alpha_\eta\} \times P_\eta, \{\alpha_\eta\} \times Q_\eta \rangle : \eta < \xi \rangle$ is a partial game, then put

$$S_0(\bar{T}, \{\alpha\} \times P_\xi) = \{\alpha\} \times S(\langle \langle P_\eta, Q_\eta \rangle : \eta < \xi \rangle, P_\xi).$$

By Theorem 3.6.2 there is a game $\langle \langle \{\alpha_\xi\} \times P_\xi, \{\alpha_\xi\} \times Q_\xi \rangle : \xi < \omega_1 \rangle$ played according to S_0 in which Player I wins, that is, $\bigcup_{\xi < \omega_1} \{\alpha_\xi\} \times Q_\xi = \omega_1 \times X$.

It is easy to see that $\langle\langle P_\xi, Q_\xi\rangle\colon \xi < \omega_1\rangle$ is a game in $\mathrm{GAME}_{\mathrm{prism}}(X)$ played according to S in which every $x \in X$ belongs to ω_1 many Q_ξ's.   ∎

Remark 3.6.4 The following weaker version of $\mathrm{CPA}_{\mathrm{prism}}$ implies that continuum is a cardinal successor:

$\mathrm{CPA}_{\mathrm{prism}}^{\mathfrak{c}}$: For every Polish space X and every $\mathcal{F}_{\mathrm{prism}}$-dense family $\mathcal{E} \subset \mathrm{Perf}(X)$ there is an $\mathcal{E}_0 \subset \mathcal{E}$ such that $|\mathcal{E}_0| < \mathfrak{c}$ and $|X \backslash \bigcup \mathcal{E}_0| < \mathfrak{c}$.

PROOF. First note that $\mathrm{CPA}_{\mathrm{prism}}^{\mathfrak{c}}$ implies the following version of part (b) from Theorem 4.1.1:

There exists an $\mathcal{F} \subset \mathcal{C}^1$ such that $|\mathcal{F}| < \mathfrak{c}$ and $\mathbb{R}^2 = \bigcup_{f \in \mathcal{F}}(f \cup f^{-1})$.

The proof is identical to that for Theorem 4.1.1(b). On the other hand, for a cardinal number κ we have

$\mathfrak{c} = \kappa^+$ if and only if there exists an $\mathcal{F} \subset \mathbb{R}^{\mathbb{R}}$ such that

$$|\mathcal{F}| = \kappa < \mathfrak{c} \quad \text{and} \quad \mathbb{R}^2 = \bigcup_{f \in \mathcal{F}}(f \cup f^{-1}).$$

This is an easy generalization of the property P_1 from [120]. (See also [25, thm. 6.1.8].) It is easy to see that the above two conditions imply that $\mathfrak{c}$ is a cardinal successor.   ∎

Note also that $\mathrm{CPA}_{\mathrm{prism}}^{\mathfrak{c}}$ can be considered as a prism version of property (ii) from Remark 1.8.2. Note also that, similarly as for Remarks 2.5.2 and 2.5.3, we can argue that the part of axiom $\mathrm{CPA}_{\mathrm{prism}}^{\mathrm{game}}$ without $\mathfrak{c} = \omega_2$ follows trivially from CH. However, the version of $\mathrm{CPA}_{\mathrm{prism}}^{\mathrm{game}}$ with $\mathfrak{c} = \omega_2$ replaced by $\mathfrak{c} \geq \omega_2$ is equivalent to the original form of $\mathrm{CPA}_{\mathrm{prism}}^{\mathrm{game}}$.

4

CPA$_{\text{prism}}$ and coverings with smooth functions

This chapter is based on K. Ciesielski and J. Pawlikowski [38]. Below we will use standard notation for the classes of differentiable partial functions from $\mathbb{R}$ into $\mathbb{R}$. Thus, if X is an arbitrary subset of $\mathbb{R}$ without isolated points, we will write $\mathcal{C}^0(X)$ or $\mathcal{C}(X)$ for the class of all continuous functions $f\colon X \to \mathbb{R}$ and $D^1(X)$ for the class of all differentiable functions $f\colon X \to \mathbb{R}$, that is, those for which the limit

$$f'(x_0) = \lim_{x \to x_0,\ x \in X} \frac{f(x) - f(x_0)}{x - x_0}$$

exists and is finite for all $x_0 \in X$. Also, for $0 < n < \omega$ we will write $D^n(X)$ to denote the class of all functions $f\colon X \to \mathbb{R}$ that are n times differentiable with all derivatives being finite and $\mathcal{C}^n(X)$ for the class of all $f \in D^n(X)$ whose n-th derivative $f^{(n)}$ is continuous. The symbol $\mathcal{C}^\infty(X)$ will be used for all infinitely many times differentiable functions from X into $\mathbb{R}$. In addition, we say that a function $f\colon X \to \mathbb{R}$ is in the class "$D^n(X)$" if $f \in \mathcal{C}^{n-1}(X)$ and it has the n-th derivative, which can be infinite; f is in the class "$\mathcal{C}^n(X)$" when f is in "$D^n(X)$" and its n-th derivative is continuous when its range $[-\infty, \infty]$ is considered with the standard topology. "$\mathcal{C}^\infty(X)$" will stand for all functions $f\colon X \to \mathbb{R}$ that are either in $\mathcal{C}^\infty(X)$ or, for some $0 < n < \omega$, are in "$\mathcal{C}^n(X)$" and $f^{(n)}$ is constant and equal to ∞ or $-\infty$. (Thus, in general, "$\mathcal{C}^\infty(X)$" is not a subclass of "$\mathcal{C}^n(X)$.") In addition, we assume that functions defined on a singleton are in the $\mathcal{C}^\infty$ class, that is, $\mathcal{C}^\infty(\{x\}) = \mathbb{R}^{\{x\}}$. We will use these symbols mainly for X's that are either in the class $\mathrm{Perf}(\mathbb{R})$ or are the singletons. In particular, $\mathcal{C}^n_{\mathrm{perf}}$ will stand for the union of all $\mathcal{C}^n(P)$ for which $P \subset \mathbb{R}$ is either in $\mathrm{Perf}(\mathbb{R})$ or is a `singleton`. The classes D^n_{perf}, $\mathcal{C}^\infty_{\mathrm{perf}}$, and "$\mathcal{C}^\infty_{\mathrm{perf}}$" are defined in a similar way. We will drop the parameter X if $X = \mathbb{R}$. In particular, $D^n = D^n(\mathbb{R})$ and $\mathcal{C}^n = \mathcal{C}^n(\mathbb{R})$. The relations between these classes for $n < \omega$ are given in the following chart, where the arrows $\longrightarrow$ indicate the strict inclusions $\subsetneq$.

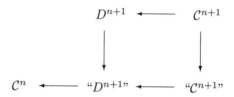

Chart 1.

In addition, for $F \subset \mathbb{R}^2$ we define $F^{-1} = \{\langle y, x \rangle : \langle x, y \rangle \in F\}$ and for $\mathcal{F} \subset \mathcal{P}(\mathbb{R}^2)$ we put $\mathcal{F}^{-1} = \{F^{-1} : F \in \mathcal{F}\}$.

4.1 Chapter overview; properties (H*) and (R)

The main result of this chapter is the following theorem.

Theorem 4.1.1 *The following facts follow from* CPA$_{\text{prism}}$.

(a) *For every Borel measurable function $g : \mathbb{R} \to \mathbb{R}$ there exists a family of functions $\{f_\xi \in \text{``}C^\infty_{\text{perf}}\text{''} : \xi < \omega_1\}$ such that*

$$g = \bigcup_{\xi < \omega_1} f_\xi.$$

Moreover, for each $\xi < \omega_1$ there exists an extension $\bar{f}_\xi : \mathbb{R} \to \mathbb{R}$ of f_ξ such that

 (i) *$\bar{f}_\xi \in \text{``}C^1\text{''}$ and*
 (ii) *either $\bar{f}_\xi \in C^1$ or $\bar{f}_\xi$ is a homeomorphism from $\mathbb{R}$ onto $\mathbb{R}$ such that $\bar{f}_\xi^{-1} \in C^1$.*

(b) *There exists a sequence $\{f_\xi \in \mathbb{R}^{\mathbb{R}} : \xi < \omega_1\}$ of C^1 functions such that*

$$\mathbb{R}^2 = \bigcup_{\xi < \omega_1} (f_\xi \cup f_\xi^{-1}).$$

Clearly parts (a) and (b) of the theorem imply properties (R) and (H*), respectively. In particular, we have

Corollary 4.1.2 CPA$_{\text{prism}}$ *implies properties (R) and (H*).*

Note also that, by Corollary 5.0.2, under $\text{CPA}_{\text{prism}}^{\text{game}}$, the functions f_ξ in Theorem 4.1.1(a) may be chosen to have disjoint graphs. Also, $\mathbb{R}^2$ can be covered by ω_1 pairwise disjoint sets P such that either P or P^{-1} is a function in the class $\mathcal{C}_{\text{perf}}^1 \cap$ "$\mathcal{C}_{\text{perf}}^\infty$ ".

The essence of Theorem 4.1.1 lies in the following real analysis fact. Its proof is combinatorial in nature and uses no extra set-theoretical assumptions.

Proposition 4.1.3 *Let* $g: \mathbb{R} \to \mathbb{R}$ *be Borel and* $0 < \alpha < \omega_1$.

(a) *For every continuous injection* $h: \mathfrak{C}^\alpha \to \mathbb{R}$ *there exists an* $E \in \mathbb{P}_\alpha$ *such that* $g \restriction h[E] \in$ "$\mathcal{C}_{\text{perf}}^\infty$" *and there is an extension* $f: \mathbb{R} \to \mathbb{R}$ *of* $g \restriction h[E]$ *such that* $f \in$ "$\mathcal{C}^1$" *and either* $f \in \mathcal{C}^1$ *or* f *is an autohomeomorphism of* $\mathbb{R}$ *with* $f^{-1} \in \mathcal{C}^1$.

(b) *For every continuous injection* $h: \mathfrak{C}^\alpha \to \mathbb{R}^2$ *there exists an* $E \in \mathbb{P}_\alpha$ *such that either* $F = h[E] \subset \mathbb{R}^2$ *or its inverse,* F^{-1}, *is a function that can be extended to a* $\mathcal{C}^1$ *function* $f: \mathbb{R} \to \mathbb{R}$.

With Proposition 4.1.3 in hand the proof of Theorem 4.1.1 becomes an easy exercise.

PROOF OF THEOREM 4.1.1. (a) Let $g: \mathbb{R} \to \mathbb{R}$ be a Borel function and let $\mathcal{E}$ be the family of all $P \in \text{Perf}(\mathbb{R})$ such that

$g \restriction P \in$ "$\mathcal{C}_{\text{perf}}^\infty$" and there is an extension $f: \mathbb{R} \to \mathbb{R}$ of $g \restriction P$ such that $f \in$ "$\mathcal{C}^1$" and either $f \in \mathcal{C}^1$ or f is an autohomeomorphism of $\mathbb{R}$ with $f^{-1} \in \mathcal{C}^1$.

By Proposition 4.1.3(a), the family $\mathcal{E}$ is $\mathcal{F}_{\text{prism}}$-dense: If $P \in \text{Perf}(\mathbb{R})$ is a prism and $h: \mathfrak{C}^\alpha \to \mathbb{R}$ from $\mathcal{F}_{\text{prism}}$ witnesses it, then $Q = h[E]$ as in the proposition is a subprism of P with $Q \in \mathcal{E}$. So, by $\text{CPA}_{\text{prism}}$, there exists an $\mathcal{E}_0 \in [\mathcal{E}]^{\leq \omega_1}$ such that $|\mathbb{R} \backslash \bigcup \mathcal{E}_0| \leq \omega_1$. Let $\mathcal{E}_1 = \mathcal{E}_0 \cup \{\{r\}: r \in \mathbb{R} \backslash \bigcup \mathcal{E}_0\}$. Then the family $\{g \restriction P: P \in \mathcal{E}_1\}$ satisfies the theorem.

(b) Let $\mathcal{E}$ be the family of all $P \in \text{Perf}(\mathbb{R}^2)$ such that either P or P^{-1} is a function that can be extended to a $\mathcal{C}^1$ function $f: \mathbb{R} \to \mathbb{R}$. By Proposition 4.1.3(b) the family $\mathcal{E}$ is $\mathcal{F}_{\text{prism}}$-dense, so there exists an $\mathcal{E}_0 \in [\mathcal{E}]^{\leq \omega_1}$ such that $|\mathbb{R} \backslash \bigcup \mathcal{E}_0| \leq \omega_1$. Let $\mathcal{E}_1 = \mathcal{E}_0 \cup \{\{x\}: x \in \mathbb{R}^2 \backslash \bigcup \mathcal{E}_0\}$. For every $P \in \mathcal{E}_1$ let $f_P: \mathbb{R} \to \mathbb{R}$ be a $\mathcal{C}^1$ function that extends either P or P^{-1}. Then the family $\{f_P: P \in \mathcal{E}_1\}$ is as desired. ∎

The proof of Proposition 4.1.3 will be left to the next sections of this chapter. Meanwhile, we want to present a discussion of Theorem 4.1.1.

First we reformulate Theorem 4.1.1 in a language of a *covering number* cov defined below, where X is an infinite set (in our case $X \subset \mathbb{R}^2$ with $|X| = \mathfrak{c}$) and $\mathcal{A}, \mathcal{F} \subset \mathcal{P}(X)$:

$$\mathrm{cov}(\mathcal{A}, \mathcal{F}) = \min \left(\left\{ \kappa \colon (\forall A \in \mathcal{A})(\exists \mathcal{G} \in [\mathcal{F}]^{\leq \kappa}) \ A \subset \bigcup \mathcal{G} \right\} \cup \{|X|^+\} \right).$$

If $A \subset X$, we will write $\mathrm{cov}(A, \mathcal{F})$ for $\mathrm{cov}(\{A\}, \mathcal{F})$. Notice the following monotonicity of the cov operator: For every $A \subset B \subset X$, $\mathcal{A} \subset \mathcal{B} \subset \mathcal{P}(X)$, and $\mathcal{F} \subset \mathcal{G} \subset \mathcal{P}(X)$,

$$\mathrm{cov}(\mathcal{A}, \mathcal{G}) \leq \mathrm{cov}(\mathcal{B}, \mathcal{G}) \leq \mathrm{cov}(\mathcal{B}, \mathcal{F}) \ \& \ \mathrm{cov}(\mathcal{A}, \mathcal{G}) \leq \mathrm{cov}(\mathcal{B}, \mathcal{G}) \leq \mathrm{cov}(\mathcal{B}, \mathcal{F}).$$

In terms of the cov operator, Theorem 4.1.1 can be expressed in the following form, where Borel stands for the class of all Borel functions $f \colon \mathbb{R} \to \mathbb{R}$.

Corollary 4.1.4 CPA$_{\mathrm{prism}}$ *implies that*

(a) $\mathrm{cov}\left(\mathrm{Borel}, \text{``}\mathcal{C}_{\mathrm{perf}}^{\infty}\text{''}\right) = \omega_1 < \mathfrak{c}$;

(b) $\mathrm{cov}\left(\mathrm{Borel}, \text{``}\mathcal{C}^1\text{''}\right) = \omega_1 < \mathfrak{c}$;

(c) $\mathrm{cov}\left(\mathrm{Borel}, \mathcal{C}^1 \cup (\mathcal{C}^1)^{-1}\right) = \omega_1 < \mathfrak{c}$;

(d) $\mathrm{cov}\left(\mathbb{R}^2, \mathcal{C}^1 \cup (\mathcal{C}^1)^{-1}\right) = \omega_1 < \mathfrak{c}$.

PROOF. The fact that all numbers $\mathrm{cov}(\mathcal{A}, \mathcal{G})$ listed above are $\leq \omega_1$ follows directly from Theorem 4.1.1. The other inequalities follow from Examples 4.5.6 and 4.5.8. ∎

Theorem 4.1.1(b) and Corollary 4.1.4(d) can be treated as generalizations of a result of J. Steprāns [124] who proved that in the iterated perfect set model we have $\mathrm{cov}\left(\mathbb{R}^2, \left(\text{``}D^1\text{''}\right) \cup \left(\text{``}D^1\text{''}\right)^{-1}\right) \leq \omega_1$. This clearly follows from Corollary 4.1.4(d) since $\mathcal{C}^1 \subsetneq \text{``}D^1\text{''}$. (See survey article [15]. For more information on how to "locate" Steprāns' result in [124], see also [32, cor. 9].)

The following proposition shows that Theorem 4.1.1 is, in a way, the best possible. (Parts (i), (ii), and (iii) relate, respectively, to items (b), (c) together with (d), and (a) from Corollary 4.1.4.)

Proposition 4.1.5 *The following are true in ZFC.*

(i) $\mathrm{cov}\left(\mathrm{Borel}, \mathcal{C}^1\right) = \mathrm{cov}\left(\text{``}\mathcal{C}^1\text{''}, \mathcal{C}^1\right) = \mathrm{cov}\left(\text{``}\mathcal{C}^1\text{''}, D^1_{\mathrm{perf}}\right) = \mathfrak{c}$. *Moreover,*
$\mathrm{cov}(\text{``}\mathcal{C}^n\text{''}, \mathcal{C}^n) = \mathrm{cov}(\text{``}\mathcal{C}^n\text{''}, D^n_{\mathrm{perf}}) = \mathfrak{c}$ *for every* $0 < n < \omega$.

(ii) $\mathrm{cov}\left(\mathrm{Borel}, \mathcal{C}^2 \cup (\mathcal{C}^2)^{-1}\right) = \mathrm{cov}\left(\text{``}\mathcal{C}^2\text{''}, D^2_{\mathrm{perf}} \cup (D^2_{\mathrm{perf}})^{-1}\right) = \mathfrak{c}$ *and*
$\mathrm{cov}\left(\mathbb{R}^2, \mathcal{C}^2 \cup (\mathcal{C}^2)^{-1}\right) = \mathrm{cov}\left(\text{``}\mathcal{C}^2\text{''}, D^2_{\mathrm{perf}} \cup (D^2_{\mathrm{perf}})^{-1}\right) = \mathfrak{c}$.

(iii) $\mathrm{cov}\left(\mathrm{Borel}, \mathcal{C}^\infty_{\mathrm{perf}}\right) = \mathrm{cov}\left(\text{``}\mathcal{C}^1\text{''}, \mathcal{C}^\infty_{\mathrm{perf}}\right) = \mathrm{cov}\left(\text{``}\mathcal{C}^1\text{''}, D^1_{\mathrm{perf}}\right) = \mathfrak{c}$ *and*
$\mathrm{cov}\left(\mathrm{Borel}, \text{``}\mathcal{C}^\infty\text{''}\right) = \mathrm{cov}\left(\mathcal{C}^1, \text{``}\mathcal{C}^\infty\text{''}\right) = \mathrm{cov}\left(\mathcal{C}^1, \text{``}D^2\text{''}\right) = \mathfrak{c}$. *Moreover,*
$\mathrm{cov}\left(\mathcal{C}^n, \text{``}D^{n+1}\text{''}\right) = \mathfrak{c}$ *for every* $0 < n < \omega$.

PROOF. Part (i) follows immediately from Examples 4.5.2 and 4.5.3.

Part (ii) follows from the monotonicity of the cov operator and Example 4.5.1.

The first part of (iii) follows from (i). The remaining two parts follow, respectively, from Examples 4.5.4 and 4.5.5. ∎

Corollary 4.1.4 and Proposition 4.1.5 establish the values of the cov operator for all classes in Chart 1 except for $\mathrm{cov}\left(D^n, \mathcal{C}^n\right)$ and $\mathrm{cov}\left(\text{``}D^n\text{''}, \text{``}\mathcal{C}^n\text{''}\right)$. These are established in the following theorem, the proof of which will be left to the latter sections of this chapter.

Theorem 4.1.6 *If* $\mathrm{CPA}_{\mathrm{prism}}$ *holds, then, for every* $0 < n < \omega$,

$$\mathrm{cov}\left(D^n, \mathcal{C}^n\right) = \mathrm{cov}\left(\text{``}D^n\text{''}, \text{``}\mathcal{C}^n\text{''}\right) = \omega_1 < \mathfrak{c}.$$

Note also that, by Corollary 5.0.2, under $\mathrm{CPA}^{\mathrm{game}}_{\mathrm{prism}}$, covering functions in Theorem 4.1.6 may be chosen to have disjoint graphs.

With this theorem in hand we can summarize the values of the cov operator between the classes from Chart 1 in the following graphical form. Here the mark "$\mathfrak{c}$" next to the arrow means that the covering of the larger class by the functions from the smaller class is equal to $\mathfrak{c}$ and that this can be proved in ZFC. The mark "$< \mathfrak{c}$" next to the arrow means that it is consistent with ZFC (and it follows from $\mathrm{CPA}_{\mathrm{prism}}$) that the appropriate cov number is $< \mathfrak{c}$. (From Examples 4.5.6, 4.5.7, and 4.5.8 it follows that all these numbers are greater than or equal to $\min\{\mathrm{cov}(\mathcal{M}), \mathrm{cov}(\mathcal{N})\} > \omega$. So, under the continuum hypothesis CH or Martin's axiom, all these numbers are equal to $\mathfrak{c}$.)

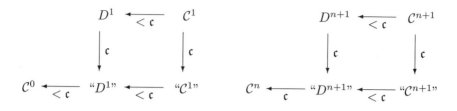

Chart 2. Values of the cov operator for $n = 0$ (left) and $n > 0$ (right).

The values of cov operator next to the vertical arrows are justified by $\text{cov}(\text{``}\mathcal{C}^n\text{''}, D^n) = \mathfrak{c}$ (Proposition 4.1.5(i)), while the marks "$< \mathfrak{c}$" below the upper horizontal arrows and that directly below them follow from Theorem 4.1.6. The remaining arrow in the right part of the chart is the restatement of the last part of Proposition 4.1.5(iii), while its counterpart in the left part of the chart follows from Corollary 4.1.4(b): $\text{cov}\left(\mathcal{C}, \text{``}\mathcal{C}^1\text{''}\right) = \text{cov}\left(\text{Borel}, \text{``}\mathcal{C}^1\text{''}\right) < \mathfrak{c}$ is a consequence of CPA$_{\text{prism}}$. Finally, let us mention that in Corollary 4.1.4(b) there is no chance of increasing the Borel family in any essential way and keep the result. This follows from the following fact:

$$\text{cov}(\text{Sc}, \mathcal{C}) = \text{cov}\left(\mathbb{R}^{\mathbb{R}}, \mathcal{C}\right) \geq \text{cof}(\mathfrak{c}), \tag{4.1}$$

where the symbol Sc stands for the family of all symmetrically continuous functions $f \colon \mathbb{R} \to \mathbb{R}$ that are, in particular, continuous outside of some set of measure zero and the first category. (See K. Ciesielski [28, cor. 1.1] and the remarks below on the operator dec.)

The number $\text{cov}(\mathcal{A}, \mathcal{F})$ is very closely related to the decomposition number $\text{dec}(\mathcal{A}, \mathcal{F})$ defined as

$$\min\left(\{\kappa \geq \omega \colon (\forall A \in \mathcal{A})(\exists \mathcal{G} \in [\mathcal{F}]^\kappa)\ \mathcal{G} \text{ is a partition of } A\} \cup \{|X|^+\}\right),$$

which was first studied by J. Cichoń, M. Morayne, J. Pawlikowski, and S. Solecki [24] for the Baire class α functions. (More information on $\text{dec}(\mathcal{F}, \mathcal{G})$ can be found in a survey article [26, sec. 4].) It is easy to see that if $\mathcal{A}$ and $\mathcal{F}$ are some classes of partial functions and $\mathcal{F}_r$ denotes all possible restrictions of functions from $\mathcal{F}$, then $\text{cov}(\mathcal{A}, \mathcal{F}) = \text{dec}(\mathcal{A}, \mathcal{F}_r)$. In particular, for all situations relevant to our discussion above, the operators cov and dec have the same values.

Our number cov is also related to the following general class of problems. We say that the families $\mathcal{A}, \mathcal{F} \subset \mathcal{P}(X)$ satisfy the *Intersection Theorem*, which we denote by

$$\mathrm{IntTh}(\mathcal{A}, \mathcal{F}),$$

if for every $A \in \mathcal{A}$ there exists an $F \in \mathcal{G}$ such that $|A \cap F| = |X|$. If $\mathcal{A} = \{A\}$, we write $\mathrm{IntTh}(A, \mathcal{F})$ in place of $\mathrm{IntTh}(\mathcal{A}, \mathcal{F})$. This kind of theorem has been studied for a large part of the twentieth century. In particular, in the early 1940s S. Ulam asked in the *Scottish Book* [90, problem 17.1] if $\mathrm{IntTh}(\mathcal{C}, \mathrm{Analytic})$ holds, that is, whether for every $f \in \mathcal{C}$ there exists a real analytic function $g \colon \mathbb{R} \to \mathbb{R}$ that agrees with f on a perfect set. (See [127].) In 1947, Z. Zahorski [130] gave a negative answer to this question by proving that the proposition $\mathrm{IntTh}(\mathcal{C}^\infty, \mathrm{Analytic})$ is false. In the same paper he also raised a natural question, which has become known as the Ulam-Zahorski Problem: *Does* $\mathrm{IntTh}(\mathcal{C}, \mathcal{G})$ *hold for* $\mathcal{G} = \mathcal{C}^\infty$ *(or* $\mathcal{G} = \mathcal{C}^n$ *or* $\mathcal{G} = D^n$*)?* Here is a quick summary of what is known about this problem. (See [15].)

Proposition 4.1.7

(a) $\neg\mathrm{IntTh}(\mathcal{C}^\infty, \mathrm{Analytic})$ (Z. Zahorski [130])

(b) $\mathrm{IntTh}(\mathcal{C}, \mathcal{C}^1)$ (S. Agronsky, A. M. Bruckner, M. Laczkovich, and D. Preiss [1])

(c) $\mathrm{IntTh}(\mathcal{C}^1, \mathcal{C}^2)$ (A. Olevskiĭ [109])

(d) $\neg\mathrm{IntTh}(\mathcal{C}, \mathcal{C}^2)$ *and* $\neg\mathrm{IntTh}(\mathcal{C}^n, \mathcal{C}^{n+1})$ *for* $n \geq 2$. (A. Olevskiĭ [109])

We are interested in these problems because for the families $\mathcal{A}, \mathcal{F} \in \mathcal{P}(\mathbb{R}^n)$ of uncountable Borel sets

$$\neg\mathrm{IntTh}(\mathcal{A}, \mathcal{F}) \implies \mathrm{cov}(\mathcal{A}, \mathcal{F}) = \mathfrak{c}, \qquad (4.2)$$

as, in this situation, if $\neg\mathrm{IntTh}(\mathcal{A}, \mathcal{F})$, then there exists an $A \in \mathcal{A}$, $|A| = \mathfrak{c}$, such that $|A \cap F| \leq \omega$ for every $F \in \mathcal{F}$. Thus in the examples relevant to Proposition 4.1.5, instead of proving $\mathrm{cov}(\mathcal{A}, \mathcal{F}) = \mathfrak{c}$ we will in fact be showing a stronger fact that $\neg\mathrm{IntTh}(A_0, \mathcal{F})$ for an appropriate choice of $A_0 \subset A \in \mathcal{A}$.

4.2 Proof of Proposition 4.1.3

Proposition 4.1.3 will be deduced from the following fact, which is a generalization of a theorem of M. Morayne [99]. (Morayne proved his results for E and E_1 being perfect sets, that is, for $\alpha = 1$.) For a set X we will use the symbol Δ_X to denote the diagonal in $X \times X$, that is, $\Delta_X = \{\langle x, x \rangle \colon x \in X\}$. We will usually write simply Δ in place of Δ_X, since X is always clear from the context.

Proposition 4.2.1 *Let* $0 < \alpha < \omega_1$, $E \in \mathbb{P}_\alpha$, $h\colon E \to \mathbb{R}$ *be a continuous injection, and* G *be a function from* $(h[E])^2 \setminus \Delta$ *into* $[0, 1]$ *that is continuous and symmetric, that is, such that* $G(x, y) = G(y, x)$ *for all* $x, y \in (h[E])^2 \setminus \Delta$. *Then there exists an* $E_1 \in \mathbb{P}_\alpha$, $E_1 \subset E$, *such that* G *is uniformly continuous on* $(h[E_1])^2 \setminus \Delta$.

The proof of Proposition 4.2.1 will be presented in the next section. In the proof of Proposition 4.1.3 we will also use the following lemma.

Lemma 4.2.2 *Let* $g\colon \mathbb{R} \to \mathbb{R}$ *be Borel,* $0 < \alpha < \omega_1$, *and* $E \in \mathbb{P}_\alpha$. *For every continuous injection* $h\colon E \to \mathbb{R}$ *there exist a subset* $E_1 \in \mathbb{P}_\alpha$ *of* E *and a "*C^1*" function* $f\colon \mathbb{R} \to \mathbb{R}$ *such that* f *extends* $g \restriction h[E_1]$.

In addition, we can require that either $f \in C^1$ *or*

($\star$) $f' \restriction h[E_1]$ *is constant and equal to* ∞ *or* $-\infty$ *and* f *is an autohomeomorphism of* $\mathbb{R}$ *such that* $f^{-1} \in C^1$.

PROOF. First note that there exists an $E' \in \mathbb{P}_\alpha$, $E' \subset E$, such that

$$g \restriction h[E'] \text{ is continuous.} \tag{4.3}$$

Indeed, let $h_0 \in \Phi_{\mathrm{prism}}$ be such that $E = h_0[\mathfrak{C}^\alpha]$ and let U be a comeager subset of $h[E] = (h \circ h_0)[\mathfrak{C}^\alpha]$ such that the restriction $g \restriction U$ is continuous. Then $(h \circ h_0)^{-1}(U)$ is comeager in $\mathfrak{C}^\alpha$ and, by Claim 1.1.5, there is a perfect cube $Q \subset (h \circ h_0)^{-1}(U)$. The set $E' = h_0[Q] \in \mathbb{P}_\alpha$ has the desired property since $h[E'] = h[h_0[Q]] \subset U$.

Now let $k\colon [-\infty, \infty] \to [0, 1]$ be a homeomorphism and let G be defined on $(h[E'])^2 \setminus \Delta$ by

$$G(x, y) = k\left(\frac{g(x) - g(y)}{x - y}\right).$$

Then, by Proposition 4.2.1, there exists an $E_1' \in \mathbb{P}_\alpha$, $E_1' \subset E'$, such that G is uniformly continuous on $(h[E_1'])^2 \setminus \Delta$. So, there exists a uniformly continuous extension of $G \restriction (h[E_1'])^2 \setminus \Delta$ to $\hat{G} \restriction (h[E_1'])^2$. Clearly $k^{-1}(\hat{G}(x, x))$ is the derivative (possibly infinite) of $g_0 = g \restriction h[E_1']$ for every $x \in h[E_1']$, so $g_0 \in$ "$C^1(h[E_1'])$".

Now, if $(g_0')^{-1}(\mathbb{R})$ is nonempty, then, as in the argument for (4.3), we can find an $E_1 \in \mathbb{P}_\alpha$, $E_1 \subset E_1'$, such that $h[E_1] \subset (g_0')^{-1}(\mathbb{R})$. This obviously implies $g \restriction h[E_1] \in C^1_{\mathrm{perf}}$. But we also know that the difference quotient function $\frac{g(x) - g(y)}{x - y}$ is uniformly continuous on $(h[E_1])^2 \setminus \Delta$. Therefore, by Whitney's extension theorem [129] (see also Lemma 4.4.1), we can find a C^1 extension $f\colon \mathbb{R} \to \mathbb{R}$ of $g \restriction h[E_1]$.

So, assume that $(g_0')^{-1}(\mathbb{R}) = \emptyset$. Then either $(g_0')^{-1}(\infty)$ or $(g_0')^{-1}(-\infty)$ is nonempty and open in $h[E_1']$. Assume the former case. Similarly as above we can find an $E_1'' \in \mathbb{P}_\alpha$, $E_1'' \subset E_1'$, such that $g_0'[h[E_1'']] = \{\infty\}$. Then, by a version of Whitney's extension theorem from [14, thm. 2.1], we can find a "$\mathcal{C}^1$" extension $f_0 \colon \mathbb{R} \to \mathbb{R}$ of $g \upharpoonright h[E_1'']$.

But then there exists an open interval J in $\mathbb{R}$ intersecting $h[E_1'']$ on the closure of which f_0' is positive. So $f_1 = f_0 \upharpoonright \mathrm{cl}(J)$ is strictly increasing and the derivative of f_1^{-1} is continuous, nonnegative, and bounded. Thus there exists a homeomorphism $f_2 \colon \mathbb{R} \to \mathbb{R}$ extending f_1^{-1} with $f_2 \in \mathcal{C}^1$. Now put $f = f_2^{-1}$ and take an $E_1 \in \mathbb{P}_\alpha$ with $E_1 \subset E_1'' \cap h^{-1}(J)$. It is easy to see that E_1 and f are as required. ∎

PROOF OF PROPOSITION 4.1.3(a). By Lemma 4.2.2 we can find an $E_0 \in \mathbb{P}_\alpha$ for which there is an extension $f \colon \mathbb{R} \to \mathbb{R}$ of $g \upharpoonright h[E_0]$ such that $f \in$ "$\mathcal{C}^1$" and either $f \in \mathcal{C}^1$ or f is an autohomeomorphism of $\mathbb{R}$ with $f^{-1} \in \mathcal{C}^1$. Thus, it is enough to find a subset $E \in \mathbb{P}_\alpha$ of E_0 for which $g \upharpoonright h[E] \in$ "$\mathcal{C}^\infty_{\mathrm{perf}}$".

If there exist a subset $E \in \mathbb{P}_\alpha$ of E_0 and $n < \omega$ such that

$$f = g \upharpoonright h[E] \in \text{``}\mathcal{C}^n_{\mathrm{perf}}\text{,''} \text{ and } f^{(n)} \text{ has a constant value } \infty \text{ or } -\infty, \quad (4.4)$$

then this E is as desired. So assume that there is no such E. We will use Corollary 3.1.3 to find a subprism E of E_0 for which $g \upharpoonright h[E] \in \mathcal{C}^\infty_{\mathrm{perf}}$.

First notice that we can assume that $E_0 = \mathfrak{C}^\alpha$, since we can replace h with $h \circ h_0$, where $h_0 \in \Phi_{\mathrm{prism}}$ is such that $E_0 = h_0[\mathfrak{C}^\alpha]$. Next, for every $k < \omega$ put

$$D_k = \{E \in \mathbb{P}_\alpha \colon g \upharpoonright h[E] \in \mathcal{C}^k_{\mathrm{perf}}\}$$

and notice that each D_k is dense and open in $\mathbb{P}_\alpha$. This is the case, since for every $\bar{E} \in \mathbb{P}_\alpha$ applying Lemma 4.2.2 k times and using the fact that (4.4) is false we can find a sequence $\bar{E} = P_0 \supset \cdots \supset P_k$ from $\mathbb{P}_\alpha$ such that $g \upharpoonright h[P_i] \in \mathcal{C}^i_{\mathrm{perf}}$ for each $i \le k$.

Notice also that, in the notation of Corollary 3.1.3, we have $D_k^* = D_k$ for every $k < \omega$. So, the set $\bar{D} = \bigcap_{k<\omega} D_k$ is dense in $\mathbb{P}_\alpha$ and clearly $g \upharpoonright h[E] \in \mathcal{C}^\infty_{\mathrm{perf}}$ for such an E. ∎

PROOF OF PROPOSITION 4.1.3(b). Let π_x and π_y be the projections of $\mathbb{R}^2$ onto the x-axis and y-axis, respectively, and consider the functions $h_x = \pi_x \circ h$ and $h_y = \pi_y \circ h$. Applying Lemma 3.2.5 two times we can find $\beta_x, \beta_y \le \alpha$ and $E = P_y \subset P_x$ from $\mathbb{P}_\alpha$ such that $h_x \circ \pi_{\beta_x}^{-1}$ is a function on $\pi_{\beta_x}[P_x] \in \mathbb{P}_{\beta_x}$, $h_y \circ \pi_{\beta_y}^{-1}$ is a function on $\pi_{\beta_y}[P_y] \in \mathbb{P}_{\beta_y}$, and each of these functions is either one to one or constant. Notice that

$$\text{either } h_x \text{ or } h_y \text{ is one to one on } E. \tag{4.5}$$

To see this first note that for every $z \in E$ we have

$$h(z) = \langle \pi_x \circ h(z), \pi_y \circ h(z) \rangle = \langle (h_x \circ \pi_{\beta_x}^{-1})(\pi_{\beta_x}(z)), (h_y \circ \pi_{\beta_y}^{-1})(\pi_{\beta_y}(z)) \rangle.$$

Since h is one to one, this implies that $\max\{\beta_x, \beta_y\} = \alpha$. By symmetry, we can assume that $\alpha = \beta_x$. Thus, $h_x = h_x \circ \pi_{\beta_x}^{-1}$ is either one to one or constant on $P_x = \pi_{\beta_x}[P_x]$. If h_x is one to one on P_x, then (4.5) holds. So, assume that h_x is constant on P_x. Then $\pi_x \circ h = h_x$ is constant on $E \subset P_x$, and so $h_y = \pi_y \circ h$ must be one to one on E, since h is one to one. Thus, (4.5) holds.

By symmetry, we can assume that h_x is one to one on E. Therefore $\pi_x \circ h$ is a one to one function from E onto $\pi_x[h[E]] \subset \mathbb{R}$. In particular, $F_0 = h[E] \subset \mathbb{R}^2$ is a function from $\pi_x[h[E]]$ into $\mathbb{R}$. Then, by Lemma 4.2.2 used with $g = F_0$ and $h = \pi_x \circ h \upharpoonright E$, we can find a subset $E_1 \in \mathbb{P}_\alpha$ of E and a function $f \colon \mathbb{R} \to \mathbb{R}$ extending $h[E_1] = g \upharpoonright h[E_1]$ such that either f or f^{-1} belongs to $\mathcal{C}^1$. ∎

4.3 Proposition 4.2.1: a generalization of a theorem of Morayne

Our proof of Proposition 4.2.1 is based on the following lemmas, the first of which is a version of a theorem of F. Galvin [60, 61]. (For the proof see [74, thm. 19.7] or [19]. Galvin proved his results for $\alpha = 1$.)

Lemma 4.3.1 *For every $0 < \alpha < \omega_1$ and every continuous symmetric function h from $(\mathcal{C}^\alpha)^2 \setminus \Delta$ into $2 = \{0, 1\}$ there exists a $P \in \mathbb{P}_\alpha$ such that h is constant on $P^2 \setminus \Delta$.*

PROOF. For $j < 2$, let G_j be the set of all $s \in \mathcal{C}^\alpha$ such that

$$(\forall \beta < \alpha)(\forall \varepsilon > 0)(\exists t \in \mathcal{C}^\alpha)\ 0 < \rho(s, t) < \varepsilon\ \&\ s \upharpoonright \beta = t \upharpoonright \beta\ \&\ h(s, t) = j$$

and notice that

$$\text{each } G_j \text{ is a } G_\delta\text{-set} \quad \text{and} \quad \mathcal{C}^\alpha = G_0 \cup G_1. \tag{4.6}$$

Indeed, to see that G_j is a G_δ-set it is enough to note that for every $\beta < \alpha$ and $\varepsilon > 0$ the set

$$G_j^{\beta, \varepsilon} = \{s \in \mathcal{C}^\alpha \colon (\exists t \in \mathcal{C}^\alpha)\ 0 < \rho(s, t) < \varepsilon\ \&\ s \upharpoonright \beta = t \upharpoonright \beta\ \&\ h(s, t) = j\}$$

is open in $\mathfrak{C}^\alpha$. So let $s \in G_j^{\beta,\varepsilon}$ and take $t \in \mathfrak{C}^\alpha$ witnessing it, that is, such that $0 < \rho(s,t) < \varepsilon$, $s \restriction \beta = t \restriction \beta$, and $h(s,t) = j$. We can choose basic open neighborhoods U and V of s and t, respectively, such that $U \times V \setminus \Delta \subset h^{-1}(j)$. In addition, we can assume that $\pi_\beta[U] = \pi_\beta[V]$ and that each of the sets U and V has diameter less than $\delta = (\varepsilon - \rho(s,t))/3$. Then $s \in U \subset G_i^{\beta,\varepsilon}$ since for every $s' \in U$ there exists a $t' \in V$, $t' \neq s'$, with $s' \restriction \beta = t' \restriction \beta$ (since $\pi_\beta[U] = \pi_\beta[V]$), $h(s',t') \in h[U \times V \setminus \Delta] = \{j\}$ and

$$0 < \rho(s',t') \leq \rho(s',s) + \rho(s,t) + \rho(t,t') \leq \delta + \rho(s,t) + \delta < \varepsilon.$$

Thus each $G_j^{\beta,\varepsilon}$ is open and G_j is a G_δ-set.

To see the second part of (4.6) assume, by way of contradiction, that there exists an $s \in \mathfrak{C}^\alpha \setminus (G_0 \cup G_1)$. Let β_0, ε_0 and β_1, ε_1 witness that $s \notin G_0$ and $s \notin G_1$, respectively. Define $\varepsilon = \min\{\varepsilon_0, \varepsilon_1\} > 0$ and $\beta = \max\{\beta_0, \beta_1\} < \alpha$ and find $t \in \mathfrak{C}^\alpha$ such that $t \restriction \beta = s \restriction \beta$, $\rho(s,t) < \varepsilon$, and $t(\beta) \neq s(\beta)$. Then there exists a $j < 2$ such that $h(s,t) = j$, and this, together with $t \restriction \beta_j = s \restriction \beta_j$ and $\rho(s,t) < \varepsilon_j$, contradicts the choice of β_j and ε_j. This finishes the proof of (4.6).

Next find a $j < 2$ and a basic clopen set U in $\mathfrak{C}^\alpha$ such that G_j is comeager in U. Replacing $\mathfrak{C}^\alpha$ with U, if necessary, we can assume that G_j is comeager in $\mathfrak{C}^\alpha$. Define a binary relation R on $\mathfrak{C}^\alpha$ by

$$R = \{\langle x_0, x_1 \rangle : h(x_0, x_1) = 1 - j\} \cup \Delta.$$

Clearly, if $P \in \mathbb{P}_\beta$ is R-independent, then $P^2 \setminus \Delta \subset h^{-1}(j)$. Thus, it is enough to find an R-independent $P \in \mathbb{P}_\beta$. But this follows immediately from Proposition 3.2.4, since G_j and R clearly satisfy its condition (bin). ∎

We will also need the following simple fact, which must be well known.

Lemma 4.3.2 *There exists a continuous function $h: \mathfrak{C} \to [0,1]$ with the following property. If X is a zero-dimensional Polish space, then for every continuous function $f: X \to [0,1]$ there exists a continuous $g: X \to \mathfrak{C}$ such that $f = h \circ g$.*

PROOF. Let $\{U_\sigma : \sigma \in 2^{<\omega}\}$ be an open basis for $[0,1]$ such that $U_\emptyset = [0,1]$ and, for every $\sigma \in 2^k$, $U_\sigma = U_{\sigma^\frown 0} \cup U_{\sigma^\frown 1}$ and $\mathrm{diam}(U_\sigma) \leq 2^{1-k}$. For every $s \in 2^\omega$, let $h(s) \in [0,1]$ be such that $\{h(s)\} = \bigcap_{n<\omega} \mathrm{cl}(U_{s \restriction n})$. It is clear that h is continuous.

To see that h is as required take X and f as in the lemma. For every $\sigma \in 2^{<\omega}$ choose an open set $V_\sigma \subset f^{-1}(U_\sigma)$ such that $V_\emptyset = X$, $V_{\sigma^\frown 0}$ and $V_{\sigma^\frown 1}$

are disjoint, and $V_{\sigma^\frown 0} \cup V_{\sigma^\frown 1} = V_\sigma$. This can be easily done by induction on the length of σ using the zero-dimensionality of X.[1] Thus for every $n < \omega$ the sets $\{V_\sigma : \sigma \in 2^n\}$ form a clopen partition of X.

Define $g(x)$ as the unique $s \in \mathfrak{C}$ for which $x \in \bigcap_{n<\omega} V_{s \restriction n}$. Clearly g is continuous. Moreover, if $g(x) = s$, then

$$x \in \bigcap_{n<\omega} V_{s \restriction n} \subset f^{-1}\left(\bigcap_{n<\omega} \mathrm{cl}(U_{s\restriction n})\right) = f^{-1}(\{h(s)\}) = f^{-1}(\{h(g(x))\})$$

so that $f(x) \in \{h(g(x))\}$. Hence $f = h \circ g$. ∎

The next lemma is already a very close approximation of Proposition 4.2.1.

Lemma 4.3.3 *If $\alpha < \omega_1$ and H is a continuous symmetric function from the set $(\mathfrak{C}^\alpha)^2 \setminus \Delta$ into $\mathfrak{C}$, then there exists an $E \in \mathbb{P}_\alpha$ such that H is uniformly continuous on $E^2 \setminus \Delta$.*

PROOF. For $k < \omega$ define $h_k : (\mathfrak{C}^\alpha)^2 \setminus \Delta \to 2$ by $h_k(s,t) = H(s,t)(k)$. It is enough to find an $E \in \mathbb{P}_\alpha$ such that each h_k is uniformly continuous on $E^2 \setminus \Delta$.

For this, first note that each h_k satisfies the assumptions of Lemma 4.3.1. In particular, each set

$$D_k = \{P \in \mathbb{P}_\alpha : h \text{ is constant on } P^2 \setminus \Delta\}$$

is dense and open in $\mathbb{P}_\alpha$. So, by Corollary 3.1.3, we can find an $E \in \mathbb{P}_\alpha$ that belongs to all sets

$$D_k^* = \left\{\bigcup \mathcal{D} : \mathcal{D} \in [D]^{<\omega} \text{ and sets in } \mathcal{D} \text{ are pairwise disjoint}\right\}.$$

To finish the proof it is enough to show that each h_k is uniformly continuous on $P^2 \setminus \Delta$ for $P \in D_k^*$. So, let $\{P_i : i < m\} \subset D_k$ be pairwise disjoint such that $P = \bigcup_{i<m} P_i$. Then h_k is constant (so, uniformly continuous) on $P_i \times P_i \setminus \Delta$ for each $i < m$. Compactness also guarantees that h_k is uniformly continuous on $P_i \times P_j$ for all distinct $i, j < m$. So h_k is uniformly continuous on

$$P^2 \setminus \Delta = \left(\bigcup_{i<m} P_i\right)^2 \setminus \Delta = \left(\bigcup_{i,j<m,\ i \neq j} P_i \times P_j\right) \cup \bigcup_{i<m} (P_i \times P_i \setminus \Delta).$$

This finishes the proof. ∎

---

[1] Recall that every second countable zero-dimensional space X is strongly zero-dimensional, see, e.g., [53, thm. 6.2.7]. In particular, for every open cover $\{W_0, W_1\}$ of X there are disjoint clopen sets $V_0 \subset W_0$ and $V_1 \subset W_1$ such that $V_0 \cup V_1 = X$.

PROOF OF PROPOSITION 4.2.1. Pick $f_0 \in \Phi_{\mathrm{prism}}(\alpha)$ such that $E = f_0 [\mathfrak{C}^\alpha]$ and put

$$F = G \circ \langle h \circ f_0, h \circ f_0 \rangle \colon (\mathfrak{C}^\alpha)^2 \setminus \Delta \to [0,1]. \tag{4.7}$$

Note also that

$$F = h_0 \circ H \tag{4.8}$$

for some continuous symmetric function from $H \colon (\mathfrak{C}^\alpha)^2 \setminus \Delta \to \mathfrak{C}$ and continuous $h_0 \colon \mathfrak{C} \to [0,1]$. This follows immediately from Lemma 4.3.2 used with a function $f = F \upharpoonright \{\langle x, y \rangle \in \mathfrak{C}^\alpha \times \mathfrak{C}^\alpha \colon x < y\}$, where $<$ is the lexicographical order on $\mathfrak{C}^\alpha$. (We use the lexicographical order in which $\mathfrak{C}^\alpha$ is identified with $2^{\alpha \times \omega}$ and $\alpha \times \omega$ is ordered in type ω. Then the set $\{\langle x, y \rangle \in \mathfrak{C}^\alpha \times \mathfrak{C}^\alpha \colon x < y\}$ is open in $\mathfrak{C}^\alpha \times \mathfrak{C}^\alpha$.)

Now, by Lemma 4.3.3, there exists an $E_0 \in \mathbb{P}_\alpha$ such that H is uniformly continuous on $(E_0)^2 \setminus \Delta$. So H can be extended to a uniformly continuous function $\hat{H}$ on $(E_0)^2$. Then the function

$$\hat{G} = h_0 \circ \hat{H} \circ \langle h \circ f_0, h \circ f_0 \rangle^{-1} = h_0 \circ \hat{H} \circ \langle (f_0)^{-1} \circ h^{-1}, (f_0)^{-1} \circ h^{-1} \rangle$$

is also uniformly continuous on $(h[f_0[E_0]])^2$. Put $E_1 = f_0[E_0]$ and notice that it is as desired.

Indeed, clearly $E_1 \in \mathbb{P}_\alpha$ and $E_1 \subset E$. Moreover, it is not difficult to see that $G \upharpoonright (h[E_1])^2 \setminus \Delta = \hat{G} \upharpoonright (h[E_1])^2 \setminus \Delta$. So G is uniformly continuous on $(h[E_1])^2 \setminus \Delta$. ∎

4.4 Theorem 4.1.6: on cov $(D^n, C^n) < \mathfrak{c}$

In the proof we will use the following lemma.

Lemma 4.4.1 *For $n < \omega$ let $f \in C^n$ and let $P \subset \mathbb{R}$ be a perfect set for which the function $F \colon P^2 \setminus \Delta \to \mathbb{R}$ defined by*

$$F(x,y) = \frac{f^{(n)}(x) - f^{(n)}(y)}{x - y}$$

is uniformly continuous and bounded. Then $f \upharpoonright P$ can be extended to a C^{n+1} function.

PROOF. This follows from the fact that $f \upharpoonright P$ satisfies the assumptions of Whitney's extension theorem. To see this, notice first that F naturally extends to a continuous function on P^2 with $F(a, a) = f^{(n+1)}(a)$. Next, for $q = 1, 2, 3, \ldots$ and $a \in P$, let

$$\eta_q(a) = \sup \left\{ \left| \frac{f^{(n)}(x) - f^{(n)}(a)}{x - a} - f^{(n+1)}(a) \right| : 0 < |x - a| < \frac{1}{q} \right\}.$$

In the second part of the proof of [56, thm. 3.1.15] it is shown that if

$$\lim_{q \to \infty} \sup\{\eta_q(a) : a \in P\} = 0, \tag{4.9}$$

then $f \upharpoonright P$ satisfies the assumptions of Whitney's extension theorem. However, we have

$$\frac{f^{(n)}(x) - f^{(n)}(a)}{x - a} - f^{(n+1)}(a) = F(x, a) - F(a, a),$$

so the uniform continuity of F clearly implies (4.9).   ∎

PROOF OF THEOREM 4.1.6. The inequalities $\mathrm{cov}\,(\text{``}D^n\text{''}, \text{``}\mathcal{C}^n\text{''}) > \omega$ and $\mathrm{cov}\,(D^n, \mathcal{C}^n) > \omega$ follow from Example 4.5.7. So it is enough to prove only that these numbers are $\leq \omega_1$.

To prove $\mathrm{cov}\,(D^n, \mathcal{C}^n) \leq \omega_1$, take an $f \in D^n$ and note that, by CPA_prism, it is enough to show that the set

$$\mathcal{E} = \{E \in \mathrm{Perf}(\mathbb{R}) : (\exists h \in \mathcal{C}^n(\mathbb{R}))\, h \upharpoonright E = f \upharpoonright E\}$$

is $\mathcal{F}_{\mathrm{prism}}$-dense. So fix a prism P in $\mathbb{R}$. Let $k : [-\infty, \infty] \to [0, 1]$ be a homeomorphism. Applying Proposition 4.2.1 n times in the same way as in the proof of Lemma 4.2.2, we find a subprism E of P such that for each $i < n$ the function $k \circ F_i : E^2 \setminus \Delta \to [0, 1]$ is uniformly continuous, where $F_i : E^2 \setminus \Delta \to \mathbb{R}$ is defined by

$$F_i(x, y) = \frac{f^{(i)}(x) - f^{(i)}(y)}{x - y}.$$

So each F_i can be extended to a continuous function $\bar{F}_i : E^2 \to [-\infty, \infty]$. Note also that since $\bar{F}_i(x, x) = f^{(i+1)}(x) \in \mathbb{R}$, as $f \in D^n$, we in fact have $\bar{F}_i[E^2] \subset \mathbb{R}$.

Next, starting with $f_0 = f$, we use Lemma 4.4.1 to prove by induction that for every $i < n$ there exists an $f_{i+1} \in \mathcal{C}^{i+1}(\mathbb{R})$ extending $f_i \upharpoonright E$. Then the function $h = f_n \in \mathcal{C}^n(\mathbb{R})$ witnesses that $E \in \mathcal{E}$.

To prove $\mathrm{cov}\,(\text{``}D^n\text{''}, \text{``}\mathcal{C}^n\text{''}) \leq \omega_1$, take an $f \in \text{``}D^n\text{''}$. As before, it is enough to show that

$$\mathcal{E}' = \{E' \in \mathrm{Perf}(\mathbb{R}) : (\exists h \in \text{``}\mathcal{C}^n(\mathbb{R})\text{''})\, h \upharpoonright E' = f \upharpoonright E'\}$$

is $\mathcal{F}_{\text{prism}}$-dense. So fix a prism P in $\mathbb{R}$ and find E, F_i's, and $\bar{F}_i$'s as above. Note that the F_i's are well defined since $f \in$ "D^n" $\subset \mathcal{C}^{n-1}$. By the same reason, we have that $\bar{F}_i[P^2] \subset \mathbb{R}$ for all $i < n-1$. However, $\bar{F}_{n-1}$ can have infinite values.

Proceeding as in the proof of Lemma 4.2.2, decreasing E if necessary, we can assume that either the range of $\bar{F}_{n-1}$ is bounded or $\bar{F}_{n-1} \upharpoonright P^2 \cap \Delta$ is constant and equal to ∞ or $-\infty$. If $\bar{F}_{n-1}$ is bounded, then, taking $E' = E$, we are done as previously. So, assume that $\bar{F}_{n-1}[P^2 \cap \Delta] = \{\infty\}$. (The case of $-\infty$ is handled by replacing f with $-f$.) Then $f^{(n-1)}$ and E satisfy the assumptions of Brown's version of Whitney's extension theorem [14, thm. 2.1]. So, we can find a "$\mathcal{C}^1$" extension $g \colon \mathbb{R} \to \mathbb{R}$ of $f^{(n-1)} \upharpoonright E$ such that $g'[E] = (f^{(n-1)})'[E] = \{\infty\}$ and $g'[\mathbb{R} \setminus E] \subset \mathbb{R}$. By integrating g $n-1$ times we can find a $G \colon \mathbb{R} \to \mathbb{R}$ such that $G^{(n-1)} = g$. Then $G \in$ "$\mathcal{C}^n$". Next notice that $G - f \in \mathcal{C}^n(E)$, since $(G - f)^{(n-1)} = g - f^{(n-1)} \equiv 0$ on E. Now, proceeding as above for the case of $f \in \mathcal{C}^n$, we can find a subprism E' of E and a function $\hat{h} \in \mathcal{C}^n(\mathbb{R})$ extending $G - f \upharpoonright E'$. Then function $h = G - \hat{h}$ belongs to "$\mathcal{C}^n$" as a difference of functions from "$\mathcal{C}^n$" and $\mathcal{C}^n$. Moreover, h extends $f \upharpoonright E'$ since $h = G - \hat{h} = G - (G - f) = f$ on E'. So, h witnesses $E' \in \mathcal{E}'$. ∎

4.5 Examples related to the cov operator

We will start with the examples needed for the proof of Proposition 4.1.5 that give $\mathfrak{c}$ as a lower bound for the appropriate numbers $\text{cov}(\mathcal{A}, \mathcal{F})$.

Example 4.5.1 *There exist a homeomorphism $h \colon \mathbb{R} \to \mathbb{R}$ and a perfect set $P \subset \mathbb{R}$ such that $h, h^{-1} \in$ "$\mathcal{C}^2$", $h'' \upharpoonright P \equiv \infty$, and $(h^{-1})'' \upharpoonright h[P] \equiv -\infty$. In particular, $\neg \text{IntTh}\left(h \upharpoonright P, D^2_{\text{perf}} \cup (D^2_{\text{perf}})^{-1}\right)$ and*

$$\text{cov}\left(\text{``}\mathcal{C}^2\text{''}, D^2_{\text{perf}} \cup (D^2_{\text{perf}})^{-1}\right) = \text{cov}\left(h, D^2_{\text{perf}} \cup (D^2_{\text{perf}})^{-1}\right) = \mathfrak{c}.$$

PROOF. First notice that there exist a strictly increasing homeomorphism h_0 from $\mathbb{R}$ onto $(0, \infty)$ and a perfect set $P \subset \mathbb{R}$ such that

$$h_0 \in \text{``}\mathcal{C}^1\text{''} \quad \text{and} \quad h_0' \upharpoonright P \equiv \infty. \tag{4.10}$$

Indeed, let C be an arbitrary nowhere dense, perfect subset of $[2, 3]$ with $2 \in C$ and let $d(x)$ denote the distance between $x \in \mathbb{R}$ and C. Let $f_0 \colon (0, \infty) \to [0, \infty)$ be defined by a formula $f_0(x) = x^{-2}$ for $x \in (0, 1]$ and $f_0(x) = d(x)$ for $x \in [1, \infty)$. Then f_0 is continuous and $f_0(x) = 0$ precisely when $x \in C$. Define a strictly increasing function f from $(0, \infty)$ onto $\mathbb{R}$ by a formula $f(x) = \int_1^x f_0(t)\,dt$. Then $f' = f_0$ and $f(x) = 1 - \frac{1}{x}$ on $(0,1)$. It is easy to see that $h_0 = f^{-1}$ and $P = f[C] \subset (0,\infty)$ satisfy (4.10).

Now put $h(x) = \int_0^x h_0(t)\,dt$. Then clearly h is strictly increasing since h_0 is positive. Also, h is onto $\mathbb{R}$ as on $(-\infty, 0)$ we have $h_0(x) = \frac{1}{1-x}$ and so $h(x) = -\ln(1-x)$. It is easy to see that $h' = h_0$; so, by condition (4.10), $h \in$ "$\mathcal{C}^2$" and $h'' \upharpoonright P \equiv \infty$. Also, if $g = h^{-1}$, then we obtain that $g'(x) = 1/h'(g(x)) = 1/h_0(g(x)) > 0$ is strictly decreasing and $h^{-1} = g \in \mathcal{C}^1$. Therefore, to see that $h^{-1} = g \in$ "$\mathcal{C}^2$" and $(h^{-1})'' \equiv -\infty$ on $h[P] = h[f[C]]$, it is enough to differentiate $g'(x)$ (note that the differentiation formulas are valid, if just one of the terms is infinite) to get $g''(x) = -[h'(g(x))]^{-2}h''(g(x))g'(x) = -h''(g(x))(g'(x))^3$. Thus, h and P have the desired properties.

To see the additional part note first that for every $f \in D^2_{\text{perf}}$ the functions f and $h \upharpoonright P$ may agree on at most a countable set S since at any point x of a perfect subset Q of S we would have

$$(h \upharpoonright Q)''(x) = \infty \neq (f \upharpoonright Q)''(x).$$

Similarly, $|f \cap (h \upharpoonright P)| \leq \omega$ for every $f \in (D^2_{\text{perf}})^{-1}$. This clearly implies the additional part. ∎

Example 4.5.2 *There exists a perfect set $P \subset \mathbb{R}$ and a function $f \in$ "$\mathcal{C}^1$" such that $f'(x) = \infty$ for every $x \in P$. In particular, $\neg\text{IntTh}\left(f \upharpoonright P, D^1_{\text{perf}}\right)$ and*

$$\text{cov}\left(\text{Borel}, \mathcal{C}^1\right) = \text{cov}\left(\text{"}\mathcal{C}^1\text{"}, \mathcal{C}^1\right) = \text{cov}\left(\text{"}\mathcal{C}^1\text{"}, D^1_{\text{perf}}\right) = \text{cov}\left(f, D^1_{\text{perf}}\right) = \mathfrak{c}.$$

PROOF. If f is a function h_0 from (4.10), then it has the desired properties.

For such an f and any function $g \in D^1_{\text{perf}}$, the intersection $f \cap g$ must be finite. So

$$\mathfrak{c} \geq \text{cov}\left(\text{Borel}, \mathcal{C}^1\right) \geq \text{cov}\left(\text{"}\mathcal{C}^1\text{"}, D^1_{\text{perf}}\right) \geq \text{cov}\left(f, D^1_{\text{perf}}\right) \geq \mathfrak{c}.$$

Monotonicity of the cov operator gives the other equations. ∎

Example 4.5.3 *For every $0 < n < \omega$ there exists an $f \in$ "$\mathcal{C}^n$" and a perfect set $P \subset \mathbb{R}$ such that $\neg\text{IntTh}\left(f \upharpoonright P, D^n_{\text{perf}}\right)$ so that*

$$\text{cov}(\text{"}\mathcal{C}^n\text{"}, \mathcal{C}^n) = \text{cov}(\text{"}\mathcal{C}^n\text{"}, D^n_{\text{perf}}) = \text{cov}(f, D^n_{\text{perf}}) = \mathfrak{c}.$$

PROOF. For $n = 1$, this is a restatement of Example 4.5.2. The general case can be done by induction: If f is good for some n and F is a definite integral of f, then $F \in$ "$\mathcal{C}^{n+1}$" and $\neg\text{IntTh}(F \upharpoonright P, D^{n+1}_{\text{perf}}) = \mathfrak{c}$. ∎

Example 4.5.4 *There exists an $f \in \mathcal{C}^1$ and a perfect set $P \subset \mathbb{R}$ such that $|(f \upharpoonright P) \cap g| \leq \omega$ for every $g \in \text{``}D^2\text{''}$. In particular, $\neg \text{IntTh}\left(f \upharpoonright P, \text{``}D^2\text{''}\right)$ and*

$$\text{cov}\left(\mathcal{C}^1, \text{``}D^2\text{''}\right) = \text{cov}\left(f, \text{``}D^2\text{''}\right) = \mathfrak{c}.$$

PROOF. In [1, thm. 22] the authors construct a perfect set $P \subset [0, 1]$ and a function $f \in \mathcal{C}^1$ that have the desired properties. The argument for this is implicitly included in the proof of [1, thm. 22] and goes like that.

Function f has the property that $f'(x) = 0$ for all $x \in P$. Now, assume that some $g \in \text{``}D^2\text{''}$ agrees with f on a perfect set $Q \subset P$. Then clearly we would have $(g \upharpoonright Q)'' \equiv [(g \upharpoonright Q)']' \equiv [(f \upharpoonright Q)']' \equiv [0]' \equiv 0$. On the other hand, in [1, thm. 22] it is shown[1] that for such a g we would have $g''(x) \in \{\pm\infty\}$ for every $x \in Q$, which is a contradiction. ∎

Example 4.5.5 *For every $0 < n < \omega$ there exist an $f \in \mathcal{C}^n$ and a perfect set $P \subset \mathbb{R}$ such that $\neg \text{IntTh}\left(f \upharpoonright P, \text{``}D^{n+1}\text{''}\right)$ and*

$$\text{cov}\left(\mathcal{C}^n, \text{``}D^{n+1}\text{''}\right) = \text{cov}\left(f, \text{``}D^{n+1}\text{''}\right) = \mathfrak{c}.$$

PROOF. For $n = 1$ this is a restatement of Example 4.5.4. The general case can be done by induction: If f is good for some n and F is a definite integral of f, then $F \in \mathcal{C}^{n+1}$ and $\neg \text{IntTh}(F \upharpoonright P, D_{\text{perf}}^{n+1}) = \mathfrak{c}$. ∎

Next we will describe the examples showing that the $\text{cov}(\mathcal{A}, \mathcal{F})$ numbers considered in Corollary 4.1.4 and Theorem 4.1.6 have values greater than ω. In what follows, $\text{cov}(\mathcal{M})$ ($\text{cov}(\mathcal{N})$, respectively) will stand for the smallest cardinality of a family $\mathcal{F} \subset \mathcal{P}(\mathbb{R})$ of measure zero sets (nowhere dense, respectively) such that $\mathbb{R} = \bigcup \mathcal{F}$.

Example 4.5.6 *There exists a function $f \in D^1$ such that*

$$\text{cov}\left(f, \text{``}\mathcal{C}^1\text{''} \cup (D^1)^{-1}\right) \geq \text{cov}(\mathcal{M}) > \omega.$$

In particular,

$$\text{cov}\left(\text{Borel}, \text{``}\mathcal{C}^1\text{''}\right) \geq \text{cov}\left(D^1, \text{``}\mathcal{C}^1\text{''}\right) \geq \text{cov}(\mathcal{M}) > \omega$$

and

$$\text{cov}\left(\text{Borel}, \mathcal{C}^1 \cup (\mathcal{C}^1)^{-1}\right) \geq \text{cov}\left(D^1, \mathcal{C}^1 \cup (\mathcal{C}^1)^{-1}\right) \geq \text{cov}(\mathcal{M}) > \omega.$$

---

[1] Actually, the calculation in [1, thm. 22] is done under the assumption that $g \in \mathcal{C}^2$, but it also works under our weaker assumption that $g \in \text{``}D^2\text{''}$.

PROOF. We will construct the function f only on $[0,1]$. It can be easily modified to a function defined on $\mathbb{R}$.

Let $E \subset [0,1]$ be an F_σ-set of full measure such that $E^c = [0,1] \setminus E$ is dense in $[0,1]$. It is well known that there exists a derivative $g: [0,1] \to [0,1]$ such that $g[E] \subset (0,1]$ and $g[E^c] = \{0\}$. (See, e.g., [17, p. 24].) Let $f: [0,1] \to \mathbb{R}$ be such that $f' = g$. We claim that this f is as desired.

Indeed, by way of contradiction, assume that for some $\kappa < \text{cov}(\mathcal{M})$ there exists a family $\{h_\xi \in \mathbb{R}^\mathbb{R} : \xi < \kappa\} \subset \text{``}\mathcal{C}^1\text{''} \cup (D^1)^{-1}$ such that $f \subset \bigcup_{\xi < \kappa} h_\xi$. Since h_ξ are closed subsets of $\mathbb{R}^2$ and the graph of f is compact, we see that the x-coordinate projections $P_\xi = \pi_x[f \cap h_\xi]$ are closed. So, $[0,1]$ is covered by less than $\text{cov}(\mathcal{M})$ closed sets P_ξ. Thus, there exists an $\eta < \kappa$ such that P_η has the nonempty interior $U = \text{int}(P_\eta)$.

Now, if $h_\eta \in \text{``}\mathcal{C}^1\text{''}$, then $h'_\eta = f' = g$ on U, which is impossible, since h'_η is continuous, while g is not continuous on any nonempty open set. So, assume that $h_\eta \in (D^1)^{-1}$. Note that f is strictly increasing as an integral of function g that is strictly positive a.e. So f^{-1} is strictly increasing and agrees with $h = h_\eta^{-1} \in D^1$ on an open set $f[U]$. But then, if $x \in U \setminus E$, then $h'(f(x)) = (f^{-1})'(f(x)) = \frac{1}{f'(x)} = \infty$, which contradicts $h \in D^1$. $\blacksquare$

Note also that if f from Example 4.5.6 is replaced by its $(n-1)$-st antiderivative, then we also get the following example.

Example 4.5.7 *For any $0 < n < \omega_1$ there exists an $f \in D^n$ such that*

$$\text{cov}(D^n, \text{``}\mathcal{C}^n\text{''}) \geq \text{cov}(f, \text{``}\mathcal{C}^n\text{''}) \geq \text{cov}(\mathcal{M}) > \omega.$$

Example 4.5.8 *There exists an $f \in \mathcal{C}^0$ such that*

$$\text{cov}(\mathcal{C}^0, \text{``}D^1_{\text{perf}}\text{''}) \geq \text{cov}(f, \text{``}D^1_{\text{perf}}\text{''}) \geq \text{cov}(\mathcal{N}) > \omega.$$

Moreover, for every $n < \omega$ if $F \in \mathcal{C}^n$ is such that $F^{(n)} = f$, then

$$\text{cov}(\mathcal{C}^n, \text{``}D^{n+1}_{\text{perf}}\text{''}) \geq \text{cov}(F, \text{``}D^{n+1}_{\text{perf}}\text{''}) \geq \text{cov}(\mathcal{N}) > \omega$$

and

$$\text{cov}(\text{Borel}, \text{``}\mathcal{C}^\infty_{\text{perf}}\text{''}) \geq \text{cov}(\mathcal{C}^n, \text{``}\mathcal{C}^\infty_{\text{perf}}\text{''}) \geq \text{cov}(F, \text{``}\mathcal{C}^\infty_{\text{perf}}\text{''}) \geq \text{cov}(\mathcal{N}) > \omega.$$

PROOF. A continuous function f justifying $\text{cov}(f, \text{``}D^1_{\text{perf}}\text{''}) \geq \text{cov}(\mathcal{N})$ was pointed out by M. Morayne: Just take any $f \in \mathcal{C}$ for which there is a set $A \subset \mathbb{R}$ of positive measure for which $|f^{-1}(a)| = \mathfrak{c}$ for all $a \in A$. (See [124, thm. 6.1].)

To see the additional part, let $\mathcal{G} = \{g_\xi : \xi < \kappa\}$ be an infinite subset of $\text{``}D^{n+1}_{\text{perf}}\text{''} \cup \text{``}\mathcal{C}^\infty_{\text{perf}}\text{''}$ such that $F \subset \bigcup \mathcal{G}$. We will show that $\kappa \geq \text{cov}(\mathcal{N})$. For this first note that for every $\xi < \kappa$ the domain of $F \cap g_\xi$ can be represented as a union of a perfect set P_ξ (which can be empty) and a countable (scattered) set S_ξ. Let $S = \bigcup_{\xi < \kappa} S_\xi$ and note that it has cardinality at most κ. Since $F \upharpoonright P_\xi = g_\xi \upharpoonright P_\xi$, by an easy induction on $i \leq n$ we can prove that

$$F^{(i)} \upharpoonright P_\xi = (g_\xi \upharpoonright P_\xi)^{(i)} \quad \text{provided} \quad g_\xi \in \text{``}D^i_{\text{perf}}\text{''} \text{ and } P_\xi \neq \emptyset. \quad (4.11)$$

Thus, if $g_\xi \in \text{``}D^{n+1}_{\text{perf}}\text{''}$ and $P_\xi \neq \emptyset$, then

$$f \upharpoonright P_\xi = F^{(n)} \upharpoonright P_\xi = (g_\xi \upharpoonright P_\xi)^{(n)} \in \text{``}D^1_{\text{perf}}\text{''}.$$

On the other hand,

$$\text{if } g_\xi \in \text{``}C^\infty_{\text{perf}}\text{''} \setminus \text{``}D^{n+1}_{\text{perf}}\text{''}, \text{ then } P_\xi = \emptyset.$$

Indeed, otherwise there is an $i \leq n$ such that $g_\xi \in \text{``}D^i_{\text{perf}}\text{''}$ and $g_\xi^{(i)}$ is constant and equal to ∞ or $-\infty$. So, by (4.11), for any $x \in P_\xi$ a real number $F^i(x)$ belongs to $\{-\infty, \infty\}$, which is a contradiction.

Thus $\mathcal{F} = \{f \upharpoonright P_\xi : \xi < \kappa \ \& \ P_\xi \neq \emptyset\} \cup \{f \upharpoonright \{x\} : x \in S\} \subset \text{``}D^1_{\text{perf}}\text{''}$ has cardinality at most κ and it covers f. So, by the first part, $\kappa \geq \text{cov}(\mathcal{N})$. $\blacksquare$

5

Applications of CPA$_{\text{prism}}^{\text{game}}$

First notice that the proof that is identical to that for Theorem 2.1.1 also gives its prism version which reads as follows.

Theorem 5.0.1 *Assume that* CPA$_{\text{prism}}^{\text{game}}$ *holds and let* X *be a Polish space. If* $\mathcal{D} \subset \text{Perf}(X)$ *is* $\mathcal{F}_{\text{prism}}$*-dense and it is closed under perfect subsets, then there exists a partition of* X *into* ω_1 *disjoint sets from* $\mathcal{D} \cup \{\{x\}: x \in X\}$.

Notice that, by using Theorem 5.0.1, we can obtain the following generalizations of Theorems 4.1.1 and 4.1.6.

Corollary 5.0.2 *The graphs of covering functions in Theorems 4.1.1 and 4.1.6 can be chosen as pairwise disjoint.*

5.1 Nice Hamel bases

The results presented in this section are based on K. Ciesielski and J. Pawlikowski [40].

In the next two sections we will consider $\mathbb{R}$ as a linear space over $\mathbb{Q}$. For $Z \subset \mathbb{R}$, its linear span with respect to this structure will be denoted by $\text{LIN}(Z)$. Notice that if L_m, for $0 < m < \omega$, is the collection of all functions $\ell: \mathbb{R}^m \to \mathbb{R}$ given by a formula

$$\ell(x_0, \ldots, x_{m-1}) = \sum_{i<m} q_i x_i, \text{ where } q_i \in \mathbb{Q} \setminus \{0\} \text{ for all } i < m, \quad (5.1)$$

then

$$\text{LIN}(Z) = \bigcup_{0 < m < \omega} \bigcup_{\ell \in L_m} \ell[Z^m].$$

Also, $Z \subset \mathbb{R}$ is linearly independent (over $\mathbb{Q}$) provided $\ell(x_0, \ldots, x_{m-1}) \neq 0$ for every $\ell \in L_m$, $0 < m < \omega$, and $\{x_0, \ldots, x_{m-1}\} \in [Z]^m$. It should be clear that the linear independence over $\mathbb{Q}$ is an $\mathcal{F}$-independence for the family $\mathcal{F}$ of closed symmetric relations F_ℓ, with $\ell \in L_n$, defined by

$$F_\ell(x_0, \ldots, x_{n-1}) \Leftrightarrow \ell(x_{\pi(0)}, \ldots, x_{\pi(n-1)}) = 0 \text{ for some permutation } \pi \text{ of } n.$$

We also have $\mathrm{LIN}(Z) = \mathrm{cl}_{\mathcal{F}}(Z)$ for every $Z \subset \mathbb{R}$. Recall also that the subset H of $\mathbb{R}$ is a *Hamel basis* provided it is a linear basis of $\mathbb{R}$ over $\mathbb{Q}$, that is, it is linearly independent and $\mathrm{LIN}(H) = \mathbb{R}$.

The first result we prove in this section is that $\mathrm{CPA}_{\mathrm{prism}}^{\mathrm{game}}$ implies the existence of a Hamel basis that is a union of ω_1 pairwise disjoint perfect sets. This can be viewed as a generalization of Theorem 5.0.1. To prove this we will need, as usual, some prism density results.

For a Polish space X and a family $\mathcal{F}$ of finitary relations on X, we say that $\mathcal{F}$ has *countable character* provided for every n-ary relation $F \in \mathcal{F}$ and every F-independent set $\{x_1, \ldots, x_n\} \in [X]^n$ the set

$$\{x \in X : F(x, x_1, \ldots, x_{n-1})\}$$

is countable. It should be clear that the linear independence family $\mathcal{F}$ defined above is of countable character. Similarly, algebraic independence can be expressed in this language. (See page 65.)

The next fact can be considered a first approximation of what we will need for finding our Hamel basis. However, it is not strong enough for what we need: Both its assumptions and its conclusion are too strong.

Proposition 5.1.1 *Let X be a Polish space and $\mathcal{F}$ be a countable family of closed finitary relations on X such that $\mathcal{F}$ has countable character. Then for every prism P in X there is a subprism Q of P such that Q is $\mathcal{F}$-independent.*

PROOF. This follows easily from Proposition 3.2.3. Indeed, pick an $h \in \Phi_{\mathrm{prism}}(\alpha)$ such that $P = h[\mathfrak{C}^\alpha]$, for every n-ary relation F on X let

$$F^* = \{\langle p_1, \ldots, p_n \rangle \in (\mathfrak{C}^\alpha)^n : F(h(p_1), \ldots, h(p_n)) \text{ holds}\},$$

and put $\mathcal{F}^* = \{F^* : F \in \mathcal{F}\}$. The countable character of $\mathcal{F}$ implies that (ex) holds with $G = \mathfrak{C}^\alpha$ for every $F^* \in \mathcal{F}^*$. So, by Proposition 3.2.3, there exists an $\mathcal{F}^*$-independent $E \in \mathbb{P}_\alpha$. But this means that $Q = h[E]$ is an $\mathcal{F}$-independent subprism of P. ∎

From Proposition 5.1.1 and the remark preceding its statement we immediately conclude that

Corollary 5.1.2 *For every prism P in $\mathbb{R}$ there is a subprism Q of P such that Q is algebraically (so linearly) independent.*

From Theorem 5.0.1 and Corollary 5.1.2 we can also easily deduce the following fact.

Corollary 5.1.3 CPA$_{\text{prism}}^{\text{game}}$ *implies that $\mathbb{R}$ is a union of ω_1 disjoint, closed, algebraically independent sets.*

Remark 5.1.4 Note that Corollary 5.1.2 is false if we replace prisms with cubes. In particular, there is a cube P in $\mathbb{R}$ without a linearly independent subcube.

PROOF. Indeed, let P_1 and P_2 be disjoint perfect subsets of $\mathbb{R}$ such that $P_1 \cup P_1$ is linearly independent over $\mathbb{Q}$. Let $f \colon P_1 \times P_2 \to \mathbb{R}$ be defined by a formula $f(x_1, x_2) = x_1 + x_2$. Then $P = f[P_1 \times P_2]$ is a cube in $\mathbb{R}$. To see that P has no linearly independent subcube, let $Q = Q_1 \times Q_2$ be a subcube of P and choose different $a_1, b_1 \in Q_1$ and $a_2, b_2 \in Q_2$. Then the set $\{a_1 + a_2, a_1 + b_2, b_1 + a_2, b_1 + a_2\} \subset Q$ is clearly linearly dependent. ∎

It seems that the conclusion from Corollary 5.1.3 is already close to the existence of a Hamel basis that is union of ω_1 disjoint closed sets. However, the sets from Corollary 5.1.3 can be pairwise highly linearly dependent. Thus, in order to prove the Hamel basis result, we need a density result that is considerably stronger than that from Corollary 5.1.2.

Lemma 5.1.5 *Let $M \subset \mathbb{R}$ be a σ-compact and linearly independent. Then for every prism P in $\mathbb{R}$ there exist a subprism Q of P and a compact subset R of $P \setminus M$ such that $M \cup R$ is a maximal linearly independent subset of $M \cup Q$.*

PROOF. Let $\mathcal{F}$ be the linear independent family defined at the beginning of this section and let $\bar{M} = \langle M_n \colon n < \omega \rangle$ be an increasing family of compact sets such that $M = \bigcup_{n < \omega} M_n$. Let $\mathcal{F}_{\bar{M}} = \bigcup_{n < \omega} \mathcal{F}_{M_n}$, where each $\mathcal{F}_{M_n}$ is defined as on page 65, that is, $\mathcal{F}_{M_n}$ is the the collection of all possible projections of the relations from $\mathcal{F}$ along M_n.

If $M \cap P$ is of second category in P, then we can choose a subprism Q of P with $Q \subset M$. Then Q and $R = \emptyset$ have the desired properties. On the other hand, if $M \cap P$ is of the first category in P, then, by Claim 1.1.5, we can find a subprism P_1 of P disjoint with M.

Now, applying Lemma 3.2.7 we can find a subprism Q of P_1 and a compact $\mathcal{F}_{\bar{M}}$-independent set $R \subset P_1 \subset P \setminus M$ such that $Q \subset \mathrm{cl}_{\mathcal{F}_{\bar{M}}}(R)$. But then $M \cup R$ is $\mathcal{F}$-independent; see (3.18). Moreover,

$$Q \subset \mathrm{cl}_{\mathcal{F}_{\bar{M}}}(R) = \mathrm{cl}_{\mathcal{F}}(M \cup R) = \mathrm{LIN}(M \cup R).$$

So, $M \cup Q \subset \mathrm{LIN}(M \cup R)$, proving that Q and R are as desired. ∎

Remark 5.1.6 In Lemma 5.1.5 we cannot require $R = Q$.

PROOF. Let P_1, P_2, and f be as in Remark 5.1.4. If $M = P_2$, then P has no subprism Q such that $M \cup Q$ is linearly independent, since any vertical section of Q is a translation of a portion of M. ∎

The next theorem represents a generalization of Proposition 5.1.1 and Theorem 5.0.1.

Theorem 5.1.7 $\mathrm{CPA}_{\mathrm{prism}}^{\mathrm{game}}$ *implies that there exists a family $\mathcal{H}$ of ω_1 pairwise disjoint perfect subsets of $\mathbb{R}$ such that $H = \bigcup \mathcal{H}$ is a Hamel basis.*

PROOF. For a linearly independent σ-compact set $M \subset \mathbb{R}$ and a prism P in $\mathbb{R}$, let $Q(M, P) = Q$ and $R(M, P) = R \subset P \setminus M$ be as in Lemma 5.1.5. Consider Player II strategy S given by

$$S(\langle\langle P_\eta, Q_\eta\rangle : \eta < \xi\rangle, P_\xi) = Q\left(\bigcup\{R_\eta : \eta < \xi\}, P_\xi\right),$$

where the R_η's are defined inductively by $R_\eta = R(\bigcup\{R_\zeta : \zeta < \eta\}, P_\eta)$.

By $\mathrm{CPA}_{\mathrm{prism}}^{\mathrm{game}}$, strategy S is not a winning strategy for Player II. So there exists a game $\langle\langle P_\xi, Q_\xi\rangle : \xi < \omega_1\rangle$ played according to S in which Player II loses, that is, $\mathbb{R} = \bigcup_{\xi < \omega_1} Q_\xi$.

Let $\mathcal{H} = \{R_\xi : \xi < \omega_1\}$ and notice that $\bigcup \mathcal{H}$ is a Hamel basis. Indeed, clearly $\bigcup \mathcal{H}$ is linearly independent. To see that it spans $\mathbb{R}$ it is enough to notice that $\mathrm{LIN}\left(\bigcup_{\eta < \xi} Q_\eta\right) \subset \mathrm{LIN}\left(\bigcup_{\eta < \xi} R_\eta\right)$ for every $\xi < \omega_1$.

Although sets in $\mathcal{H}$ need not be perfect, they are clearly pairwise disjoint and compact. Thus, the theorem follows immediately from the following remark. ∎

Remark 5.1.8 If there exists a family $\mathcal{H}$ of ω_1 pairwise disjoint compact subsets of $\mathbb{R}$ such that $\bigcup \mathcal{H}$ is a Hamel basis, then there exists such an $\mathcal{H}$ with $\mathcal{H} \subset \mathrm{Perf}(\mathbb{R})$.

PROOF. Let $\mathcal{H}_0$ be a family of ω_1 pairwise disjoint compact subsets of $\mathbb{R}$ such that $\bigcup \mathcal{H}_0$ is a Hamel basis. Partitioning each $H \in \mathcal{H}_0$ into its perfect part and singletons from its scattered part we can assume that

$\mathcal{H}_0$ contains only perfect sets and singletons. To get $\mathcal{H}$ as required, fix a perfect set $P_0 \in \mathcal{H}_0$ and an $x \in P_0$ and notice that if we replace each $P \in \mathcal{H}_0 \setminus \{P_0\}$ with $px + qP$ for some $p, q \in \mathbb{Q} \setminus \{0\}$, then the resulting family will still be pairwise disjoint with union being a Hamel basis. Thus, without loss of generality, we can assume that every open interval in $\mathbb{R}$ contains ω_1 perfect sets from $\mathcal{H}_0$. Now, for every singleton $\{x\}$ in $\mathcal{H}_0$, we can choose a sequence $P_1^x > P_2^x > P_3^x > \cdots$ from $\mathcal{H}_0$ converging to x and replace a family $\{x\} \cup \{P_n^x : n < \omega\}$ with its union. (We assume that we choose different sets P_n^x for different singletons.) If $\mathcal{H}$ is such a modification of $\mathcal{H}_0$, then $\mathcal{H}$ is as desired. ∎

Recall that a subset T of $\mathbb{R}$ is a *transcendental basis* of $\mathbb{R}$ over $\mathbb{Q}$ provided T is a maximal algebraically independent subset of $\mathbb{R}$. The proof that is identical to that for Theorem 5.1.7 also gives the following result.

Theorem 5.1.9 CPA$_{\text{prism}}^{\text{game}}$ *implies that there exists a family $\mathcal{T}$ of ω_1 pairwise disjoint perfect subsets of $\mathbb{R}$ such that $T = \bigcup \mathcal{T}$ is a transcendental basis of $\mathbb{R}$ over $\mathbb{Q}$.*

Next, we will present two interesting consequences of the existence of a Hamel basis described in Theorem 5.1.7. Let $\mathcal{I}$ be a translation invariant ideal on $\mathbb{R}$. We say that a subset X of $\mathbb{R}$ is $\mathcal{I}$-*rigid* provided $X, \mathbb{R} \setminus X \notin \mathcal{I}$ but $X \triangle (r + X) \in \mathcal{I}$ for every $r \in \mathbb{R}$. An easy inductive construction gives a nonmeasurable subset X of $\mathbb{R}$ without the Baire property, which is $[\mathbb{R}]^{<\mathfrak{c}}$-rigid. (The first such construction, under CH, can be found in a paper [119] by W. Sierpiński. Compare also [68].) Thus, under CH or MA there are $\mathcal{N} \cap \mathcal{M}$-rigid sets. Recently these sets have been studied by M. Laczkovich [83] and J. Cichoń, A. Jasiński, A. Kamburelis, and P. Szczepaniak [23]. In particular, M. Laczkovich's result from [83, thm. 2] implies that there is no $\mathcal{N} \cap \mathcal{M}$-rigid set in the random and Cohen models. The next corollary shows that the existence of such sets follows from CPA$_{\text{prism}}^{\text{game}}$.

Corollary 5.1.10 CPA$_{\text{prism}}^{\text{game}}$ *implies there exists an $\mathcal{N} \cap \mathcal{M}$-rigid set X that is neither measurable nor has the Baire property.*

PROOF. Let $\mathcal{H} = \{Q_\xi : \xi < \omega_1\}$ be from Theorem 5.1.7 and for every $\xi < \omega_1$ let $L_\xi = \text{LIN}\left(\bigcup_{\eta < \xi} Q_\eta\right)$. Then $\mathbb{R}$ is an increasing union of L_ξ's and each L_ξ belongs to $\mathcal{N} \cap \mathcal{M}$, since it is a proper Borel subgroup of $\mathbb{R}$.

Since CPA$_{\text{prism}}^{\text{game}}$ implies that $\text{cof}(\mathcal{N}) = \text{cof}(\mathcal{M}) = \omega_1$, there exists a family $\{C_\xi : \xi < \omega_1\} \subset \mathcal{M} \cup \mathcal{N}$ such that every $S \in \mathcal{M} \cup \mathcal{N}$ is a subset of some C_ξ. By induction choose $X_0 = \{x_\xi : \xi < \omega_1\} \subset \mathbb{R}$ such that

$$x_\xi \notin C_\xi \cup \mathrm{LIN}(L_\xi \cup \{x_\zeta : \zeta < \xi\}).$$

Then X_0 intersects the complement of every set from $\mathcal{M} \cup \mathcal{N}$. Define

$$X = \bigcup_{\xi < \omega_1} (x_\xi + L_\xi)$$

and notice that $X_0 \subset X$ and $2X_0 \subset \mathbb{R} \setminus X$. Therefore, both X and $\mathbb{R} \setminus X$ intersect the complement of every set from $\mathcal{M} \cup \mathcal{N}$. In particular, $X, \mathbb{R} \setminus X \notin \mathcal{M} \cup \mathcal{N}$.

Next notice that for every $r \in L_\zeta$

$$X \triangle (r + X) \subset \bigcup_{\xi < \zeta} [(x_\xi + L_\xi) \cup (r + x_\xi + L_\xi)] \in \mathcal{N} \cap \mathcal{M}.$$

Thus, X is $\mathcal{N} \cap \mathcal{M}$-rigid, but also $\mathcal{N}$-rigid and $\mathcal{M}$-rigid. These last two facts imply that X is neither measurable nor does it have the Baire property. ∎

Our next application of Theorem 5.1.7 is the following.

Corollary 5.1.11 $\mathrm{CPA}_{\mathrm{prism}}^{\mathrm{game}}$ *implies there exists a function* $f : \mathbb{R} \to \mathbb{R}$ *such that for every* $h \in \mathbb{R}$ *the difference function* $\Delta_h(x) = f(x + h) - f(x)$ *is Borel; however, for every* $\alpha < \omega_1$ *there is an* $h \in \mathbb{R}$ *such that* Δ_h *is not of the Borel class* α.

Note that, answering a question of M. Laczkovich from [82], R. Filipów and I. Recław [57] gave an example of such an f under CH. I. Recław also asked (private communication) whether such a function can be constructed in the absence of CH. Corollary 5.1.11 gives an affirmative answer to this question. It is an open question whether such a function exists in ZFC.

PROOF. The proof is quite similar to that for Corollary 5.1.10.

Let $\mathcal{H} = \{Q_\xi : \xi < \omega_1\}$ be from Theorem 5.1.7. For every $\xi < \omega_1$ define $L_\xi = \mathrm{LIN}\left(\bigcup_{\eta < \xi} Q_\eta\right)$ and choose a Borel subset B_ξ of Q_ξ of Borel class greater than ξ. Define

$$X = \bigcup_{\xi < \omega_1} (B_\xi + L_\xi)$$

and let f be the characteristic function χ_X of X.

To see that f is as required note that

$$\Delta_{-h}(x) = \left[\chi_{(h+X) \setminus X} - \chi_{X \setminus (h+X)}\right](x).$$

So, it is enough to show that each of the sets $(h + X) \setminus X$ and $X \setminus (h + X)$

is Borel, though they can be of arbitrary high class. For this, notice that for every $h \in L_{\alpha+1} \setminus L_\alpha$ we have

$$h + X = h + \bigcup_{\xi < \omega_1} (B_\xi + L_\xi) = \bigcup_{\xi \leq \alpha} (h + B_\xi + L_\xi) \cup \bigcup_{\alpha < \xi < \omega_1} (B_\xi + L_\xi)$$

and that the sets $\bigcup_{\xi \leq \alpha} (h + B_\xi + L_\xi) \subset L_{\alpha+1}$ and $\bigcup_{\alpha < \xi < \omega_1} (B_\xi + L_\xi)$ are disjoint. So

$$(h + X) \setminus X = \bigcup_{\xi \leq \alpha} (h + B_\xi + L_\xi) \setminus X = \bigcup_{\xi \leq \alpha} (h + B_\xi + L_\xi) \setminus \bigcup_{\xi \leq \alpha} (B_\xi + L_\xi)$$

is Borel, since each set $B_\xi + L_\xi$ is Borel. (It is a subset of $Q_\xi + L_\xi$, which is homeomorphic to $Q_\xi \times L_\xi$ via the addition function.) Similarly, set $X \setminus (h + X)$ is Borel.

Finally notice that for $h \in Q_\alpha \setminus B_\alpha$ the set

$$(h + X) \setminus X = \bigcup_{\xi \leq \alpha} (h + B_\xi + L_\xi)$$

is of the Borel class greater than α, since $(h + Q_\alpha) \cap [(h + X) \setminus X] = h + B_\alpha$ has the same property. Thus, $\Delta_h(x)$ can be of an arbitrarily high Borel class. ∎

5.2 Some additive functions and more on Hamel bases

The results presented in this section come from K. Ciesielski and J. Pawlikowski [42].

The proof of the next application is essentially more involved than those presented so far and requires considerably more preparation. However, it can be viewed as a "model example" of how some CH proofs can be modified to the proofs from CPA$_{\text{prism}}^{\text{game}}$. Recall that a function $f: \mathbb{R} \to \mathbb{R}$ is *almost continuous* provided any open subset U of $\mathbb{R}^2$ that contains the graph of f also contains a graph of a continuous function from $\mathbb{R}$ to $\mathbb{R}$. It is known that if f is almost continuous, then its graph is connected in $\mathbb{R}^2$ (i.e., f is a connectivity function) and that f has the intermediate value property (i.e., f is Darboux). (See, e.g., [105] or [29].) Recall also that a function $f: \mathbb{R} \to \mathbb{R}$ is *additive* provided $f(x + y) = f(x) + f(y)$ for every $x, y \in \mathbb{R}$. It is well known that every function defined on a Hamel basis can be uniquely extended to an additive function. (See, e.g., [25, thm. 7.3.2].)

Our next goal will be to construct an additive discontinuous, almost continuous function $f: \mathbb{R} \to \mathbb{R}$ whose graph is of measure zero. In fact, we will show that, under CPA$_{\text{prism}}^{\text{game}}$, such an f can be found inside a set

$(\mathbb{R}\times G)\cup(G\times\mathbb{R})$ for every G_δ subset G of $\mathbb{R}$ with $0\in G$. A first construction of such a function, under Martin's axiom, was given by K. Ciesielski in [27]. Although it can be shown that such a function (i.e., with graph being a subset of $(\mathbb{R}\times G)\cup(G\times\mathbb{R})$) does not exist in the Cohen model, it is unknown whether the existence of an additive discontinuous, almost continuous function with graph of measure zero can be proved in ZFC alone. Note also that, under $\mathrm{CPA}_{\mathrm{prism}}^{\mathrm{game}}$, it is also possible to find an f as above with G being a meager set of measure zero. However, this construction is even more technical and will not be presented in this text.

Recall also that a function $f\colon\mathbb{R}\to\mathbb{R}$ is almost continuous if and only if it intersects every *blocking set*, that is, a closed set $K\subseteq\mathbb{R}^2$ that meets every continuous function from $\mathcal{C}(\mathbb{R})$ and is disjoint with at least one function from $\mathbb{R}^{\mathbb{R}}$. The domain of every blocking set contains a nondegenerate connected set. (See [75] or [105].) It is important for us to note that every blocking set contains a graph of a continuous function $g\colon G\to\mathbb{R}$, where G is a dense G_δ subset of some nontrivial interval. (See [76]. This follows from the fact that for every closed bounded set B with domain I, the mapping $I\ni x\mapsto\inf\{y\colon\langle x,y\rangle\in B\}$ is of the first Baire class, so it is continuous when restricted to a dense G_δ subset.) Thus, in order to make sure that a function is almost continuous, it is enough to ensure that its graph intersects every function from the family

$$\mathcal{K}=\bigcup\{\mathcal{C}(G)\colon G\text{ is a }G_\delta\text{ second category subset of }\mathbb{R}\}.\tag{5.2}$$

Now we are ready to state our next theorem.

Theorem 5.2.1 $\mathrm{CPA}_{\mathrm{prism}}^{\mathrm{game}}$ *implies that for every dense G_δ set $G\subset\mathbb{R}$ such that $0\in G$ there exists an additive discontinuous, almost continuous function $f\colon\mathbb{R}\to\mathbb{R}$ whose graph is a subset of $(\mathbb{R}\times G)\cup(G\times\mathbb{R})=(G^c\times G^c)^c$.*

Using Theorem 5.2.1 with G of measure zero, we immediately obtain the following corollary.

Corollary 5.2.2 $\mathrm{CPA}_{\mathrm{prism}}^{\mathrm{game}}$ *implies that there exists a discontinuous, almost continuous, additive function $f\colon\mathbb{R}\to\mathbb{R}$ whose graph is of measure zero.*

In what follows we will use the following notation for $G,P\subset\mathbb{R}$:

$$G[P]=\{x\in\mathbb{R}\colon x-P\subset G\}=\bigcap_{p\in P}(p+G).\tag{5.3}$$

It is also convenient to note that

$$G[P]^c = G^c + P.$$

We start with noticing some simple properties of this operation.

Fact 5.2.3 *Let* $P, S, G, G' \subset \mathbb{R}$.

(a) *If* $P \subset S$ *and* $G' \subset G$, *then* $G[P] \supset G'[S]$.
(b) *If* P *is compact and* G *is open, then* $G[P]$ *is open.*
(c) *If* $P = \bigcup_{i<\omega} P_i$ *and* $G = \bigcap_{n<\omega} G_n$, *then* $G[P] = \bigcap_{i,n<\omega} G_n[P_i]$.
(d) *If* P *is* σ-*compact and* G *is a* G_δ *set, then* $G[P]$ *is also a* G_δ *set.*
(e) *If* $G[P_n]$ *is a dense* G_δ *set for every* $n < \omega$, *then so is* $G[\bigcup_{n<\omega} P_n]$.
(f) $G[P][S] = G[P + S]$.

PROOF. (a) follows immediately from the second part of (5.3) while (b) follows from its first part. To see (c) notice that, by (5.3),

$$G[P] = \bigcap_{i<\omega} \{x \in \mathbb{R} : x - P_i \subset G\} = \bigcap_{i,n<\omega} \{x \in \mathbb{R} : x - P_i \subset G_n\} = \bigcap_{i,n<\omega} G_n[P_i].$$

So, (d) follows immediately from (b), while (e) is an easy consequence of (c). Note also that

$$G[P][S]^c = G[P]^c + S = G^c + P + S = G[P + S]^c,$$

so (f) holds.   ∎

Recall that for a Polish space X the space $\mathcal{C}(X)$ of continuous functions from X into $\mathbb{R}$ is considered with the metric of uniform convergence.

Lemma 5.2.4 *Let* X *be a Polish space and* $\bar{x} \in \bar{K} \in \text{Perf}(X)$. *For every dense* G_δ-*set* $G \subset \mathbb{R}$ *and a prism* P *in* $\mathcal{C}(X)$ *there exist a subprism* Q *of* P *and a* $K \in \text{Perf}(\bar{K})$ *with* $\bar{x} \in K$ *such that* $G[\text{LIN}(R_K(Q))]$ *is a dense* G_δ *subset of* $\mathbb{R}$, *where* $R_K(Q) = \{h(x) : h \in Q \ \& \ x \in K\}$.

PROOF. Let $\mathcal{U}$ be a countable family of open subsets of $\mathbb{R}$ with the property that $G = \bigcap \mathcal{U}$ and fix a countable basis $\mathcal{B}$ for $\mathbb{R}$. For $0 < m < \omega$, let L_m be the set of all functions ℓ defined as in (5.1) and put $L = \bigcup_{0<m<\omega} L_m$. In what follows, for $\ell : X^m \to \mathbb{R}$ from L and $Z \subset X$ we will write $\ell[Z]$ in place of $\ell[Z^m]$.

Let $h \in \mathcal{F}_{\text{prism}}(\mathcal{C}(X))$ be such that $P = h[\mathfrak{C}^\alpha]$ and fix an enumeration $\{\langle U_k, \ell_k, B_k \rangle : k < \omega\}$ of $\mathcal{U} \times L \times \mathcal{B}$. By induction on $k < \omega$ we will construct the sequences $\langle \mathcal{E}_k : k < \omega \rangle$ and $\langle \mathcal{K}_k : k < \omega \rangle$ such that, for every $k < \omega$:

(a) $\mathcal{K}_k$ is a family $\{K_t \in \mathrm{Perf}(\bar{K}) : t \in 2^k\}$ of pairwise disjoint sets such that $\bar{x} \in \bigcup \mathcal{K}_k$.

(b) $K_s \subset K_t$ for each $t \in 2^k$ and $t \subset s \in 2^{k+1}$.

(c) $\mathcal{E}_k = \{E_s \in \mathbb{P}_\alpha : s \in 2^{A_k}\}$.

(d) $\mathcal{E}_k$ and $\mathcal{E}_{k+1}$ satisfy conditions (i), (ii), (ag), and (sp) from Lemma 3.1.1 for every $s, t \in 2^{A_{k+1}}$ and $r \in 2^{A_k}$.

(e) If $R_k = \{h(g)(x) : g \in \bigcup \mathcal{E}_k \ \& \ x \in \bigcup \mathcal{K}_k\}$, then $U_k(\ell_k[R_k]) \cap B_k \neq \emptyset$.

Before we construct such sequences, note how this will complete the proof. Clearly, by (a) and (b), sequence $\langle \mathcal{E}_k : k < \omega \rangle$ satisfies the assumptions of Lemma 3.1.1. Thus, $E = \bigcap_{k<\omega} \bigcup \mathcal{E}_k$ belongs to $\mathbb{P}_\alpha$, and so $Q = h[E]$ is a subprism of P. Also, if $K = \bigcap_{k<\omega} \bigcup \mathcal{K}_k$, then $\bar{x} \in K \in \mathrm{Perf}(\bar{K})$. To see that $G[\mathrm{LIN}(R_K(Q))]$ is a dense G_δ, notice that $R_K \subset R_k$ for all $k < \omega$. So, by (e), we have $U_k(\ell_k[R_K]) \cap B_k \neq \emptyset$. In particular, $U(\ell[R_K])$ is dense and open for every $U \in \mathcal{U}$ and $\ell \in L$. Thus, for every, $U \in \mathcal{U}$ the set

$$\bigcap_{\ell \in L} U[\ell[R_K(Q)]] = U\left[\bigcup_{\ell \in L} \ell[R_K(Q)]\right] = U\left[\mathrm{LIN}(R_K(Q))\right]$$

is a dense G_δ-set, and so is $G[\mathrm{LIN}(R_K(Q))] = \bigcap_{U \in \mathcal{U}} U[\mathrm{LIN}(R_K(Q))]$, as desired.

To choose $\mathcal{E}_0 = \{E_\emptyset\}$ and $\mathcal{K}_0 = \{K_\emptyset\}$, pick $g_0 \in \mathfrak{C}^\alpha$, put $y = h(g_0)(\bar{x})$, and let $\{z\} = \ell_0[\{y\}]$. Clearly $U_0[\{z\}] = \{x \in \mathbb{R} : x - \{z\} \subset U_0\}$ is open and dense, so there is a $b_0 \in B_0$ such that $b_0 - \{z\} \subset U_0$. Let $\varepsilon > 0$ be such that $b_0 - (z - \varepsilon, z + \varepsilon) \subset U_0$. Find a number $\delta > 0$ such that $\ell_0[(y - 2\delta, y + 2\delta)] \subset (z - \varepsilon, z + \varepsilon)$ and a clopen subset $K_\emptyset$ of $\bar{K}$ containing $\bar{x}$ for which $h(g_0)[K_\emptyset] \subset (y - \delta, y + \delta)$. Also, let $\delta_0 > 0$ be such that the diameter of $h[B_\alpha(g, \delta_0)]$ is less than δ and put $E_\emptyset = B_\alpha(g_0, \delta_0)$. We just need to check condition (e). But for every $g \in E_\emptyset$ and $x \in K_\emptyset$ we have $|h(g)(x) - y| \leq |h(g)(x) - h(g_0)(x)| + |h(g_0)(x) - h(g_0)(\bar{x})| < 2\delta$. So, $R_0 \subset (y - 2\delta, y + 2\delta)$ and $b_0 - \ell_0[R_0] \subset b_0 - (z - \varepsilon, z + \varepsilon) \subset U_0$. Thus, $b_0 \in U_0(\ell_0[R_0]) \cap B_0$.

To make an inductive step assume that, for some $k < \omega$, families $\mathcal{E}_k$ and $\mathcal{K}_k$ are already constructed. We will find the appropriate $\mathcal{E}_{k+1}$ and $\mathcal{K}_{k+1}$. First use Lemma 3.1.2(A) to pick an $\mathcal{E}'_{k+1} = \{E'_s \in \mathbb{P}_\alpha : s \in 2^{A_{k+1}}\}$ such that (d) holds. For any $s \in 2^{A_{k+1}}$ choose a $g_s \in E'_s$ such that the family $\{\{g_s\} : s \in 2^{A_{k+1}}\}$ satisfies condition (ag). Also, for every $r \in 2^{k+1}$ choose an $x_r \in K_{r \upharpoonright k}$ such that all points in $\bar{X} = \{x_r : r \in 2^{k+1}\}$ are distinct and $\bar{x} \in \bar{X}$. Put $Y = \bigcup \{h(g_s)[\bar{X}] : s \in 2^{A_{k+1}}\}$ and $Z = \ell_{k+1}[Y]$. Clearly $U_{k+1}[Z] = \{x \in \mathbb{R} : x - Z \subset U_{k+1}\} = \bigcap_{z \in Z}(z + U_{k+1})$ is open and dense since Z is finite. Thus there is a $b_{k+1} \in B_{k+1}$ such that $b_{k+1} - Z \subset U_{k+1}$. Let $\varepsilon > 0$ be such that $b_{k+1} - B(Z, \varepsilon) \subset U_{k+1}$, where $B(Z, \varepsilon)$ is the set of all $x \in \mathbb{R}$ with distance from Z less than ε. Since Y is finite, ℓ_{k+1} is continuous, and $Z = \ell_{k+1}[Y]$, we can find a $\delta > 0$ such that $\ell_{k+1}[B(Y, 2\delta)] \subset B(Z, \varepsilon)$. Also, for every $r \in 2^{k+1}$ find a clopen subset K_r of $K_{r\restriction k}$ containing x_r such that $h(g_s)[K_r] \subset B(Y, \delta)$ for every $s \in 2^{A_{k+1}}$ and $\mathcal{K}_{k+1} = \{K_r : r \in 2^{k+1}\}$ is pairwise disjoint. This ensures (a) and (b). Let $\delta_0 > 0$ be such that for every $s \in 2^{A_{k+1}}$ the diameter of $h[B_\alpha(g_s, \delta_0)]$ is less than δ and put $E_s = B_\alpha(g_s, \delta_0) \cap E_s'$. It is easy to see that with $\mathcal{E}_{k+1} = \{E_s : s \in 2^{A_{k+1}}\}$ conditions (c) and (d) are satisfied. We just need to check (e). To see it notice that $R_{k+1} \subset B(Y, 2\delta)$, since for every $h(g)(x) \in R_{k+1}$ there are $s \in 2^{A_{k+1}}$ and $r \in 2^{k+1}$ such that

$$|h(g)(x) - h(g_s)(x_r)| \leq |h(g)(x) - h(g_s)(x)| + |h(g_s)(x) - h(g_s)(x_r)| < 2\delta,$$

while $h(g_s)(x_r) \in Y$. So, $b_{k+1} - \ell_{k+1}[R_{k+1}] \subset b_{k+1} - B(Z, \varepsilon) \subset U_{k+1}$. Thus, $b_{k+1} \in U_{k+1}(\ell_{k+1}[R_{k+1}]) \cap B_{k+1}$. ∎

As a corollary needed in the proof but also interesting on its own, we conclude with the following.

Lemma 5.2.5 *For every dense G_δ subset G of $\mathbb{R}$ and for every prism P in $\mathbb{R}$ there exists a subprism Q of P such that $G[\text{LIN}(Q)]$ is a dense G_δ subset of $\mathbb{R}$.*

PROOF. Let $f \in \Phi_{\text{prism}}(\alpha)$ be such that $P = f[\mathfrak{C}^\alpha]$ and let $h : \mathbb{R} \to \mathcal{C}(\mathbb{R})$ be given by $h(r)(x) = r + x$. Then $h[P]$ is a prism in $\mathcal{C}(\mathbb{R})$ witnessed by $h \circ f$. By Lemma 5.2.4 there exist a subprism $Q_0 = h \circ f[E]$ of $h[P]$ and a $K \in \text{Perf}(\mathbb{R})$ with $0 \in K$ such that $Z = G[\text{LIN}(\{g(x) : g \in Q_0 \ \& \ x \in K\})]$ is dense in $\mathbb{R}$. But then $Q = f[E] = h^{-1}(Q)$ is a subprism of P and, since $0 \in K$,

$$
\begin{aligned}
Z &= G[\text{LIN}(\{h(r)(x) : r \in Q \ \& \ x \in K\})] \\
&= G[\text{LIN}(\{r + x : r \in Q \ \& \ x \in K\})] \\
&\subset G[\text{LIN}(\{r : r \in Q\})] \\
&= G[\text{LIN}(Q)].
\end{aligned}
$$

So, $G[\text{LIN}(Q)]$ is dense. It is G_δ by Fact 5.2.3(d) since $\text{LIN}(Q)$ is σ-compact. ∎

We will also need the following fact about perfect sets.

Lemma 5.2.6 *Let G be a proper dense G_δ subset of $\mathbb{R}$, W a second category G_δ subset of $\mathbb{R}$, and M an F_σ subset of $\mathbb{R}$ such that $G[\text{LIN}(M)]$ is a dense G_δ subset of $\mathbb{R}$. Then there exists a linearly independent set $K \in \text{Perf}(W)$ such that $G[\text{LIN}(M \cup K)]$ is dense, $\text{LIN}(M) \cap \text{LIN}(K) = \{0\}$, and $\text{LIN}(M \cup K) \setminus \text{LIN}(M) \subset G$.*

In particular, if M is linearly independent, then so is $M \cup K$.

PROOF. First note that the density of $G[\text{LIN}(M)]$ implies $\text{LIN}(M) \neq \mathbb{R}$. So, $\text{LIN}(M)$ must be of the first category.

Replacing G with $\bigcap\{q\,G : q \in \mathbb{Q} \setminus \{0\}\}$, if necessary, we can assume that $q\,G = G$ for every $q \in \mathbb{Q} \setminus \{0\}$. Notice that then for every $q \in \mathbb{Q} \setminus \{0\}$ and linear subspace V of $\mathbb{R}$ we also have

$$q\,G[V] = \{q\,x : x - V \subset G\} = \{y : (y/q) - V \subset G\} = \{y : y - q\,V \subset q\,G\} = G[V].$$

Let J be a nonempty open interval such that W is dense in J, and let $\langle G_k : k < \omega \rangle$ and $\langle W_k : k < \omega \rangle$ be the decreasing sequences of open subsets of $\mathbb{R}$ such that $G = \bigcap_{k < \omega} G_k$ and $W \cap J = \bigcap_{k < \omega} W_k$. Choose an increasing sequence $\langle M_k : k < \omega \rangle$ of compact sets such that $\text{LIN}(M) = \bigcup_{k < \omega} M_k$; let $\mathcal{R}$ be a family of all triples $\langle \ell, m, n \rangle$ such that $m, n < \omega$, $n > 0$, and $\ell \in L_{m+n}$, where the L_i's are as in (5.1); and fix a sequence $\langle \langle \ell_k, m_k, n_k \rangle \in \mathcal{R} : k < \omega \rangle$ with each triple appearing infinitely many times. We will construct, by induction on $k < \omega$, a fusion sequence $\langle U_s : s \in 2^k \ \& \ k < \omega \rangle$ of nonempty open subsets of $\mathbb{R}$ such that $U_\emptyset = J$ and for every $0 < k < \omega$ and $s \in 2^{k-1}$ the following inductive conditions hold.

(a) $\text{cl}(U_{s^\frown 0})$ and $\text{cl}(U_{s^\frown 1})$ are disjoint subsets of $U_s \cap W_k$.
(b) $\ell_k(\bar{a}, x_1, \ldots, x_{n_k}) \in G_k \setminus M_k$ for every $\bar{a} \in (M_k)^{m_k}$ and x_j chosen from a different U_t with $t \in 2^k$.

To see that such a sequence can be built, assume that for some $0 < k < \omega$ the sets $\{U_s : s \in 2^k\}$ have already been constructed. Let $\{t_i : i < 2^k\}$ be an enumeration of 2^k and by induction on i choose

$$x_{t_i} \in U_{t_i \upharpoonright k-1} \cap (W \setminus \text{LIN}(M \cup \{x_{t_j} : j < i\}) \cap \bigcap_{y \in \text{LIN}\{x_{t_j} : j < i\}} (y + G[\text{LIN}(M)]).$$

The choice can be made since $U_{t_i \upharpoonright k-1}$ is nonempty and open while the remaining sets are dense G_δ's in $U_{t_i \upharpoonright k-1} \subset J$. Notice that the choice guarantees that

$\qquad$ (b) holds for x_j chosen as different elements of $\{x_t : t \in 2^k\}$. $\qquad$ (5.4)

To see it first notice that clearly $\{x_{t_i} : i < 2^k\}$ is linearly independent and that $\text{LIN}(M) \cap \text{LIN}(\{x_{t_i} : i < 2^k\}) = \{0\}$. Also, it is easy to see that our choice ensures that $q\, x_{t_i} - \text{LIN}\{x_{t_j} : j < i\} \subset G[\text{LIN}(M)]$, that means that $q\, x_{t_i} - \text{LIN}\{x_{t_j} : j < i\} - \text{LIN}(M) \subset G$ for every $q \in \mathbb{Q} \setminus \{0\}$ and $i < 2^k$. But if $\bar{a} \in (M_k)^{m_k}$ and $\{x_1, \ldots, x_{n_k}\} \in [\{x_{t_i} : i < 2^k\}]^{n_k}$, then for appropriate $q \in \mathbb{Q} \setminus \{0\}$ and $i < 2^k$ we have

$$\ell_k(\bar{a}, x_1, \ldots, x_{n_k}) \in q\, x_{t_i} - \text{LIN}\{x_{t_j} : j < i\} - \text{LIN}(M) \subset G \setminus \text{LIN}(M).$$

So, (5.4) is proved.

Now, by the compactness of M_k and the continuity of ℓ_k, the set

$$Z = \{\langle x_1, \ldots, x_{n_k}\rangle : (\exists \bar{a} \in (M_k)^{m_k})\ \ell_k(\bar{a}, x_1, \ldots, x_{n_k}) \in G_k \setminus M_k\}$$

is open and, by (5.4), contains all one to one sequences $\bar{s}$ of points from the set $\{x_t : t \in 2^k\}$. Since there is only finitely many such sequences $\bar{s}$, we can find disjoint basic clopen neighborhoods U_t of x_t such that (a) and (b) hold. This finishes the inductive construction.

Let $K_0 = \bigcap_{k<\omega} \bigcup_{t \in 2^k} U_t$. By (a), K_0 is a perfect subset of W. Notice also that, by condition (b),

$$T = \bigcup_{m,n<\omega} \{\ell(\bar{a}, x_0, \ldots, x_n) : \bar{a} \in M^m\ \&\ \{x_0, \ldots, x_n\} \in [K_0]^{n+1}\ \&\ \ell \in L_{n+m}\}$$

is a subset of $G \setminus \text{LIN}(M)$. Clearly $0 \in \text{LIN}(M)$, and so $0 \notin T$. Therefore K_0 is linearly independent and $\text{LIN}(M) \cap \text{LIN}(K_0) = \{0\}$. In particular, we have $\text{LIN}(M \cup K_0) \setminus \text{LIN}(M) = T \subset G$.

Now fix an $x \in K_0$ and $K \in \text{Perf}(K_0 \setminus \{x\})$. Then for every $q \in \mathbb{Q} \setminus \{0\}$ and $v \in \text{LIN}(M \cup K)$ we have $q\, x - v \in \text{LIN}(M \cup K_0) \setminus \text{LIN}(M) \subset G$. Thus, $q\, x - \text{LIN}(M \cup K) \subset G$, and so $G[\text{LIN}(M \cup K)]$ contains a set $\{q\, x : q \in \mathbb{Q} \setminus \{0\}\}$, which is clearly dense. Thus, K is as desired. ∎

We will also need the following strengthening of Theorem 5.1.7.

Proposition 5.2.7 CPA$^{\text{game}}_{\text{prism}}$ *implies that for every dense* G_δ *subset* G *of* $\mathbb{R}$ *there exists a family* $\mathcal{H}$ *of compact pairwise disjoint sets such that* $H = \bigcup \mathcal{H}$ *is a Hamel basis, and for every nonmeager* G_δ *subset* B *of* $\mathbb{R}$ *and every countable* $\mathcal{H}_0 \subset \mathcal{H}$ *there exists an uncountable* $H \in \mathcal{H} \setminus \mathcal{H}_0$ *such that* $H \subset B$ *and* $\text{LIN}(H \cup \bigcup \mathcal{H}_0) \setminus \text{LIN}(\bigcup \mathcal{H}_0) \subset G$.

PROOF. First notice that if $\mathcal{G}_\delta$ stands for the family of all G_δ second category subsets of $\mathbb{R}$, then, assuming CPA$^{\text{game}}_{\text{prism}}$, there exists a $\mathcal{B} \in [\mathcal{G}_\delta]^{\omega_1}$ coinitial with $\mathcal{G}_\delta$, that is, such that

$$\text{for every } G \in \mathcal{G}_\delta \text{ there exists a } B \in \mathcal{B} \text{ such that } B \subset G. \qquad (5.5)$$

Indeed, since $\mathrm{CPA}_{\mathrm{prism}}^{\mathrm{game}}$ implies $\mathrm{cof}(\mathcal{M}) = \omega_1$ (see Corollary 1.3.3), there exists a decreasing sequence $\langle G_\xi : \xi < \omega_1 \rangle$ of dense G_δ subsets of $\mathbb{R}$ such that for every dense G_δ-set $W \subset \mathbb{R}$ there exists a $\xi < \omega_1$ with $G_\xi \subset W$. It is easy to see that $\mathcal{B} = \{G_\xi \cap (p_0, p_1) : \xi < \omega_1 \ \& \ p_0, p_1 \in \mathbb{Q} \ \& \ p_0 < p_1\}$ satisfies (5.5).

Decreasing G, if necessary, we can assume that $G \neq \mathbb{R}$. Fix a sequence $\langle B_\xi \in \mathcal{B} : \xi < \omega_1 \rangle$ in which each $B \in \mathcal{B}$ is listed ω_1 times. For a sequence $\langle P_\xi : \xi < \omega_1 \rangle$ of prisms in $\mathbb{R}$ representing the potential play of Player I, construct a sequence $\langle \langle Q_\xi, R_\xi^0, R_\xi^1 \rangle : \xi < \omega_1 \rangle$ such that the following inductive conditions hold for every $\xi < \omega_1$, where $R_\xi = \bigcup_{\eta < \xi} (R_\eta^0 \cup R_\eta^1)$.

(i) The sets $\{R_\eta^i : \eta \leq \xi \ \& \ i < 2\}$ are compact and pairwise disjoint.
(ii) $R_{\xi+1} = \bigcup \{R_\eta^i : \eta \leq \xi \ \& \ i < 2\}$ is linearly independent over $\mathbb{Q}$.
(iii) Q_ξ is a subprism of P_ξ and $Q_\xi \subset \mathrm{LIN}(R_{\xi+1})$.
(iv) $R_\xi^0 \in \mathrm{Perf}(B_\xi)$ and $\mathrm{LIN}(R_\xi \cup R_\xi^0) \setminus \mathrm{LIN}(R_\xi) \subset G$.
(v) $G[\mathrm{LIN}(R_{\xi+1})]$ is a dense G_δ in $\mathbb{R}$.

To make an inductive step assume that for some $\xi < \omega_1$ the required sequence $\langle \langle Q_\zeta, R_\zeta^0, R_\zeta^1 \rangle : \zeta < \xi \rangle$ is already constructed. So, R_ξ is already defined and, by the inductive assumption, R_ξ is clearly linearly independent. Next notice that

$$G[\mathrm{LIN}(R_\xi)] \text{ is a dense } G_\delta.$$

If $\xi = \eta + 1$, then it follows from (v) for η. On the other hand, if ξ is a limit ordinal, then $G[\mathrm{LIN}(R_\xi)] = G\left[\bigcup_{\eta < \xi} \mathrm{LIN}(R_{\eta+1})\right] = \bigcap_{\eta < \xi} G[\mathrm{LIN}(R_{\eta+1})]$ so it follows from the inductive assumption as well.

We define R_ξ^0 as a K from Lemma 5.2.6 applied to $W = B_\xi$ and $M = R_\xi$. This guarantees (iv), $R_\xi \cap R_\xi^0 = \emptyset$, the density of $G[\mathrm{LIN}(R_\xi \cup R_\xi^0)]$, and the linear independence of $R_\xi \cup R_\xi^0$.

Next use Lemma 5.2.5 to prism P_ξ and $G[\mathrm{LIN}(R_\xi \cup R_\xi^0)]$ to find a subprism Q' of P_ξ such that

$$
\begin{aligned}
G[\mathrm{LIN}(R_\xi \cup R_\xi^0)][\mathrm{LIN}(Q')] &= G[\mathrm{LIN}(R_\xi \cup R_\xi^0) + \mathrm{LIN}(Q')] \\
&= G[\mathrm{LIN}(R_\xi \cup R_\xi^0 \cup Q')]
\end{aligned}
$$

is a dense G_δ, where the first equation follows from Fact 5.2.3(f). Further, apply Lemma 5.1.5 to $M = R_\xi \cup R_\xi^0$ and prism $P = Q'$ to find a subprism Q_ξ of Q' and a compact R_ξ^1 subset of $Q' \setminus M$ such that $M \cup R_\xi^1$ is a maximal linearly independent subset of $M \cup Q_\xi$.

The maximality immediately implies $Q_\xi \subset \mathrm{LIN}(M \cup R_\xi^1) = \mathrm{LIN}(R_{\xi+1})$; so (iii) holds. We also clearly have (i) and (ii). Condition (v) follows from the density of $G[\text{LIN}(R_\xi \cup R_\xi^0 \cup Q')]$ and the fact that $R_\xi^1 \subset Q'$. This finishes the inductive construction.

Now, if S is a Player II strategy associated with our construction, then by CPA$_{\text{prism}}^{\text{game}}$ there exists a game $\langle\langle P_\xi, Q_\xi\rangle : \xi < \omega_1\rangle$ played according to S in which $\mathbb{R} = \bigcup_{\xi<\omega_1} Q_\xi$. Let $\langle\langle R_\xi^0, R_\xi^1\rangle : \xi < \omega_1\rangle$ be a sequence associated with this game. Then $\mathcal{H} = \{R_\xi^i : \xi < \omega_1 \ \& \ i < 2\}$ is as desired. $\blacksquare$

PROOF OF THEOREM 5.2.1. Let $\mathcal{X} = \{\mathcal{C}(B) : B \in \mathcal{B}\}$, where $\mathcal{B}$ is as in (5.5). We will play GAME$_{\text{prism}}(\mathcal{X})$ in which, by Theorem 3.6.2, Player II has no winning strategy. Notice that since each $g \in \mathcal{K}$, where $\mathcal{K}$ is defined as in (5.2), contains some function from $\bigcup\mathcal{X}$, every function f intersecting each $g \in \bigcup\mathcal{X}$ is almost continuous.

Let $\mathcal{H} = \{H_\xi : \xi < \omega_1\}$ be as in Proposition 5.2.7. We also fix a sequence $\bar{P} = \langle P_\xi : \xi < \omega_1\rangle$ such that each P_ξ represents a prism in some $\mathcal{C}(B) \in \mathcal{X}$. The sequence $\bar{P}$ represents the potential play for Player I in GAME$_{\text{prism}}(\mathcal{X})$, and we will construct, by induction, a strategy S for Player II that will describe a game played according to S in response to $\bar{P}$. To make S a legitimate strategy, its value at stage $\xi < \omega_1$ will depend only on $\bar{P}_\xi = \langle P_\eta : \eta \leq \xi\rangle$.

So, construct a sequence $\langle\langle H_\xi^0, H_\xi^1, Q_\xi, K_\xi, R_\xi, Y_\xi\rangle : \xi < \omega_1\rangle$ of subsets of $\mathbb{R}$ such that for every $\xi < \omega_1$ the following inductive conditions are satisfied, where $B_\xi \in \mathcal{B}$ is such that $P_\xi \subset \mathcal{C}(B_\xi)$ and $\mathcal{H}_\xi = \{H_\eta^i : \eta < \xi \ \& \ i < 2\}$.

(I) H_ξ^0 and H_ξ^1 are distinct elements of $\mathcal{H} \setminus \bigcup\mathcal{H}_\xi$.

(II) $H_\xi^0 \in [B_\xi]^{\mathfrak{c}}$ and $\text{LIN}(H_\xi^0 \cup \bigcup\mathcal{H}_\xi) \setminus \text{LIN}(\bigcup\mathcal{H}_\xi) \subset G$.

(III) $H_\xi \in \{H_\eta^i : \eta \leq \xi \ \& \ i < 2\}$.

We can choose such H_ξ^0 and H_ξ^1 since $\mathcal{H}$ was taken from Proposition 5.2.7. Also, if $F_\xi = \bigcup_{\eta<\xi}(R_\eta \cup Y_\eta)$ and $U_\xi = G[\text{LIN}(F_\xi)]$, then:

(IV) U_ξ is a dense G_δ in $\mathbb{R}$.

(V) $K_\xi \in \text{Perf}(H_\xi^0)$, Q_ξ is a subprism of P_ξ, $R_\xi = \{h(x) : h \in Q_\xi, \ x \in K_\xi\}$, and $U_\xi[\text{LIN}(R_\xi)]$ is dense G_δ in $\mathbb{R}$.

(VI) $Y_\xi \in \text{Perf}(\mathbb{R})$ is linearly independent such that $G[\text{LIN}(F_{\xi+1})]$ is dense, $\text{LIN}(F_\xi \cup R_\xi) \cap \text{LIN}(Y_\xi) = \{0\}$, and $\text{LIN}(F_\xi \cup R_\xi \cup Y_\xi) \setminus \text{LIN}(F_\xi \cup R_\xi)$ is a subset of G.

Assuming that (IV) holds, the possibility of a choice of Q_ξ, K_ξ, and R_ξ as in (V) follows directly from Lemma 5.2.4. Next, since by Fact 5.2.3(f)

$$U_\xi[\mathrm{LIN}(R_\xi)] \;=\; G[\mathrm{LIN}(F_\xi)][\mathrm{LIN}(R_\xi)]$$
$$=\; G[\mathrm{LIN}(F_\xi) + \mathrm{LIN}(R_\xi)]$$
$$=\; G[\mathrm{LIN}(F_\xi \cup R_\xi)],$$

we can apply Lemma 5.2.6 to our G, $W = \mathbb{R}$, and $M = F_\xi \cup R_\xi$ to find a linearly independent $K \in \mathrm{Perf}(\mathbb{R})$ for which $G[\mathrm{LIN}(F_\xi \cup R_\xi \cup K)]$ is a dense G_δ subset of $\mathbb{R}$, $\mathrm{LIN}(F_\xi \cup R_\xi \cup K) \setminus \mathrm{LIN}(F_\xi \cup R_\xi) \subset G$, and $\mathrm{LIN}(F_\xi \cup R_\xi) \cap \mathrm{LIN}(K) = \{0\}$. Then put $Y_\xi = K$ and notice that (VI) is satisfied, since $F_{\xi+1} = F_\xi \cup R_\xi \cup Y_\xi$.

To finish the construction it is enough to argue that (IV) is preserved. But if $\xi = \eta+1$ is a successor ordinal, then it follows immediately from (VI) for η. But if ξ is a limit ordinal, then (IV) follows easily from the density of sets U_η for $\eta < \xi$ since $U_\xi = G\left[\bigcup_{\eta<\xi} \mathrm{LIN}\left(\bigcup_{\zeta<\eta}(R_\zeta \cup \{y_\zeta\})\right)\right] = \bigcap_{\eta<\xi} U_\eta$. This finishes the inductive construction of the sequence.

We define a strategy S for Player II by $S(\langle\langle P_\eta, Q_\eta\rangle : \eta < \xi\rangle, P_\xi) = Q_\xi$. By Theorem 3.6.2, this is not a winning strategy, so there exists a game $\langle\langle P_\xi, Q_\xi\rangle : \xi < \omega_1\rangle$ played according to S in which $\bigcup \mathcal{X} = \bigcup_{\xi<\omega_1} Q_\xi$. We will use the sequence $\langle\langle H_\xi^0, H_\xi^1, Q_\xi, K_\xi, R_\xi, y_\xi\rangle : \xi < \omega_1\rangle$ associated with this game to construct the desired function f.

Since, by (I) and (III), $\{H_\xi^i : \xi < \omega_1 \ \& \ i < 2\} = \mathcal{H}$, it is enough to define f on each H_ξ^i and extend it to a unique additive function. So, for each $\xi < \omega$ define f on H_ξ^1 as a one to one function with values in Y_ξ. On each H_ξ^0 we define f such that $f[H_\xi^0] \subset R_\xi$ and f intersects every $g \in Q_\xi$ on a set K_ξ. It remains to prove that f is as advertised.

Certainly f is additive. It is also not difficult to see that f defined this way cannot be continuous. To see that it is almost continuous it is enough to notice that every $g \in \bigcup \mathcal{X}$ belongs to some Q_ξ, so it is intersected by f. To finish the proof it is enough to show that $f \subset (\mathbb{R} \times G) \cup (G \times \mathbb{R})$. So, define f_ξ as $f \upharpoonright \mathrm{LIN}(\bigcup \mathcal{H}_\xi)$. Since $f = \bigcup_{\xi<\omega_1} f_\xi$, it is enough to prove that

$$f_\eta \subset (\mathbb{R} \times G) \cup (G \times \mathbb{R}) \tag{5.6}$$

for every $\eta < \omega_1$. This will be proved by induction.

Clearly $f_0 = \{\langle 0,0\rangle\} \subset (\mathbb{R} \times G) \cup (G \times \mathbb{R})$ since $0 \in G$. So assume that, for some $0 < \eta < \omega_1$, condition (5.6) holds for every $\zeta < \eta$. If η is a limit ordinal, then $f_\eta = \bigcup_{\zeta<\eta} f_\zeta$ so (5.6) clearly holds. So assume that $\eta = \xi+1$ and notice that

$$f_\eta^* = f_\eta \upharpoonright \mathrm{LIN}(H_\xi^0 \cup \bigcup \mathcal{H}_\xi) \text{ is a subset of } (\mathbb{R} \times G) \cup (G \times \mathbb{R}).$$

This is so since $f_\xi \subset (\mathbb{R} \times G) \cup (G \times \mathbb{R})$ by the inductive assumption while

$f^*_\eta \setminus f_\xi \subset (G \times \mathbb{R})$ since $\operatorname{dom}(f^*_\eta \setminus f_\xi) = \operatorname{LIN}(H^0_\xi \cup \bigcup \mathcal{H}_\xi) \setminus \operatorname{LIN}(\bigcup \mathcal{H}_\xi) \subset G$ is guaranteed by (II).

Thus, to finish the proof, it is enough to show that

$$f_\eta \setminus f^*_\eta \subset (\mathbb{R} \times G).$$

To see it first note that, from our construction, $\operatorname{range}(f^*_\eta) \subset \operatorname{LIN}(F_\xi \cup R_\xi)$. Now, if $x \in \operatorname{dom}(f_\eta \setminus f^*_\eta) = \operatorname{LIN}(\bigcup \mathcal{H}_{\xi+1}) \setminus \operatorname{LIN}(H^0_\xi \cup \bigcup \mathcal{H}_\xi)$, then $x = v + w$ for some $v \in \operatorname{LIN}(H^1_\xi) \setminus \{0\}$ and $w \in \operatorname{LIN}(H^0_\xi \cup \bigcup \mathcal{H}_\xi)$. Hence, by the definition of f and condition (VI),

$$
\begin{aligned}
f_\eta(x) = f_\eta(v) + f_\eta(w) \ &\in \ (\operatorname{LIN}(Y_\xi) \setminus \{0\}) + \operatorname{LIN}(F_\xi \cup R_\xi) \\
&= \ \operatorname{LIN}(F_\xi \cup R_\xi \cup Y_\xi) \setminus \operatorname{LIN}(F_\xi \cup R_\xi) \subset G.
\end{aligned}
$$

This completes the proof. ∎

For a subset A of $\mathbb{R}$ we define $E^+(A)$ as

$$E^+(A) = \left\{ \sum_{i=0}^{k} q_i a_i : k < \omega \ \& \ a_i \in A \ \& \ q_i \in \mathbb{Q} \cap [0,\infty) \text{ for every } i \le k \right\}.$$

In [54], P. Erdős proved that under the continuum hypothesis there exists a Hamel basis H for which $E^+(H)$ is a Luzin set. In particular, such an $E^+(H)$ is of measure zero. K. Muthuvel [100], answering a question of H. Miller [98], generalized Erdős' result by proving that, under Martin's axiom, there exists a Hamel basis H for which $E^+(H)$ is simultaneously of measure zero and the first category. However, it is unknown whether there is a ZFC example of a Hamel basis H for which $E^+(H)$ is of measure zero. In what follows we show that the existence of such a Hamel basis is a consequence of CPA$^{\text{game}}_{\text{prism}}$.

Theorem 5.2.8 CPA$^{\text{game}}_{\text{prism}}$ *implies that for every dense G_δ subset G of $\mathbb{R}$ with $0 \in G$ there exists an $A \subset \mathbb{R}$ such that $\operatorname{LIN}(A) = \mathbb{R}$ and $E^+(A) \subset G$.*

Using Theorem 5.2.8 with G of measure zero and the fact that every set A spanning $\mathbb{R}$ contains a Hamel basis, we immediately obtain the following corollary.

Corollary 5.2.9 CPA$^{\text{game}}_{\text{prism}}$ *implies that there exists a Hamel basis H such that $E^+(H)$ has measure zero.*

PROOF OF THEOREM 5.2.8. Decreasing G, if necessary, we can assume that $qG = G$ for every nonzero $q \in \mathbb{Q}$. Since G has a Polish metric, we can use CPA$^{\text{game}}_{\text{prism}}$ for $\operatorname{GAME}_{\text{prism}}(X)$ with $X = G$.

Fix a sequence $\bar{P} = \langle P_\xi : \xi < \omega_1 \rangle$ such that each P_ξ represents a prism in X. Sequence $\bar{P}$ represents a potential play for Player I. We will construct, by induction, a strategy S for Player II that will describe a game played according to S in response to $\bar{P}$. The value of S at stage $\xi < \omega_1$ will depend only on $\bar{P}_\xi = \langle P_\eta : \eta \leq \xi \rangle$.

For this, we will construct a sequence $\langle \langle Q_\xi, A_\xi \rangle : \xi < \omega_1 \rangle$ of pairs of σ-compact subsets of $\mathbb{R}$ such that, for every $\zeta \leq \xi < \omega_1$:

(I) Q_ξ is a subprism of P_ξ.
(II) $A_\zeta \subset A_\xi$ and $\bigcup_{\eta \leq \xi} Q_\eta \subset \mathrm{LIN}(A_\xi)$.
(III) The set $G[E^+(A_\xi)]$ is dense and $E^+(A_\xi) \subset G$.

Assume that for some $\xi < \omega_1$ the desired sequence $\langle \langle Q_\eta, A_\eta \rangle : \eta < \xi \rangle$ is already constructed. Let $B_\xi = \bigcup_{\eta < \xi} A_\eta$. Then $E^+[B_\xi] = \bigcup_{\eta < \xi} E^+[A_\eta]$ is σ-compact and $G_\xi = G[E^+(B_\xi)] = \bigcap_{\eta < \xi} G[E^+(A_\eta)]$ is a dense G_δ. Thus, by Lemma 5.2.5, we can find a subprism Q_ξ of P_ξ such that $G_\xi[\mathrm{LIN}(Q_\xi)]$ is a dense G_δ subset of $\mathbb{R}$. Since

$$G_\xi[\mathrm{LIN}(Q_\xi)] = G[E^+(B_\xi)][\mathrm{LIN}(Q_\xi)] = G[E^+(B_\xi) + \mathrm{LIN}(Q_\xi)],$$

there exists an $x \in \mathbb{R}$ such that $x + E^+(B_\xi) + \mathrm{LIN}(Q_\xi) \subset G$. Thus, we also have $qx + E^+(B_\xi) + \mathrm{LIN}(Q_\xi) \subset G$ for every nonzero $q \in \mathbb{Q}$. Let us define $C_\xi = x + \mathrm{LIN}(Q_\xi)$ and put $A_\xi = B_\xi \cup C_\xi$. This clearly ensures (II). To see $E^+(A_\xi) \subset G$, notice that every element of $E^+(A_\xi)$ belongs to either $E^+(B_\xi) \subset G$ or to $qx + E^+(B_\xi) + \mathrm{LIN}(Q_\xi) \subset G$ for some positive $q \in \mathbb{Q}$. The density of $G[E^+(A_\xi)]$ follows from

$$
\begin{aligned}
G[E^+(A_\xi)] &= G[E^+(B_\xi) + E^+(C_\xi)] \\
&= G[E^+(B_\xi)][E^+(C_\xi)] \\
&= G_\xi[E^+(C_\xi)] \\
&= G_\xi\left[\bigcup_{q \in \mathbb{Q}^+}(qx + \mathrm{LIN}(Q_\xi))\right] \\
&= \bigcap_{q \in \mathbb{Q}^+} G_\xi[qx + \mathrm{LIN}(Q_\xi)] \\
&= \bigcap_{q \in \mathbb{Q}^+}(qx + G_\xi[\mathrm{LIN}(Q_\xi)]),
\end{aligned}
$$

where $\mathbb{Q}^+ = \mathbb{Q} \cap (0, \infty)$, since $G_\xi[\mathrm{LIN}(Q_\xi)]$ is a dense G_δ. This finishes the inductive construction.

Let S be a strategy of Player II given by the above inductive construction. Since S is not winning, there is a game $\langle \langle P_\xi, Q_\xi \rangle : \xi < \omega_1 \rangle$ played according to S in which $G = X = \bigcup_{\xi < \omega_1} Q_\xi$. Therefore, for $A = \bigcup_{\xi < \omega_1} A_\xi$, condition (III) implies that $E^+(A) \subset G$, while by condition (II) we have $\mathbb{R} = \mathrm{LIN}(G) = \mathrm{LIN}\left(\bigcup_{\xi < \omega_1} Q_\xi\right) \subset \mathrm{LIN}(A)$. ∎

5.3 Selective ultrafilters and the number $\mathfrak{u}$

The next three sections are based on K. Ciesielski and J. Pawlikowski [37]. We will use here the terminology introduced in the Preliminaries chapter. Recall, in particular, that every weakly selective ultrafilter is selective and that the ideal $\mathcal{I} = [\omega]^{<\omega}$ is selective. Another example of a weakly selective ideal that we will use in what follows is given below.

Fact 5.3.1 *The ideal $\mathcal{I}$ of a nowhere dense subset of $\mathbb{Q}$ is weakly selective.*

PROOF. Let $A \in \mathcal{I}^+$ and take an $f \colon A \to \omega$. If there is a $B \in \mathcal{I}^+ \cap \mathcal{P}(A)$ such that $f \upharpoonright B$ constant, then we are done. So, assume that it is not the case and let $A_0 \subset A$ be dense on some interval. By induction on $n < \omega$, define a sequence $\{b_n \in A_0 \colon n < \omega\}$ dense in A_0 such that f restricted to $B = \{b_n \colon n < \omega\}$ is one to one. Then B is as desired. ∎

In what follows we will also need the following fact about weakly selective ideals, which can be found in S. Grigorieff [65, prop. 14].

Proposition 5.3.2 *Let $\mathcal{I}$ be a weakly selective ideal on ω and $A \in \mathcal{I}^+$. If $T \subset A^{<\omega}$ is a tree such that*

$$A \setminus \{j < \omega \colon s^\frown j \in T\} \in \mathcal{I} \quad \text{for every } s \in T,$$

then there exists a branch b of T such that $b[\omega] \in \mathcal{I}^+$.

Theorem 5.3.3 CPA$_{\text{prism}}^{\text{game}}$ *implies that for every selective ideal $\mathcal{I}$ on ω there exists a selective ultrafilter $\mathcal{F}$ on ω such that $\mathcal{F} \subset \mathcal{I}^+$. In particular, if CPA$_{\text{prism}}^{\text{game}}$ holds, then there is a selective ultrafilter on ω.*

The proof is based on the following lemma.

Lemma 5.3.4 *Let $\mathcal{I}$ be a weakly selective ideal on ω.*

(a) *For every $A \in \mathcal{I}^+$ and every prism P in ω^ω there exist a $B \in \mathcal{I}^+$, $B \subset A$, and a subprism Q of P such that either*

    (i) *$g \upharpoonright B$ is one to one for every $g \in Q$, or else*

    (ii) *there exists an $n < \omega$ such that $g \upharpoonright B$ is constant and equal to n for every $g \in Q$.*

(b) *For every $A \in \mathcal{I}^+$ and every prism P in $[\omega]^\omega$ there exist a $B \in \mathcal{I}^+$, $B \subset A$, and a subprism Q of P such that either*

    • *$x \cap B = \emptyset$ for every $x \in Q$, or else*

    • *$B \subset x$ for every $x \in Q$.*

PROOF. (a) Fix an $A \in \mathcal{I}^+$, an $f \in \mathcal{F}_{\text{prism}}(\omega^\omega)$ from $\mathfrak{C}^\alpha$ onto P, and assume that for no subprism Q of P and $B \in \mathcal{I}^+ \cap \mathcal{P}(A)$ condition (ii) holds. We will find Q satisfying (i).

For $i, n < \omega$, let $D(i, n) = \{E_0 \in \mathbb{P}_\alpha : (\forall g \in E_0) \, f(g)(i) \neq n\}$ and for $\gamma \leq \alpha$, $E \in \mathbb{P}_\alpha$, and $A' \subset \omega$, put $D_\gamma(E, i, n) = \{\pi_\gamma[E_0] : E_0 \in D(i, n) \cap \mathcal{P}(E)\}$ and

$$D_\gamma(E, A', n) = \bigcap_{i \in A'} D_\gamma(E, i, n).$$

Notice that the sets $D_\gamma(E, i, n)$ and $D_\gamma(E, A', n)$ are open in $\mathbb{P}_\gamma$. By induction on $0 < \beta \leq \alpha$ we are going to prove the following property.

ψ_β: For all $0 < \gamma \leq \beta$, $E \in \mathbb{P}_\alpha$, $n < \omega$, and $\hat{A} \in \mathcal{I}^+ \cap \mathcal{P}(A)$ there exists an $A' \in \mathcal{I}^+ \cap \mathcal{P}(\hat{A})$ such that $D_\gamma(E, A', n) \neq \emptyset$.

In what follows for $k < \omega$ and $\mathcal{E}_k = \{E_s \in \mathbb{P}_\beta : s \in 2^{A_k}\}$ satisfying (i), (ag), and (sp) from Lemma 3.1.1 and $\mathcal{E}_{k+1} = \{E_s \in \mathbb{P}_\beta : s \in 2^{A_{k+1}}\}$, we will write

$$\mathcal{E}_{k+1} \prec \mathcal{E}_k$$

provided (i), (ii), (ag), and (sp) hold for all $s, t \in 2^{A_{k+1}}$ and $r \in 2^{A_k}$. One of the main facts used in the proof of ψ_β is the following property.

(∗) If ψ_γ holds for all $\gamma < \beta$, $\bar{A} \in \mathcal{I}^+ \cap \mathcal{P}(A)$, $n < \omega$, $E \in \mathbb{P}_\alpha$, $\mathcal{E}_k$ is as above and such that $\bigcup \mathcal{E}_k \subset \pi_\beta[E]$, and

$$Z(\bar{A}, \mathcal{E}_k, n) = \left\{i \in \bar{A} : (\exists \mathcal{E}_{k+1} \prec \mathcal{E}_k) \, \bigcup \mathcal{E}_{k+1} \in D_\beta(E, i, n)\right\},$$

then $\bar{A} \setminus Z(\bar{A}, \mathcal{E}_k, n) \in \mathcal{I}$.

In order to prove (∗), fix an $\hat{A} \in \mathcal{P}(\bar{A}) \cap \mathcal{I}^+$ and note that it is enough to show that $\hat{A} \cap Z(\bar{A}, \mathcal{E}_k, n) \neq \emptyset$. Fix an $\bar{\mathcal{E}}_{k+1} = \{\bar{E}_s \in \mathbb{P}_\beta : s \in 2^{A_{k+1}}\}$ such that $\bar{\mathcal{E}}_{k+1} \prec \mathcal{E}_k$. We can find such an $\bar{\mathcal{E}}_{k+1}$ by Lemma 3.1.2(A). Let $\gamma = \max\{\delta : \langle \delta, m \rangle \in A_{k+1}\} < \beta$.

First assume that $\gamma = 0$. Then, for every $s \in 2^{A_{k+1}}$, the set

$$Z_s = \left\{i \in \hat{A} : D_\beta(E, i, n) \cap \mathcal{P}(\bar{E}_s) = \emptyset\right\}$$

belongs to $\mathcal{I}$, as otherwise $Q = f\left[\pi_\beta^{-1}(\bar{E}_s) \cap E\right]$ and $B = Z_s \in \mathcal{I}^+ \cap \mathcal{P}(A)$ would satisfy condition (ii), contradicting our assumption. Let us define $A' = \hat{A} \setminus \bigcup\{Z_s : s \in 2^{A_{k+1}}\} \in \mathcal{I}^+$ and notice that $A' \subset Z(\bar{A}, \mathcal{E}_k, n)$. Indeed, take an $i \in A'$ and for every $s \in 2^{A_{k+1}}$ choose $E_s \in D_\beta(E, i, n) \cap \mathcal{P}(\bar{E}_s)$. Then $\mathcal{E}_{k+1} \overset{\text{def}}{=} \{E_s : s \in 2^{A_{k+1}}\} \prec \mathcal{E}_k$, since (i), (ii), and (sp) hold for $\mathcal{E}_{k+1}$ as they were true for $\bar{\mathcal{E}}_{k+1}$, and (ag) is satisfied trivially, by the maximality of γ. Condition $\bigcup \mathcal{E}_{k+1} \in D_\beta(E, i, n)$ is guaranteed by the choice of E_s's; so indeed $i \in Z(\bar{A}, \mathcal{E}_k, n)$.

Next assume that $\gamma > 0$. Let $B = \{\langle \delta, m \rangle \in A_{k+1} : \delta < \gamma\}$, and define $\mathcal{E}_{k+1}^* = \{E_t^* \in \mathbb{P}_\gamma : t \in 2^B\}$, where $E_t^* = \pi_\gamma[\bar{E}_s]$ for any $s \in 2^{A_{k+1}}$ with $t \subset s$. Note that, by (ag), the definition of E_t^* is independent of the choice of s. It is easy to see that $\mathcal{E}_{k+1}^*$ satisfies (ag) and (sp), where α is replaced by γ. For $\hat{A}_0 \in \mathcal{P}(\hat{A}) \cap \mathcal{I}^+$ and $t \in 2^B$, define $D(\hat{A}_0, t)$ as the collection of all $E_0 \in \mathbb{P}_\gamma$ for which there exists an $A' \in \mathcal{P}(\hat{A}_0) \cap \mathcal{I}^+$ such that

$$E_0 \in \bigcap \left\{ D_\gamma \left(\pi_\beta^{-1}(\bar{E}_s) \cap E, A', n \right) : t \subset s \in 2^{A_{k+1}} \right\}.$$

Clearly each $D(\hat{A}_0, t)$ is open, since so is each $D_\gamma \left(\pi_\beta^{-1}(\bar{E}_s) \cap E, A', n \right)$. It is also important to notice that $D(\hat{A}_0, t)$ is dense below E_t^*. To see this, fix an $E_0 \in \mathbb{P}_\gamma \cap \mathcal{P}(E_t^*)$ and let $\{s_1, \ldots, s_m\}$ be an enumeration of the set $\{s \in 2^{A_{k+1}} : t \subset s\}$. By induction on $i \leq m$ we define two decreasing sequences $\{E_i \in \mathbb{P}_\gamma : i \leq m\}$ and $\{\hat{A}_i \in \mathcal{I}^+ : i \leq m\}$ with the property that $E_i \in D_\gamma \left(\pi_\gamma^{-1}(E_{i-1}) \cap \left(\pi_\beta^{-1}(\bar{E}_{s_i}) \cap E \right), \hat{A}_i, n \right)$ provided $0 < i \leq m$. The inductive step can be made since property ψ_γ holds. Then $E_m \in \bigcap \left\{ D_\gamma \left(\pi_\beta^{-1}(\bar{E}_s) \cap E, \hat{A}_m, n \right) : t \subset s \in 2^{A_{k+1}} \right\}$, and so we have $E_m \in D(\hat{A}_0, t) \cap \mathcal{P}(E_0)$.

Let $\mathcal{D}$ be the collection of all pairwise disjoint families $\mathcal{E} \in [\mathbb{P}_\gamma]^{<\omega}$ for which there exists an $A' \in \mathcal{P}(\hat{A}) \cap \mathcal{I}^+$ working simultaneously for all $E_0 \in \mathcal{E}$; that is, such that for all $t \in 2^B$ and $E_0 \in \mathcal{E}$, if $E_0 \subset E_t^*$, then

$$E_0 \in \bigcap \left\{ D_\gamma \left(\pi_\beta^{-1}(\bar{E}_s) \cap E, A', n \right) : t \subset s \in 2^{A_{k+1}} \right\}.$$

Notice that $\mathcal{D}$ satisfies condition (†) from Lemma 3.1.2 used with α replaced by γ. Indeed, if $\mathcal{E} \in \mathcal{D}$ is witnessed by $A' \in \mathcal{P}(\hat{A}) \cap \mathcal{I}^+$ and $E \in \mathbb{P}_\gamma$ is disjoint with $\bigcup \mathcal{E}$, choose $E' \in \mathbb{P}_\gamma$ below E, which is either disjoint with $\bigcup \mathcal{E}_{k+1}^*$ or contained in some $E_t^* \in \mathcal{E}_{k+1}^*$. If $E' \cap \bigcup \mathcal{E}_{k+1}^* = \emptyset$, then $\{E'\} \cup \mathcal{E} \in \mathcal{D}$ is witnessed by A'. If $E' \subset E_t^* \in \mathcal{E}_{k+1}^*$ by the density of $D(A', t)$ below E_t^*, we can find an $A'' \in \mathcal{P}(A') \cap \mathcal{I}^+$ and

$$E'' \in \mathcal{P}(E') \cap \bigcap \left\{ D_\gamma \left(\pi_\beta^{-1}(\bar{E}_s) \cap E, A'', n \right) : t \subset s \in 2^{A_{k+1}} \right\}.$$

Then $\{E''\} \cup \mathcal{E} \in \mathcal{D}$ is witnessed by A''.

Now, by Lemma 3.1.2(B), there is an $\hat{\mathcal{E}}_{k+1} = \left\{ \hat{E}_t \in D : t \in 2^B \right\} \in \mathcal{E}$ satisfying conditions (ag) and (sp) and such that $\hat{E}_t \subset E_t^*$ for all $t \in 2^B$. Let $A' \in \mathcal{P}(\hat{A}) \cap \mathcal{I}^+$ witness $\hat{\mathcal{E}}_{k+1} \in \mathcal{E}$. We will show that $A' \subset Z(\bar{A}, \mathcal{E}_k, n)$. So fix an $i \in A'$. Since for every $t \in 2^B$ and $t \subset s \in 2^{A_{k+1}}$ we have

$\hat{E}_t \in D_\gamma(\pi_\beta^{-1}(\bar{E}_s) \cap E, A', n)$, there is an $\hat{E}_s \in D(\pi_\beta^{-1}(\bar{E}_s) \cap E, i, n)$ with $\pi_\gamma[\hat{E}_s] = \hat{E}_t$. Let $E_s = \pi_\beta[\hat{E}_s] \subset \bar{E}_s$ and notice that

$$\mathcal{E}_{k+1} \stackrel{\text{def}}{=} \{E_s \colon s \in 2^{A_{k+1}}\} \prec \mathcal{E}_k.$$

Indeed, $\mathcal{E}_{k+1}$ satisfies (i), (ii), and (sp) since they were true for $\bar{\mathcal{E}}_{k+1}$ and $\mathcal{E}_{k+1}$ is a refinement of $\bar{\mathcal{E}}_{k+1}$. Condition (ag) is satisfied by $\mathcal{E}_{k+1}$ since, by the maximality of γ, it is nontrivial for $\hat{\beta} \leq \gamma$ and for such $\hat{\beta}$ it is guaranteed by (ag) for $\hat{\mathcal{E}}_{k+1}$. Finally, $\bigcup \mathcal{E}_{k+1} \in D_\beta(E, i, n)$ is guaranteed by our definition, so indeed $i \in Z(\bar{A}, \mathcal{E}_k, n)$. This finishes the proof of $(*)$.

To prove ψ_β assume that ψ_γ holds for all $\gamma < \beta$. Fix $E \in \mathbb{P}_\alpha$, $n < \omega$, and $\hat{A} \in \mathcal{I}^+ \cap \mathcal{P}(A)$. We need to find an $A' \in \mathcal{I}^+ \cap \mathcal{P}(\hat{A})$ such that $D_\beta(E, A', n) \neq \emptyset$, that is, $\bigcap_{i \in A'} D_\beta(E, i, n) \neq \emptyset$. We will construct a tree $T \subset \hat{A}^{<\omega}$ and the mapping $T \ni s \mapsto \mathcal{E}_s \in [\mathbb{P}_\beta]^{<\omega}$ such that $\mathcal{E}_\emptyset = \{E\}$ and for every $r \in T$ and $s = r^\frown i \in T$ we have $\mathcal{E}_s \prec \mathcal{E}_r$ and $\bigcup \mathcal{E}_s \in D_\beta(E, i, n)$. Notice that, by $(*)$, for every $r \in T$ we can define $\mathcal{E}_{r^\frown i}$ for all $i \in Z(\hat{A}, \mathcal{E}_r, n)$. So we can ensure that T satisfies the assumptions of Proposition 5.3.2. Let b be a branch of T with $A' = b[\omega] \in \mathcal{I}^+$. By Lemma 3.1.1, $E_0 = \bigcap_{k<\omega} \bigcup \mathcal{E}_{b \upharpoonright k}$ belongs to $\mathbb{P}_\beta$ and $E_0 \in \bigcap_{i \in A'} D_\beta(E, i, n)$. This concludes the proof of ψ_β.

For the conclusion of the proof we first need to refine the prism P. For every $i < \omega$, let $h_i \colon \mathfrak{C}^\alpha \to \omega \subset \mathbb{R}$ be defined by $h_i(g) = f(g)(i)$. Clearly each h_i is continuous. Hence each set $h_i^{-1}(n)$ is open in $\mathfrak{C}^\alpha$, and so

$$D_i = \{E \in \mathbb{P}_\alpha \colon h_i \text{ is constant on } E\}$$

is dense and open in $\mathbb{P}_\alpha$. Therefore, by Corollary 3.1.3, there exists an $E \in \bigcap_{i<\omega} D_i^*$, where

$$D_i^* = \left\{ \bigcup \mathcal{D} \colon \mathcal{D} \in [D_i]^{2^i} \text{ and the sets in } \mathcal{D} \text{ are pairwise disjoint} \right\}.$$

Let $P_0 = f[E]$. Then P_0 is a subprism of P. We will find a subprism Q of P_0. Notice also that, by our construction, for every $i < \omega$ there is a set $V_i \in [\omega]^{\leq 2^i}$ such that

$$f(g)(i) \in V_i \text{ for all } g \in E \text{ and } i < \omega.$$

Also, since ψ_α holds, so is the conclusion of $(*)$ for $\beta = \alpha$. In particular, for every $n < \omega$ and $\mathcal{E}_k = \{E_s \in \mathbb{P}_\alpha \colon s \in 2^{A_k}\}$ satisfying (i), (ag), and (sp) from Lemma 3.1.1 and such that $\bigcup \mathcal{E}_k \subset E$, we have

$$Z(A, \mathcal{E}_k, n) = \left\{ i \in A \colon (\exists \mathcal{E}_{k+1} \prec \mathcal{E}_k) \bigcup \mathcal{E}_{k+1} \in D(i, n) \right\}$$

and $A \setminus Z(A, \mathcal{E}_k, n) \in \mathcal{I}$.

We will construct a tree $T \subset A^{<\omega}$ as in Proposition 5.3.2 and the mapping $T \ni s \mapsto \mathcal{E}_s \in [\mathbb{P}_\alpha]^{<\omega}$. The construction is done by induction on the levels of T. We start with $\mathcal{E}_\emptyset = \{E\}$ and, for every $r \in T$ and $s = r\hat{\ }i \in T$, we ensure that $\mathcal{E}_s \prec \mathcal{E}_r$ and

$$\bigcup \mathcal{E}_s \in \bigcap \{D(i,n) : n \in V_j \text{ for some } j \in \text{range}(r)\}. \tag{5.7}$$

Notice that for $r \in T$ if $Z_r = \bigcap \{Z(A, \mathcal{E}_r, n) : n \in V_j \text{ for some } j \in \text{range}(r)\}$, then $A \setminus Z_r \in \mathcal{I}$. Moreover, for all $i \in Z_r$ we can find $\mathcal{E}_{r\hat{\ }i}$ as in (5.7). So, T as above can be constructed. Take a branch b of T with $B = b[\omega] \in \mathcal{I}^+$ and $E_0 = \bigcap_{k<\omega} \bigcup \mathcal{E}_{b\restriction k} \in \mathbb{P}_\beta$. Then B and $Q = f[E_0]$ satisfy (i). This finishes the proof of (a).

(b) Since the characteristic function $\mathcal{X}$ gives an embedding from $[\omega]^\omega$ into $2^\omega \subset \omega^\omega$, the prism P can be identified with $\mathcal{X}[P] = \{\mathcal{X}_x : x \in P\}$. Applying part (a) to $\mathcal{X}[P]$ we can find a subprism Q of P, $k < 2$, and $B \in \mathcal{I}^+$, $B \subset A$, such that $\mathcal{X}_x \restriction B \equiv k$ for every $x \in Q$. If $k = 0$, this gives $x \cap B = \emptyset$ for every $x \in Q$. If $k = 1$, we have $B \subset x$ for every $x \in Q$. ∎

PROOF OF THEOREM 5.3.3. Let $\mathcal{I}$ be a selective ideal on ω. For a countable family $\mathcal{A} \subset \mathcal{I}^+$ linearly ordered by $\subset^*$, let $C(\mathcal{A}) \in \mathcal{I}^+$ be such that $C(\mathcal{A}) \subset^* A$ for every $A \in \mathcal{A}$.

For $A \in \mathcal{I}^+$ and $f \in \mathcal{F}_{\text{prism}}(\omega^\omega)$ define $P = \text{range}(f)$ and take $B(A, P) \in [A]^\omega$ and a subprism $Q(A, P)$ of P as in Lemma 5.3.4(a). If $f \in \mathcal{C}_{\text{prism}}(\omega^\omega)$ and $P = \text{range}(f) = \{x\}$, then we put $Q(A, P) = P$ and take $B(A, P) \in [A]^\omega$ satisfying the conclusion of Lemma 5.3.4(a).

Consider the following strategy S for Player II:

$$S(\langle\langle P_\eta, Q_\eta\rangle : \eta < \xi\rangle, P_\xi) = Q(C(\{B_\eta : \eta < \xi\}), P_\xi),$$

where the sets B_η are defined inductively by $B_\eta = B(C(\{B_\zeta : \zeta < \eta\}), P_\eta)$.

By CPA$_{\text{prism}}^{\text{game}}$, strategy S is not a winning strategy for Player II. So, there exists a game $\langle\langle P_\xi, Q_\xi\rangle : \xi < \omega_1\rangle$ played according to S in which Player II loses, that is, $\omega^\omega = \bigcup_{\xi<\omega_1} Q_\xi$.

Now, let $\mathcal{F}$ be a filter generated by $\{B_\xi : \xi < \omega_1\}$ and notice that $\mathcal{F}$ is a selective ultrafilter. It is a filter, since $\{B_\xi : \xi < \omega_1\}$ is decreasing with respect to $\subset^*$. It also easy to see that for every $f \in \omega^\omega$ there exists a $B \in \mathcal{F}$ such that $f \restriction B$ is either one to one or constant. Indeed, if $f \in \omega^\omega$, then there exists a $\xi < \omega_1$ such that $f \in Q_\xi$. Then $B = B_\xi$ is as desired.

Now, to see that $\mathcal{F}$ is an ultrafilter, take an $A \subset \omega$ and let $f \in \omega^\omega$ be a characteristic function of A. Then $B \in \mathcal{F}$ as above is a subset of either A or its complement.

It is easy to see that the above two properties imply that $\mathcal{F}$ is a selective ultrafilter. ∎

Notice that $\text{CPA}_{\text{prism}}^{\text{game}}$ also implies that we have many different selective ultrafilters. The consistency of this fact, in a model obtained by adding many side by side Sacks reals, was first noticed by K. P. Hart in [66].

Remark 5.3.5 $\text{CPA}_{\text{prism}}^{\text{game}}$ implies that there are ω_2 different selective ultrafilters.

PROOF. This can be easily deduced by a simple transfinite induction from:

($*$) For every family $\mathcal{U} = \{\mathcal{F}_\xi \colon \xi < \omega_1\}$ of ultrafilters on ω there is a selective ultrafilter $\mathcal{F} \notin \mathcal{U}$.

Property ($*$) is proved as above, where we use $\mathcal{I} = [\omega]^{<\omega}$ and the operator $C(\{B_\eta \colon \eta < \xi\})$ is replaced with $C_\xi(\{B_\eta \colon \eta < \xi\}) \notin \mathcal{F}_\xi$. ∎

Since CPA holds in the iterated perfect set model, it is consistent with $2^{\omega_1} = \omega_2$ as well as with $2^{\omega_1} > \omega_2$. (See Theorem 7.2.1.) Now, from Remark 5.3.5 we obtain that $2^{\omega_1} = \omega_2 + \text{CPA}_{\text{prism}}^{\text{game}}$ implies that there are 2^{ω_1} different selective ultrafilters. It is worth noticing that the existence of 2^{ω_1} different selective ultrafilters also can be deduced from a slightly stronger version of $\text{CPA}_{\text{prism}}^{\text{game}}$ (which follows from CPA) even when we have $2^{\omega_1} > \omega_2$. (See Proposition 6.1.2.)

Recall that the number $\mathfrak{u}$ is defined as the smallest cardinality of the base for a nonprincipal ultrafilter on ω. Thus Theorem 5.3.3 and Corollaries 1.5.4 and 1.5.5 imply that

Corollary 5.3.6 $\text{CPA}_{\text{prism}}^{\text{game}}$ *implies that* $\mathfrak{u} = \mathfrak{r}_\sigma = \omega_1$.

5.4 Nonselective P-points and number i

Recall that an ultrafilter $\mathcal{F}$ on ω is a *P-point* provided for every partition $\mathcal{P}$ of ω either $\mathcal{P} \cap \mathcal{F} \neq \emptyset$ or there is an $F \in \mathcal{F}$ such that $|F \cap P| < \omega$ for all $P \in \mathcal{P}$. Clearly every selective ultrafilter is a P-point. Thus, $\text{CPA}_{\text{prism}}^{\text{game}}$ implies the existence of a P-point. On the other hand, S. Shelah proved that there are models with no P-points. (See, e.g., [4, thm. 4.4.7].) K. P. Hart in [66] proved that in a model obtained by adding many side by side Sacks reals there is a P-point that is not selective. Next, we will prove that this also follows from $\text{CPA}_{\text{prism}}^{\text{game}}$. The main idea of the proof is the same as that used in [66].

For $m < \omega$, let $P_m = \{n < \omega : 2^m - 1 \leq n < 2^{m+1} - 1\}$ and define a partition $\mathcal{P}$ of ω by $\mathcal{P} = \{P_m : m < \omega\}$. Consider the ideal $\bar{\mathcal{I}}$ on ω,

$$\bar{\mathcal{I}} = \left\{ A \subset \omega : \limsup_{m \to \infty} |A \cap P_m| < \omega \right\}, \tag{5.8}$$

and notice the following simple fact.

Fact 5.4.1 If a family $\mathcal{A} \in [\bar{\mathcal{I}}^+]^{\leq \omega}$ is linearly ordered by $\subset^*$, then there is a $C(\mathcal{A}) \in \bar{\mathcal{I}}^+$ such that $C(\mathcal{A}) \subset^* A$ for all $A \in \mathcal{A}$.

PROOF. Let $\{A_n : n < \omega\} \subset \mathcal{A}$ be a $\subset^*$-decreasing sequence coinitial with $\mathcal{A}$. For every $i < \omega$ choose $m_i < \omega$ and $C_i \in [P_{m_i}]^i$ such that $C_i \subset \bigcap_{j \leq i} A_j$. Then $C(\mathcal{A}) = \bigcup_{i < \omega} C_i$ is as desired. ∎

To construct a nonselective P-point we are going to prove the following theorem.

Theorem 5.4.2 If CPA$^{\text{game}}_{\text{prism}}$ holds, then there exists a $\subset^*$-decreasing sequence $\mathcal{B} = \{B_\xi \in \bar{\mathcal{I}}^+ : \xi < \omega_1\}$ such that the filter $\mathcal{F}$ generated by $\mathcal{B}$ is an ultrafilter on ω.

Notice that from this we will immediately deduce the required result.

Corollary 5.4.3 If CPA$^{\text{game}}_{\text{prism}}$ holds, then there exists a nonselective P-point.

PROOF. Let $\mathcal{F}$ be as in Theorem 5.4.2. Clearly $\mathcal{F}$ is nonselective, since $\mathcal{P}$ is disjoint with $\bar{\mathcal{I}}^+ \supset \mathcal{F}$ and every selector of $\mathcal{P}$ is in $\bar{\mathcal{I}} \subset P(\omega) \setminus \mathcal{F}$. The fact that $\mathcal{F}$ is a P-point follows from the fact that $\mathcal{F}$ has a base linearly ordered by $\subset^*$. Indeed, if $\{S_n : n < \omega\} \subset P(\omega) \setminus \mathcal{F}$ is a partition of ω, then for every $m < \omega$ there is $\xi_m < \omega_1$ such that $B_{\xi_m} \subset^* \omega \setminus \bigcup_{n \leq m} S_n$. Let $\beta < \omega_1$ be such that $B_\beta \subset^* B_{\xi_m}$ for all $m < \omega$. Then $F = B_\beta \in \mathcal{F}$ is such that $|F \cap S_n| < \omega$ for all $n < \omega$. ∎

The proof of Theorem 5.4.2 will be based on the following lemma, which is analogous to Lemma 5.3.4. Note that, although the statement of this lemma is identical to that of Lemma 5.3.4(b), we cannot apply this lemma here, since the ideal $\bar{\mathcal{I}}$ is not weakly selective.

Lemma 5.4.4 Let $\bar{\mathcal{I}}$ be as in (5.8). Then for every $A \in \bar{\mathcal{I}}^+$ and a prism P in 2^ω there exist a $B \in \bar{\mathcal{I}}^+$, $B \subset A$, a subprism Q of P, and a $j < 2$ such that:

(◦) $g \restriction B$ is constant and equal to j for every $g \in Q$.

PROOF. Fix an $A \in \bar{\mathcal{I}}^+$ and an $f \in \mathcal{F}_{\mathrm{prism}}(2^\omega)$ from $\mathfrak{C}^\alpha$ onto P. Since $A \in \bar{\mathcal{I}}^+$ for every $k < \omega$, we can find a number $m_k < \omega$ such that $|A \cap P_{m_k}| \geq k\, 2^{2^k}$. First we will construct a subprism Q_0 of P and a sequence $\langle A_k \in [A \cap P_{m_k}]^k : k < \omega \rangle$ such that for every $k < \omega$

$$g \restriction A_k \text{ is constant for every } g \in Q_0. \tag{5.9}$$

This will be done using Lemmas 3.1.2 and 3.1.1. So, for each $k < \omega$, let $\mathcal{D}_k$ be the collection of all pairwise disjoint families $\mathcal{E} \in [\mathbb{P}_\alpha]^{<\omega}$ such that for every $E \in \mathcal{E}$

$$f(h) \restriction P_{m_k} = f(h') \restriction P_{m_k} \text{ for all } h, h' \in E. \tag{5.10}$$

Clearly each $\mathcal{D}_k$ satisfies condition (†) from Lemma 3.1.2, so by an easy induction we can find a sequence $\langle \mathcal{E}_k \in \mathcal{D}_k : k < \omega \rangle$ satisfying the assumptions of Lemma 3.1.1. Let $E_0 = \bigcap_{k<\omega} \bigcup \mathcal{E}_k \in \mathbb{P}_\alpha$. We will show that $Q_0 = f[E_0]$ satisfies (5.9).

Indeed, fix a $k < \omega$ and notice that $\mathcal{E}_k = \{E_i : i < 2^k\}$. For each $i < 2^k$ choose an $h_i \in E_i$ and define $\varphi : A \cap P_{m_k} \to 2^{2^k}$ by $\varphi(p)(i) = f(h_i)(p)$. Since $|A \cap P_{m_k}| \geq k\, 2^{2^k}$, by the pigeon hole principle we can find an $s \in 2^{2^k}$ such that $|\varphi^{-1}(s)| \geq k$. Choose an $A_k \in [\varphi^{-1}(s)]^k$. Then for every $i < 2^k$ and $p \in A_k$ we have $f(h_i)(p) = \varphi(p)(i) = s(i)$. To see (5.9), fix a $g \in Q_0 \subset f[\bigcup \mathcal{E}_k]$ and notice that there exists an $i < 2^k$ and an $h \in E_i$ such that $g = f(h)$. Then, by (5.10), for every $p \in A_k \subset P_{m_k}$ we have $g(p) = f(h)(p) = f(h_i)(p) = s_i$. So, $g \restriction A_k$ is constant and equal to $s(i)$, proving (5.9).

To finish the proof, fix a selector $\bar{A}$ from the family $\{A_k : k < \omega\}$. Then $\bar{A} \in \mathcal{I}^+$, where $\mathcal{I}$ is the ideal of finite subsets of ω. Applying Lemma 5.3.4(a) to $\bar{A}$ and Q_0 we can find a $j < 2$, an $S \in [\bar{A}]^\omega$, and a subprism Q of Q_0 such that $g \restriction S$ is constant and equal to j for every $g \in Q$. Put $B = \bigcup \{A_k : k < \omega \ \& \ A_k \cap S \neq \emptyset\}$. Then, by (5.9), $g \restriction B$ is constant and equal to j for every $g \in Q$.

It is clear that $B \in \bar{\mathcal{I}}^+ \cap \mathcal{P}(A)$, since it is a union of infinitely many sets $A_k \in [A \cap P_{m_k}]^k$. ∎

PROOF OF THEOREM 5.4.2. The proof is almost identical to that for Theorem 5.3.3.

For $A \in \bar{\mathcal{I}}^+$ and $f \in \mathcal{F}_{\mathrm{prism}}(2^\omega)$ put $P = \mathrm{range}(f)$, and let $B(A, P) \in [A]^\omega$ and a subprism $Q(A, P)$ of P be as in Lemma 5.4.4. If $f \in \mathcal{C}_{\mathrm{prism}}(2^\omega)$ and $P = \mathrm{range}(f) = \{x\}$, then we put $Q(A, P) = P$ and take $B(A, P) \in [A]^\omega$ satisfying (∘). Consider the following strategy S for Player II:

$$S(\langle \langle P_\eta, Q_\eta \rangle : \eta < \xi \rangle, P_\xi) = Q(C(\{B_\eta : \eta < \xi\}), P_\xi),$$

where sets B_η are defined inductively by $B_\eta = B(C(\{B_\zeta : \zeta < \eta\}), P_\eta)$ and the operator C is as in Fact 5.4.1.

By CPA$^{\text{game}}_{\text{prism}}$, strategy S is not a winning strategy for Player II. So there exists a game $\langle\langle P_\xi, Q_\xi\rangle : \xi < \omega_1\rangle$ played according to S in which Player II loses, that is, $2^\omega = \bigcup_{\xi<\omega_1} Q_\xi$.

Let $\mathcal{F}$ be a filter generated by $\{B_\xi : \xi < \omega_1\}$. To see that $\mathcal{F}$ is an ultrafilter, take an $A \subset \omega$ and let $f \in 2^\omega$ be the characteristic function of A. Let $\xi < \omega_1$ be such that $f \in Q_\xi$. Then $f[B] = \{j\}$ for some $j < 2$, and so B is a subset of either A or its complement.   ∎

Note also that, similarly as for Remark 5.3.5, the conclusion of the following fact holds in a model obtained by adding many side by side Sacks reals. This was first noticed by K. P. Hart in [66].

Remark 5.4.5 CPA$^{\text{game}}_{\text{prism}}$ implies that there are ω_2 many different nonselective P-points.

The existence of 2^{ω_1} different such ultrafilters also follows from a slightly stronger version of CPA$^{\text{game}}_{\text{prism}}$. (See Proposition 6.1.2.)

Recall also that a family $\mathcal{J} \subset [\omega]^\omega$ is an *independent family* provided the set

$$\bigcap_{A \in \mathcal{A}} A \cap \bigcap_{B \in \mathcal{B}} (\omega \setminus B)$$

is infinite for every disjoint finite subsets $\mathcal{A}$ and $\mathcal{B}$ of $\mathcal{J}$. It is often convenient to express this definition in a slightly different notation. Thus, for $W \subset \omega$, let $W^0 = W$ and $W^1 = \omega \setminus W$. A family $\mathcal{J} \subset [\omega]^\omega$ is independent provided the set

$$\bigcap_{W \in \mathcal{J}_0} W^{\tau(W)}$$

is infinite for every finite subset $\mathcal{J}_0$ of $\mathcal{J}$ and $\tau \colon \mathcal{J}_0 \to \{0,1\}$.

The *independence* cardinal $\mathfrak{i}$ is defined as follows:

$$\mathfrak{i} = \min\{|\mathcal{J}| : \mathcal{J} \text{ is infinite maximal independent family}\}.$$

The fact that $\mathfrak{i} = \omega_1$ holds in the iterated perfect set model was apparently first noticed by Todd Eisworth and S. Shelah (see A. Blass [9, sec. 11.5]), though it seems that the proof of this result was never provided. The argument presented below comes from K. Ciesielski and J. Pawlikowski [37].

Theorem 5.4.6 CPA$^{\text{game}}_{\text{prism}}$ *implies that* $\mathfrak{i} = \omega_1$.

The proof of the theorem is based on the following lemma. We say that a family $\mathcal{W} \subset [\omega]^{\omega}$ *separates points* provided for every $k < \omega$ there are $U, V \in \mathcal{W}$ such that $k \in U \setminus V$.

Lemma 5.4.7 *For every countable independent family $\mathcal{W} \subset [\omega]^{\omega}$ separating points and a prism P in $[\omega]^{\omega}$ there exist $W \in [\omega]^{\omega}$ and a subprism Q of P such that $\mathcal{W} \cup \{W\}$ is independent but $\mathcal{W} \cup \{W, x\}$ is not independent for every $x \in Q$.*

PROOF. Let $\mathcal{W} = \{W_i : i < \omega\}$ and let $\varphi: \omega \to 2^{\omega}$ be a Marczewski function for $\mathcal{W}$, that is, for $i, k < \omega$

$$\varphi(k)(i) = \begin{cases} 1 & \text{for } k \in W_i \\ 0 & \text{for } k \notin W_i. \end{cases}$$

Note that φ is one to one, since $\mathcal{W}$ separates points. Notice also that for every $k, n < \omega$ and $\tau \in 2^n$

$$k \in \bigcap_{i<n} W_i^{\tau(i)} \Leftrightarrow (\forall i < n)\, k \in W_i^{\tau(i)} \Leftrightarrow (\forall i < n)\, \varphi(k)(i) = \tau(i) \Leftrightarrow \tau \subset \varphi(k).$$

Now, if $[\tau] = \{t \in 2^{\omega} : \tau \subset t\}$, then the sets $\{[\tau] : \tau \in 2^{<\omega}\}$ form a base for 2^{ω} and

$$k \in \bigcap_{i<n} W_i^{\tau(i)} \Leftrightarrow \varphi(k) \in [\tau]. \tag{5.11}$$

Thus, independence of $\mathcal{W}$ implies that $\varphi[\omega]$ is dense in 2^{ω}. We will denote $\varphi[\omega]$ by $\mathbb{Q}$, since it is homeomorphic with the set of rational numbers. Note also that from (5.11) it follows immediately that:

(a) If $W \subset \omega$ is such that $\varphi[W]$ and $\mathbb{Q} \setminus \varphi[W]$ are dense, then $\mathcal{W} \cup \{W\}$ is independent.

(b) If $W, x \subset \omega$ are such that for some $\tau \in 2^{<\omega}$ either $\varphi[x \cap W] \cap [\tau] = \emptyset$ or $\varphi[W] \cap [\tau] \subset \varphi[x]$, then $\mathcal{W} \cup \{W, x\}$ is not independent.

Let $\mathcal{I}$ be the ideal of nowhere dense subsets of $\varphi[\omega]$. Then, by Fact 5.3.1, $\mathcal{I}$ is weakly selective, since $\varphi[\omega]$ is homeomorphic to $\mathbb{Q}$. So, identifying $\varphi[\omega]$ with ω and applying Lemma 5.3.4(b), we can find a subprism Q of P and a $V \in [\omega]^{\omega} \setminus \mathcal{I}$ such that either

- $x \cap V = \emptyset$ for every $x \in Q$, or else
- $V \subset x$ for every $x \in Q$.

Since $V \notin \mathcal{I}$, there exists a $\tau \in 2^{<\omega}$ such that $\varphi[V]$ is dense in $[\tau]$. Trimming V, if necessary, we can assume that $\varphi[V] \subset [\tau]$ and that $\varphi[\omega \setminus V]$ is also dense in $[\tau]$. Now let $W \supset V$ be such that $\varphi[W] \cap [\tau] = \varphi[V]$ and both $\varphi[W]$ and $\varphi[w \setminus W]$ are dense in $\varphi[w]$. Then, by (a) and (b), $\mathcal{W} \cup \{W\}$ is independent while $\mathcal{W} \cup \{W, x\}$ is not independent for every $x \in Q$.   ∎

PROOF OF THEOREM 5.4.6. For a countable independent family $\mathcal{W} \subset [w]^\omega$ separating points and an $f \in \mathcal{F}_{\text{prism}}([w]^\omega)$ put $P = \text{range}(f)$ and let $W(\mathcal{W}, P) \in [w]^\omega$ and a subprism $Q(\mathcal{W}, P)$ of P be as in Lemma 5.4.7. If $f \in \mathcal{C}_{\text{prism}}([w]^\omega)$ and $P = \text{range}(f) = \{x\}$, then we put $Q(\mathcal{W}, P) = P$ and define $W(\mathcal{W}, P)$ as an arbitrary W such that $\mathcal{W} \cup \{W\}$ is independent while $\mathcal{W} \cup \{W, x\}$ is not.

Let $\mathcal{A}_0 \subset [w]^\omega$ be an arbitrary countable independent family separating points and consider the following strategy S for Player II:

$$S(\langle\langle P_\eta, Q_\eta\rangle: \eta < \xi\rangle, P_\xi) = Q(\mathcal{A}_0 \cup \{W_\eta: \eta < \xi\}, P_\xi),$$

where the sets W_η are defined by $W_\eta = W(\mathcal{A}_0 \cup \{W_\zeta: \zeta < \eta\}, P_\eta)$.

By CPA$_{\text{cube}}^{\text{game}}$, strategy S is not a winning strategy for Player II. So there exists a game $\langle\langle P_\xi, Q_\xi\rangle: \xi < \omega_1\rangle$ played according to S in which Player II loses, that is, $[w]^\omega = \bigcup_{\xi<\omega_1} Q_\xi$.

Now, notice that the family $\mathcal{J} = \mathcal{A}_0 \cup \{W_\xi: \xi < \omega_1\}$ is a maximal independent family. It is clear that $\mathcal{J}$ is independent, since every set W_ξ was chosen so that $\mathcal{A}_0 \cup \{W_\zeta: \zeta \leq \xi\}$ is independent. To see that the family $\mathcal{J}$ is maximal, it is enough to note that every $x \in [w]^\omega$ belongs to a Q_ξ for some $\xi < \omega_1$, and so $\mathcal{A}_0 \cup \{W_\zeta: \zeta \leq \xi\} \cup \{x\}$ is not independent.   ∎

By Theorem 5.4.6 we see that CPA$_{\text{prism}}^{\text{game}}$ implies the existence of an independent family of size ω_1. Next, answering a question of Michael Hrušák [67], we show that such a family can be simultaneously a splitting family. This is similar in flavor to Theorem 2.2.3. In the proof we will use the following lemma.

Lemma 5.4.8 *For every countable family* $\mathcal{V} \subset [w]^\omega$ *and a perfect set* P *in* $[w]^\omega$ *there exists a* $W_1 \in [w]^\omega$ *such that* $\mathcal{V} \cup \{W_1\}$ *is independent and* W_1 *splits every* $A \in P$.

PROOF. We follow the argument from [49, p. 121] that $\mathfrak{s} \leq \mathfrak{d}$.

For every $A \in [w]^\omega$, let b_A be a strictly increasing bijection from ω onto A. Then $b: [w]^\omega \to \omega^\omega$ defined by $b(A) = b_A$ is continuous. In particular, $b[P] = \{b_A: A \in P\}$ is compact, so there exists a strictly increasing $f \in \omega^\omega$ such that $b_A(n) < f(n)$ for every $A \in P$ and $n < \omega$. For $n < \omega$, let f^n denote the n-fold composition of f and let $S_n = \{m < \omega: f^n(0) \leq m < f^{n+1}(0)\}$. Then $f^n(0) \leq b_A(f^n(0)) < f(f^n(0)) = f^{n+1}(0)$ for every $A \in P$ and

$n < \omega$. In particular, for every $A \in P$,

$$S_n \cap A \neq \emptyset.$$

So, if $T \subset \omega$ be infinite and co-infinite and $W_1 = \bigcup_{n \in T} S_n$, then W_1 splits every $A \in P$. Thus, it is enough to take an infinite and co-infinite $T \subset \omega$ such that $\mathcal{V} \cup \{W_1\}$ is independent. ∎

Theorem 5.4.9 CPA$_{\text{prism}}^{\text{game}}$ *implies that there exists a family* $\mathcal{F} \subset [\omega]^\omega$ *of cardinality* ω_1 *that is simultaneously independent and splitting.*

PROOF. The proof is just a slight modification of that for Theorem 5.4.6. (Compare also Theorem 2.2.3.)

For a countable independent family $\mathcal{W} \subset [\omega]^\omega$ separating points and an $f \in \mathcal{F}_{\text{prism}}([\omega]^\omega)$, put $P = \text{range}(f)$ and let $W_0 \in [\omega]^\omega$ and a subprism Q of P be as in Lemma 5.4.7. Let W_1 be as in Lemma 5.4.8 used with $P = Q$ and $\mathcal{V} = \mathcal{W} \cup \{W_0\}$. We put $Q(\mathcal{W}, P) = Q_1$ and $\mathcal{W}(\mathcal{W}, P) = \{W_0, W_1\}$.

If $f \in \mathcal{C}_{\text{cube}}([\omega]^\omega)$ and $P = \text{range}(f) = \{x\}$, then we put $Q(\mathcal{W}, P) = P$ and $\mathcal{W}(\mathcal{W}, P) = \{W_0, W_1\}$, where W_0 and W_1 are such that $\mathcal{W} \cup \{W_0, W_1\}$ is independent and W_1 splits $P = \{x\}$.

Let $\mathcal{A}_0 \subset [\omega]^\omega$ be an arbitrary, countable, independent family separating points and consider the following strategy S for Player II:

$$S(\langle \langle P_\eta, Q_\eta \rangle : \eta < \xi \rangle, P_\xi) = Q(\mathcal{A}_0 \cup \bigcup \{\mathcal{W}_\eta : \eta < \xi\}, P_\xi),$$

where the $\mathcal{W}_\eta$'s are defined by $\mathcal{W}_\eta = \mathcal{W}(\mathcal{A}_0 \cup \bigcup \{\mathcal{W}_\eta : \eta < \xi\}, P_\eta)$.

By CPA$_{\text{prism}}^{\text{game}}$, strategy S is not a winning strategy for Player II. Therefore, there exists a game $\langle \langle P_\xi, Q_\xi \rangle : \xi < \omega_1 \rangle$ played according to the strategy S in which Player II loses, that is, $[\omega]^\omega = \bigcup_{\xi < \omega_1} Q_\xi$. Then the family $\mathcal{F} = \mathcal{A}_0 \cup \bigcup \{\mathcal{W}_\xi : \xi < \omega_1\}$ is independent and splitting. ∎

5.5 Crowded ultrafilters on $\mathbb{Q}$

Let Perf($\mathbb{Q}$) stand for the family of all closed subsets A of $\mathbb{Q}$ without isolated points, that is, such that their closures $\text{cl}_{\mathbb{R}}(A)$ in $\mathbb{R}$ are perfect sets. Recall that a filter $\mathcal{F}$ on $\mathbb{Q}$ is *crowded* provided it is generated by the sets from Perf($\mathbb{Q}$) in a sense that $\mathcal{F} \cap \text{Perf}(\mathbb{Q})$ is coinitial in $\mathcal{F}$. Crowded ultrafilters were studied by several authors (see, e.g., [50, 46]) in connection with the remainder $\beta\mathbb{Q} \setminus \mathbb{Q}$ of the Čech-Stone compactification $\beta\mathbb{Q}$ of $\mathbb{Q}$.

In what follows we will also use the following simple fact, in which a nonscattered subset of $\mathbb{Q}$ is understood as a set containing a subset dense in itself.

Fact 5.5.1 *Every nonscattered set $B \subset \mathbb{Q}$ contains a subset from* Perf$(\mathbb{Q})$.

PROOF. Since B is nonscattered, decreasing it, if necessary, we can assume that B is dense in itself. Let $\{k_n \leq n : n < \omega\}$ be an enumeration of ω with infinitely many repetitions and let $\mathbb{Q} \setminus B = \{a_n : n < \omega\}$. By induction, construct a sequence $\langle \langle p_n, U_n \rangle \in B \times \mathcal{P}(\mathbb{Q}) : n < \omega \rangle$ such that $p_n \in B \setminus \bigcup_{i<n} U_i$ and $|p_n - p_{k_n}| < 2^{-n}$ while $U_n \ni a_n$ is a clopen subset of $\mathbb{Q} \setminus \{p_i : i \leq n\}$. Then $\mathbb{Q} \setminus \bigcup_{n<\omega} U_n \subset B$ is as desired. ∎

The next theorem answers in the positive a question of M. Hrušák [67] on whether there exists a crowded ultrafilter in the iterated perfect set model.

Theorem 5.5.2 CPA$_{\text{prism}}^{\text{game}}$ *implies there exists a nonprincipal ultrafilter on $\mathbb{Q}$ that is crowded.*

Fix a $p \in \mathbb{R} \setminus \mathbb{Q}$ and for a family $\mathcal{D} \subset \mathcal{P}(\mathbb{Q})$ let $F(\mathcal{D})$ denote a filter on $\mathbb{Q}$ generated by the family $\mathcal{D} \cup \{I_n \cap \mathbb{Q} : n < \omega\}$, where $I_n = [p - 2^{-n}, p + 2^{-n}]$. The proof of the theorem is based on the following lemma, in which $[\mathbb{Q}]^\omega$ is considered with the same topology as $[\omega]^\omega$ upon natural identification.

Lemma 5.5.3 *Let $\mathcal{D} \subset$ Perf$(\mathbb{Q})$ be a countable family such that $F(\mathcal{D})$ is crowded. Then for every prism P in $[\mathbb{Q}]^\omega$ there exist a subprism Q of P and a $Z \in$ Perf$(\mathbb{Q})$ such that $F(\mathcal{D} \cup \{Z\})$ is crowded and either*

(i) $Z \cap x = \emptyset$ *for every $x \in Q$, or else*
(ii) $Z \subset x$ *for every $x \in Q$.*

PROOF. In what follows we will identify $[\mathbb{Q}]^\omega$ with $2^\mathbb{Q}$, the identification mapping given by the characteristic function. Thus, we will consider P as a prism in $2^\mathbb{Q}$. Fix an $f \in \mathcal{F}_{\text{prism}}\left(2^\mathbb{Q}\right)$ from $\mathfrak{C}^\alpha$ onto P.

Let $\{D_n \in$ Perf$(\mathbb{Q}) : n < \omega\}$ be a coinitial sequence in $F(\mathcal{D})$ with a property that $D_{n+1} \subset D_n \subset I_n$ for every $n < \omega$. Choosing a subsequence, if necessary, we can find disjoint intervals J_n with $K_n = D_n \cap J_n \in$ Perf$(\mathbb{Q})$.

First we will show that there exist a sequence $\langle B_n \subset K_n : n < \omega \rangle$ of nonscattered sets and a subprism P_0 of P such that

$$g \restriction B_n \text{ is constant for every } g \in P_0 \text{ and } n < \omega. \tag{5.12}$$

For each $n < \omega$, let $\mathcal{D}_n$ be the collection of all pairwise disjoint families $\mathcal{E} \in [\mathbb{P}_\alpha]^{<\omega}$ for which there exists a nonscattered set $B_n \subset K_n$ with the property that

$$g \restriction B_n \text{ is constant for every } g \in f\left[\bigcup \mathcal{E}\right].$$

To see that the families $\mathcal{D}_n$ satisfy condition (†) from Lemma 3.1.2, it is enough to notice that for every nonscattered set $B \subset \mathbb{Q}$ and every prism P_1 there is a subprism Q_1 of P_1 and a nonscattered subset B' of B such that $g \restriction B'$ is constant for every $g \in Q_1$. But B contains a subset W homeomorphic to $\mathbb{Q}$. So, by Fact 5.3.1, the ideal $\mathcal{I}$ of nowhere dense subsets of W is weakly selective. So, applying Lemma 5.3.4 to this ideal and the prism P_1, we can find a $B' \in \mathcal{I}^+$, that clearly is not scattered, and a Q_1 as desired.

Thus, using Lemma 3.1.2, we can find a sequence $\langle \mathcal{E}_n \in \mathcal{D}_n \colon n < \omega \rangle$ satisfying the assumptions of Lemma 3.1.1. Let $E_0 = \bigcap_{n<\omega} \bigcup \mathcal{E}_n \in \mathbb{P}_\alpha$. It is easy to see that the sets B_n witnessing $\mathcal{E}_n \in \mathcal{D}_n$ and $Q_0 = f[E_0]$ satisfy (5.12). Notice also that, by Fact 5.5.1, we can assume that $B_n \in \mathrm{Perf}(\mathbb{Q})$ for every $n < \omega$.

Now let A be a selector from the family $\{B_n \colon n < \omega\}$. Then $A \in \mathcal{I}^+$, where $\mathcal{I}$ is the ideal of finite subsets of $\mathbb{Q}$. Applying Lemma 5.3.4(a) to A and P_0, we can find $i < 2$, $S \in [A]^\omega$, and a subprism Q of P_0 with the property that $g \restriction S$ is constant and equal to i for every $g \in Q$. Define $Z = \bigcup \{B_n \colon n < \omega \ \& \ B_n \cap S \neq \emptyset\}$. Then, by (5.12), $g \restriction Z$ is constant for every $g \in Q$. Finally, note that $Z \in \mathrm{Perf}(\mathbb{Q})$ since Z is closed, as $B_n \to p \notin \mathbb{Q}$. ∎

PROOF OF THEOREM 5.5.2. For a prism P in $[\mathbb{Q}]^\omega$ and a countable family $\mathcal{D} \subset \mathrm{Perf}(\mathbb{Q})$ for which $F(\mathcal{D})$ is crowded, let $Z(\mathcal{D}, P) \in \mathrm{Perf}(\mathbb{Q})$ and a subprism $Q(\mathcal{D}, P)$ of P be as in Lemma 5.5.3. Consider the following strategy S for Player II:

$$S(\langle \langle P_\eta, Q_\eta \rangle \colon \eta < \xi \rangle, P_\xi) = Q(\{Z_\eta \colon \eta < \xi\}, P_\xi),$$

where the sets Z_η are defined inductively by $Z_\eta = Z(\{Z_\zeta \colon \zeta < \eta\}, P_\eta)$.

By $\mathrm{CPA}_{\mathrm{prism}}^{\mathrm{game}}$, strategy S is not a winning strategy for Player II. So there exists a game $\langle \langle P_\xi, Q_\xi \rangle \colon \xi < \omega_1 \rangle$ played according to S in which Player II loses, that is, $[\mathbb{Q}]^\omega = \bigcup_{\xi < \omega_1} Q_\xi$.

Now, let $\mathcal{F} = F(\{Z_\xi \colon \xi < \omega_1\})$. Then clearly $\mathcal{F}$ is a crowded nonprincipal filter. To see that it is maximal, take an $x \in [\mathbb{Q}]^\omega$. Then there is a $\xi < \omega_1$ such that $x \in Q_\xi$. Then either $Z_\xi \cap x = \emptyset$ or $Z_\xi \subset x$. Thus, either x or its complement belongs to $\mathcal{F}$. ∎

Note also that, similarly as for Remarks 5.3.5 and 5.4.5, we can argue that there are many nonprincipal crowded ultrafilters.

Remark 5.5.4 $\mathrm{CPA}_{\mathrm{prism}}^{\mathrm{game}}$ implies that there are ω_2 many different nonprincipal crowded ultrafilters.

The existence of 2^{ω_1} different such ultrafilters follows also from a slightly stronger version of CPA$_{\text{prism}}^{\text{game}}$. (See Proposition 6.1.2.)

The construction of crowded ultrafilters is quite similar to that of selective ultrafilters and of nonselective P-points. This similarity suggests that it may be possible to construct a crowded ultrafilter that is also selective. This, however, cannot be done:

Proposition 5.5.5 *There is no nonprincipal crowded ultrafilter on $\mathbb{Q}$ that is also a P-point.*

PROOF.   Let $\mathcal{F}$ be a nonprincipal crowded ultrafilter on $\mathbb{Q}$ and let $\{x\} = \bigcap_{F \in \mathcal{F}} \text{cl}_{\mathbb{R}}(F) \in \mathbb{R} \setminus \mathbb{Q}$. Then $I_n = (x - 2^{-n}, x + 2^{-n}) \cap \mathbb{Q}$ belongs to $\mathcal{F}$ for every $n < \omega$. Let $\mathcal{P} = \{I_n \setminus I_{n+1} : n < \omega\} \cup \{\mathbb{Q} \setminus I_0\}$. Then $\mathcal{P}$ is a partition of $\mathbb{Q}$ disjoint with $\mathcal{F}$. It is also easy to see that if $F \subset \mathbb{Q}$ is such that $|F \cap P| < \omega$ for every $P \in \mathcal{P}$, then $F \notin \mathcal{F}$.   ∎

It is worth mentioning that CPA$_{\text{prism}}^{\text{game}}$ also implies the existence of many other kinds of ultrafilters, like those constructed in [66]. This has been recently done by A. Millán [92, 93]. In fact, many constructions that are done under CH also can be carried out under CPA$_{\text{prism}}^{\text{game}}$. However, this always needs the use of some combinatorial lemma, such as Lemma 5.3.4, which allows replacing points with prisms.

6

CPA and properties (F*) and (G)

To express the most general version of our axiom we need yet another version of a game, which we denote by $\text{GAME}_{\text{prism}}^{\text{section}}(X)$. As before, the game is of length ω_1, has two players, Player I and Player II, and at each stage $\xi < \omega_1$ of the game Player I can play an arbitrary prism, now represented explicitly by an $f_\xi \in \mathcal{F}_{\text{prism}}^*(X)$, and Player II responds with a subprism $g_\xi \in \mathcal{F}_{\text{prism}}^*(X)$ of f_ξ. The game $\langle\langle f_\xi, g_\xi\rangle: \xi < \omega_1\rangle$ is won by Player I provided there exists a $c \in \mathfrak{C}$ such that

$$X = \bigcup_{\xi < \omega_1} g_\xi\left[\{p \in \text{dom}(g_\xi): \pi_0(p) = c\}\right],$$

where π_0 is a projection onto the first coordinate.

Now, if $\Gamma = \{\alpha < \omega_2: \text{cof}(\alpha) = \omega_1\}$, then CPA reads as follows.

CPA: $\mathfrak{c} = \omega_2$, and for any Polish space X Player II has no winning strategy in the game $\text{GAME}_{\text{prism}}^{\text{section}}(X)$. Moreover, there exist a family $\{\langle f_\xi^\alpha \in \mathcal{F}_{\text{prism}}^*(X): \xi < \omega_1\rangle: \alpha < \omega_2\}$ of (very simple) tactics for Player I, as well as the enumerations $\{c_\alpha: \alpha < \omega_2\}$ and $\{x_\alpha: \alpha < \omega_2\}$ of $\mathfrak{C}$ and X, respectively, such that for every strategy S for Player II there is a closed unbounded subset Γ_S of Γ such that for every $\alpha \in \Gamma_S$ and a game $\langle\langle f_\xi^\alpha, g_\xi\rangle: \xi < \omega_1\rangle$ played according to S we have

$$X = \bigcup_{\xi < \omega_1} g_\xi\left[\{p \in \text{dom}(g_\xi): \pi_0(p) = c_\alpha\}\right]$$

and

$$X \setminus \bigcup_{g_\xi \in \mathcal{F}_{\text{prism}}} g_\xi\left[\{p \in \text{dom}(g_\xi): \pi_0(p) = c_\alpha\}\right] \subset \{x_\xi: \xi < \alpha\}.$$

143

In what follows we will not give any application for the moreover part of the axiom. However, it is a powerful tool that we used in some earlier versions of this text and which, we believe, will find some applications in the future.

It is easy to see that CPA implies $\text{CPA}_{\text{prism}}^{\text{game}}$. We also have the following implication.

Proposition 6.0.1 *Axiom* CPA *implies the following section version of the axiom* $\text{CPA}_{\text{prism}}$:

$\text{CPA}_{\text{prism}}^{\text{sec}}$: $\mathfrak{c} = \omega_2$, and for every Polish space X, $P \in \text{Perf}(\mathfrak{C})$, and an $\mathcal{F}_{\text{prism}}^*$-dense family $\mathcal{F} \subset \mathcal{F}_{\text{prism}}^*(X)$ there exist a $c \in P$ and an $\mathcal{F}_0 \subset \mathcal{F}$ such that $|\mathcal{F}_0| \leq \omega_1$ and

$$X = \bigcup_{f \in \mathcal{F}_0} f\left[\{p \in \text{dom}(f) \colon \pi_0(p) = c\}\right].$$

PROOF. The argument is essentially identical to that used in the proof of Proposition 2.0.1. First note that the implication is true if $P = \mathfrak{C}$.

So, let $\mathcal{F} \subset \mathcal{F}_{\text{prism}}^*(X)$ be $\mathcal{F}_{\text{prism}}^*$-dense. Thus for the prism $f \in \mathcal{F}_{\text{prism}}^*(X)$ there exists a subprism $s(f) \in \mathcal{F}$ of f. Consider the following strategy S for Player II:

$$S(\langle\langle f_\eta, g_\eta \rangle \colon \eta < \xi \rangle, f_\xi) = s(f_\xi).$$

By CPA, it is not a winning strategy for Player II. So there exists a game $\langle\langle f_\xi, g_\xi \rangle \colon \xi < \omega_1 \rangle$ in which $g_\xi = s(f_\xi)$ for every $\xi < \omega_1$ and Player II loses, that is, there exists a $c \in \mathfrak{C}$ such that

$$X = \bigcup_{\xi < \omega_1} g_\xi\left[\{p \in \text{dom}(g_\xi) \colon \pi_0(p) = c\}\right].$$

Then $\mathcal{F}_0 = \{g_\xi \colon \xi < \omega_1\}$ is as desired.

To get the general version of $\text{CPA}_{\text{prism}}^{\text{sec}}$ from the version with $P = \mathfrak{C}$, it is enough to identify in every prism the first coordinate Cantor set $\mathfrak{C}$ with the set P and apply the above case when $P = \mathfrak{C}$ to the family $\mathcal{F}'$ of all restrictions $f \restriction \{x \in \text{dom}(f) \colon \pi_0(x) \in P\}$ of $f \in \mathcal{F}$. ∎

To state a cube version of $\text{CPA}_{\text{prism}}^{\text{sec}}$, we need the following additional notation: For a fixed Polish space X the symbol $\mathcal{F}_{\text{cube}}^{\text{sec}}(X)$ will stand for the family of all continuous functions f from a perfect cube $C \subset \mathfrak{C}^\eta$, where $1 \leq \eta \leq \omega$, into X such that f is either one to one or constant.

It should be clear that $\text{CPA}_{\text{prism}}^{\text{sec}}$ implies the following:

$\mathrm{CPA}^{\mathrm{sec}}_{\mathrm{cube}}$: c = ω_2, and for every Polish space X and every $\mathcal{F}^{\mathrm{sec}}_{\mathrm{cube}}$-dense family $\mathcal{F} \subset \mathcal{F}^{\mathrm{sec}}_{\mathrm{cube}}(X)$ there exist a $c \in \mathfrak{C}$ and an $\mathcal{F}_0 \subset \mathcal{F}$ such that $|\mathcal{F}_0| \le \omega_1$ and

$$X = \bigcup_{f \in \mathcal{F}_0} f\left[\{p \in \mathrm{dom}(f) : \pi_0(p) = c\}\right].$$

Note that the axiom $\mathrm{CPA}^{\mathrm{sec}}_{\mathrm{cube}}$ would be false if we use in it $\mathcal{F}^*_{\mathrm{cube}}$ in place of $\mathcal{F}^{\mathrm{sec}}_{\mathrm{cube}}$, as noticed in Remark 6.4.1. Thus, this complication is essential.

For the remainder of this chapter we will concentrate on the consequences of $\mathrm{CPA}^{\mathrm{sec}}_{\mathrm{cube}}$ and $\mathrm{CPA}^{\mathrm{sec}}_{\mathrm{prism}}$.

6.1 cov(s_0) = c and many ultrafilters

To appreciate the section part in CPA, let us note that it implies Martin's axiom for the Sacks forcing $\mathbb{P} = \langle \mathrm{Perf}(\mathfrak{C}), \subset \rangle$ in a sense that for every family $\mathcal{D}$ of less than c many dense subsets of $\mathbb{P}$ there exists a $\mathcal{D}$-generic filter in $\mathbb{P}$. This follows immediately from the following fact, since for a dense subset D of $\mathbb{P}$ we have $\mathfrak{C} \setminus \bigcup D \in s_0$. (Compare also Corollary 3.3.7.)

Proposition 6.1.1 CPA *implies that* cov(s_0) = c.

PROOF. Let $\mathcal{T} = \{T_\xi : \xi < \omega_1\} \subset s_0$. We will show that $\bigcup \mathcal{T} \ne \mathfrak{C}$. Since s_0 is a σ-ideal, we can assume that $T_\eta \subset T_\xi$ for every $\eta < \xi < \omega_1$.

Consider the following Player II strategy S in a game $\mathrm{GAME}^{\mathrm{section}}_{\mathrm{prism}}(\mathbb{R})$: If Player I plays $f_\xi \in \mathcal{F}^*_{\mathrm{prism}}(\mathbb{R})$, Player II looks at $P = \pi_0[\mathrm{dom}(f_\xi)]$, chooses $Q \in \mathrm{Perf}(P)$ disjoint with T_ξ, and replies with a subprism g_ξ of f_ξ such that $\pi_0[\mathrm{dom}(g_\xi)] = Q$.

Since, by CPA, S is not winning, there exist a game $\langle\langle f_\xi, g_\xi \rangle : \xi < \omega_1 \rangle$ played according to S and a $c \in \mathfrak{C}$ such that

$$\mathbb{R} = \bigcup_{\xi < \omega_1} g_\xi \left[\{p \in \mathrm{dom}(g_\xi) : \pi_0(p) = c\}\right].$$

We claim that $c \in \mathfrak{C} \setminus \bigcup \mathcal{T}$, so that $\bigcup \mathcal{T} \ne \mathfrak{C}$.

Indeed, if we had $c \in T_\zeta$ for some $\zeta < \omega_1$, then we would also have $c \notin \pi_0[\mathrm{dom}(g_\xi)]$ for all $\zeta \le \xi < \omega_1$ and so

$$\mathbb{R} = \bigcup_{\eta < \zeta} g_\eta \left[\{p \in \mathrm{dom}(g_\eta) : \pi_0(p) = c\}\right].$$

But this would mean that $\mathbb{R}$ is a union of countably many nowhere dense sets range(g_η), $\eta < \zeta$. This contradiction finishes the proof. ∎

Next we will show that CPA implies the following generalization of Remarks 5.3.5, 5.4.5, and 5.5.4.

Proposition 6.1.2 CPA *implies that there are* 2^{ω_1} *different selective ultrafilters,* 2^{ω_1} *different nonselective P-points, as well as* 2^{ω_1} *different nonprincipal crowded ultrafilters.*

PROOF. First notice that CPA implies the following version of CPA_{prism}^{game}:

$*$-CPA_{prism}^{game}: $\mathfrak{c} = \omega_2$, and for any Polish space X there is a collection $\mathcal{T} = \{T_\alpha : \alpha < \omega_2\}$ of Player I tactics $T_\alpha = \langle f_\xi^\alpha \in \mathcal{F}_{prism}^*(X) : \xi < \omega_1\rangle$ for $GAME_{prism}(X)$ such that for any Player II strategy S for $GAME_{prism}(X)$ there is a $\xi < \omega_2$ with the property that a game played according to T_ξ and S is won by Player I.

We will show that the conclusion follows from $*$-CPA_{prism}^{game}.

If $2^{\omega_1} = \omega_2$, this follows from Remarks 5.3.5, 5.4.5, and 5.5.4 and the fact that $*$-CPA_{prism}^{game} implies CPA_{prism}^{game}. So, assume that $2^{\omega_1} > \omega_2$.

First we will prove only that there are 2^{ω_1} different selective ultrafilters, modifying slightly the proof of Theorem 5.3.3. So, let $\mathcal{I} = [\omega]^{<\omega}$. For $i < 2$ and a countable family $\mathcal{A} \subset \mathcal{I}^+$ linearly ordered by $\subset^*$, let $C_i(\mathcal{A}) \in \mathcal{I}^+$ be such that $C_i(\mathcal{A}) \subset^* A$ for every $A \in \mathcal{A}$ and $C_0(\mathcal{A}) \cap C_1(\mathcal{A}) = \emptyset$. Let $B(A, P) \in [A]^\omega$ and $Q(A, P)$ be as in Theorem 5.3.3 and for every $x \in 2^{\omega_1}$ define a Player II strategy S_x by:

$$S_x(\langle\langle P_\eta, Q_\eta\rangle : \eta < \xi\rangle, P_\xi) = Q(C_{x(\xi)}(\{B_\eta : \eta < \xi\}), P_\xi),$$

where the sets B_η are defined by $B_\eta = B(C_{x(\xi)}(\{B_\zeta : \zeta < \eta\}), P_\eta)$.

By $*$-CPA_{prism}^{game}, for every $x \in 2^{\omega_1}$ there is an $\alpha_x < \omega_2$ such that a game $\langle\langle P_\xi^x, Q_\xi^x\rangle : \xi < \omega_1\rangle$ played according to T_{α_x} and S_x is won by Player I. As in Theorem 5.3.3, we can easily prove that the filter $\mathcal{F}_x$ generated by $\{B_\xi^x : \xi < \omega_1\}$ is a selective ultrafilter. To finish the proof, notice that if $x, y \in 2^{\omega_1}$ are different and such that $\alpha_x = \alpha_y$, then $\mathcal{F}_x \neq \mathcal{F}_y$. This is the case, since $B_\eta^x \cap B_\eta^y = \emptyset$, where $\eta < \omega_1$ is the smallest ordinal for which $x(\eta) \neq y(\eta)$.

Now, let K be the set of all regular cardinals $\kappa \leq 2^{\omega_1}$ greater than ω_2. Then for every $\kappa \in K$ there exists an $X_\kappa \in [2^{\omega_1}]^\kappa$ such that $\alpha_x = \alpha_y$ for all $x, y \in X_\kappa$. So, $\mathcal{U}_\kappa = \{\mathcal{F}_x : x \in X_\kappa\}$ is a family of κ many different selective ultrafilters. Thus, $\bigcup_{\kappa \in K} \mathcal{U}_\kappa$ consists of 2^{ω_1} different selective ultrafilters.

The proofs that $*$-CPA_{prism}^{game} implies the existence of 2^{ω_1} different nonselective P-points and of nonprincipal crowded ultrafilters consist of a similar modification of the proofs of Theorems 5.4.2 and 5.5.2, respectively. ∎

6.2 Surjections onto nice sets must be continuous on big sets

In this section we will prove that $\text{CPA}^{\text{sec}}_{\text{cube}}$ implies:

(F*) For an arbitrary function h from a subset S of a Polish space X onto a Polish space Y there exists a uniformly continuous function f from a subset of X into Y such that $|f \cap h| = \mathfrak{c}$.

Note that (F*) implies (F): If $f: \mathbb{R} \to \mathbb{R}$ is Darboux nonconstant, then its range, $Y = f[\mathbb{R}]$, is a nontrivial interval, and so f is onto a Polish space Y. We will start the proof of (F*) with the following fact.

Fact 6.2.1 *Assume that* $\text{CPA}^{\text{sec}}_{\text{cube}}$ *holds and that* X *is a Polish space. Then for every function* h *from a subset* S *of* X *onto* $\mathfrak{C}$ *there exists an* $f: E \to X$ *in* $\mathcal{F}^{\text{sec}}_{\text{cube}}(X)$ *such that the set*

$$Z_f = \{f(p) \in S : p \in E \ \& \ h(f(p)) = \pi_0(p)\}$$

intersects $f[C]$ *for every perfect subcube* C *of* E. *Moreover, function* f *is one to one.*

PROOF. By way of contradiction, assume that for every $f: E \to X$ from $\mathcal{F}^{\text{sec}}_{\text{cube}}$ there exists a perfect subcube C_f of E such that $f[C_f]$ is disjoint with Z_f. Then $\mathcal{F} = \{f \restriction C_f : f \in \mathcal{F}^{\text{sec}}_{\text{cube}}\}$ is dense in $\mathcal{F}^{\text{sec}}_{\text{cube}}$. So, by $\text{CPA}^{\text{sec}}_{\text{cube}}$, there exist a $c \in \mathfrak{C}$ and an $\mathcal{F}_0 \subset \mathcal{F}$ such that $|\mathcal{F}_0| \leq \omega_1$ and

$$X = \bigcup_{f \in \mathcal{F}_0} f\left[\{p \in \text{dom}(f) : \pi_0(p) = c\}\right].$$

Let $r \in S$ be such that $h(r) = c$. Thus, there is an $f \in \mathcal{F}_0$, $f: C_f \to X$, such that $r \in f\left[\{p \in C_f : \pi_0(p) = c\}\right]$. So, there exists a $p \in C_f$ with $\pi_0(p) = c$ for which $f(p) = r$. But this means that $h(f(p)) = h(r) = c = \pi_0(p)$. Therefore $f(p)$ belongs to Z_f and $f[C_f]$, contradicting the fact that these sets are disjoint.

Thus, there is an $f \in \mathcal{F}^{\text{sec}}_{\text{cube}}$ having the main property. To see that it is one to one, it is enough to show that it is not constant. So, by way of contradiction, assume that f is constant and find perfect subcubes C and C' of $\text{dom}(f)$ such that $\pi_0[C] \cap \pi_0[C'] = \emptyset$. So, there are $p \in C$ and $p' \in C'$ such that $f(p), f(p') \in Z_f$. Thus, since f is constant, we have $\pi_0(p) = h(f(p)) = h(f(p')) = \pi_0(p')$, contradicting the fact that $\pi_0[C]$ and $\pi_0[C']$ are disjoint. ∎

Corollary 6.2.2 $\text{CPA}^{\text{sec}}_{\text{cube}}$ *implies* (F*).

PROOF. Since Y contains a subset homeomorphic to $\mathfrak{C}$, by restricting h we can assume that $Y = \mathfrak{C}$. Let $f: E \to X$ and Z_f be as in Fact 6.2.1. Clearly there is continuum many pairwise disjoint perfect subcubes C of E, and Z_f intersects $f[C]$ for all these C. So, since f is one to one, Z_f has cardinality $\mathfrak{c}$. Note also that for every $x = f(p) \in Z_f$ we have $h(x) = \pi_0[f^{-1}(x)]$. Thus, on a set Z_f, h is equal to a continuous function $\pi_0 \circ f^{-1}$. ∎

If in (F*) we have $X, Y \subset \mathbb{R}$, then we can also get the following stronger result.

Corollary 6.2.3 $\text{CPA}^{\text{sec}}_{\text{prism}}$ *implies the following property:*

(F') *For any function h from a subset S of $\mathbb{R}$ onto a perfect subset of $\mathbb{R}$ there exists a function $f \in$ "C^∞_{perf}" such that $|f \cap h| = \mathfrak{c}$ and f can be extended to a function $\bar{f} \in$ "$C^1(\mathbb{R})$" such that either either $\bar{f} \in C^1$ or $\bar{f}$ is an autohomeomorphism of $\mathbb{R}$ with $\bar{f}^{-1} \in C^1$.*

PROOF. Let f^* be a continuous function as in condition (F*), that is, such that $|f^* \cap h| = \mathfrak{c}$, and let $g: \mathbb{R} \to \mathbb{R}$ be a Borel extension of f^*. Let $\{f_\xi: \xi < \omega_1\}$ be the functions from Theorem 4.1.1(a). Then we have $f^* \cap h \subset g \subset \bigcup_{\xi < \omega_1} f_\xi$. So, there exists a $\xi < \omega_1$ such that $|f_\xi \cap h| = \mathfrak{c}$ and $f = f_\xi$ is as desired. ∎

Notice also that function f in (F') cannot chosen from D^1_{perf}. This is evident for $h = f \upharpoonright P$, where f and P are from Example 4.5.2.

6.3 Sums of Darboux and continuous functions

In this section we show that the conjunction of properties (A) and (F*) implies the property:

(G) For every Darboux function $g: \mathbb{R} \to \mathbb{R}$ there exists a continuous nowhere constant function $f: \mathbb{R} \to \mathbb{R}$ such that $f + g$ is Darboux.

In particular, (G) follows from $\text{CPA}^{\text{sec}}_{\text{cube}}$.

The proof presented below follows the argument given by K. Ciesielski and J. Pawlikowski in [34]. Although it is more involved than many of the other proofs presented so far, it is still considerably simpler than the original proof of the consistency of (G) given in [123]. Recall that, under Martin's axiom, property (G) is false, as proved by B. Kirchheim and T. Natkaniec [78].

Our proof will be based on the following two lemmas.

Lemma 6.3.1 *Let* $d: \mathbb{R} \to \mathbb{R}$ *be Darboux and* $D \subset \mathbb{R}$ *be such that* $d \upharpoonright D$ *is dense in* d. *If* $g: \mathbb{R} \to \mathbb{R}$ *is continuous and such that*

$$[(d+g)(\alpha), (d+g)(\beta)] \subset (d+g)[[\alpha, \beta]] \quad \text{for all} \quad \alpha, \beta \in D, \qquad (6.1)$$

then $d + g$ *is Darboux.*

PROOF. An easy proof can be found in [123, lem. 4.1]. ∎

Lemma 6.3.2 *Assume that* (A) *holds and for every* $n < \omega$ *let* $A_n \in [\mathbb{R}]^\mathfrak{c}$. *Then for every* $n < \omega$ *there exists a* $B_n \in [A_n]^\mathfrak{c}$ *such that the closures of the* B_n's *are pairwise disjoint.*

PROOF. First note that (A) implies that

$$\forall \langle C_n \in [\mathbb{R}]^\mathfrak{c} : n < \omega \rangle \; \exists C \in [C_0]^\mathfrak{c} \; \forall n < \omega \; |C_n \setminus \mathrm{cl}(C)| = \mathfrak{c}. \qquad (6.2)$$

Indeed, by condition (A), we can find a continuous function $f: \mathbb{R} \to [0, 1]$ such that $|C_0 \cap f^{-1}(c)| = \mathfrak{c}$ for every $c \in [0, 1]$. (Just take a composition of a function from (A) with the Peano curve followed by a projection.) Identify $\mathfrak{C}$ with a subset of $[0, 1]$ and by induction on $n < \omega$ choose an increasing sequence $s_n \in 2^{n+1}$ such that $|C_n \setminus f^{-1}(\{c \in \mathfrak{C} : s_n \subset c\})| = \mathfrak{c}$. Put $s = \bigcup_{n < \omega} s_n$. Then $C = C_0 \cap f^{-1}(s)$ satisfies (6.2).

Next note that there exists a sequence $\langle D_n : n < \omega \rangle$ of closed subsets of $\mathbb{R}$ such that $\left| A_n \cap \left(D_n \setminus \bigcup_{i<n} D_i \right) \right| = \mathfrak{c}$. This sequence is constructed by induction on $n < \omega$ with each set D_n chosen by applying condition (6.2) to the sequence $\langle A_k \setminus \bigcup_{i<n} D_i : n \leq k < \omega \rangle$.

Finally, since each set $D_n \setminus \bigcup_{i<n} D_i$ is an F_σ-set, we can find closed subsets of them E_n with $B_n = E_n \cap A_n$ having cardinality $\mathfrak{c}$. It is easy to see that the sets B_n are as required. ∎

Remark 6.3.3 Note that Lemma 6.3.2 cannot be proved in ZFC. Indeed, if we assume that there exists a $\mathfrak{c}$-Luzin set[1] and $\langle A_n \in [\mathbb{R}]^\mathfrak{c} : n < \omega \rangle$ is a sequence of $\mathfrak{c}$-Luzin sets such that every nonempty open interval contains one of the A_n's, than for such a sequence there are no B_n's as in the lemma.

---

[1] A set $L \subset \mathbb{R}$ is a $\mathfrak{c}$-*Luzin set* if $|L| = \mathfrak{c}$ but $|L \cap N| < \mathfrak{c}$ for every nowhere dense subset N of $\mathbb{R}$. It is well known (see, e.g., [95, sec. 2]) and easy to see that no $\mathfrak{c}$-Luzin set can be mapped continuously onto $[0, 1]$. Thus (A) implies that there is no $\mathfrak{c}$-Luzin set.

Lemma 6.3.4 *Assume that* (A) *and* (F*) *hold and* $d: \mathbb{R} \to \mathbb{R}$ *is a Darboux function. If* D *is a countable subset of* $\mathbb{R}$ *such that:*

(∗) *For every* $\alpha < \beta$ *from* D *for which* $d(\alpha) = d(\beta)$ *there exist* $p, q \in \mathbb{R}$ *such that* $\alpha < p < q < \beta$ *and* $d[(p, q)] = \{f(\alpha)\}$,

then there exists a countable family $\mathcal{A} \subset [\mathbb{R}]^{\mathfrak{c}}$ *such that*

(a) *different elements of* $\mathcal{A}$ *have disjoint closures,*
(b) $d \upharpoonright A$ *is uniformly continuous for every* $A \in \mathcal{A}$, *and*
(c) *for every* $\alpha, \beta \in D$ *there exists an* $A \in \mathcal{A}$ *with the property that* $d \upharpoonright A \subset [\alpha, \beta] \times [d(\alpha), d(\beta)]$.

PROOF. First, for every $\alpha < \beta$ from D, we choose $A_\alpha^\beta \in [\mathbb{R}]^{\mathfrak{c}}$ such that the family $\mathcal{A}_0 = \{A_\alpha^\beta : \alpha, \beta \in D \ \& \ \alpha < \beta\}$ satisfies (b) and (c). For this fix $\alpha < \beta$ from D.

If $d(\alpha) = d(\beta)$, then it is enough to put $A_\alpha^\beta = (p, q)$, where p and q are from (∗). So, assume that $d(\alpha) \neq d(\beta)$. Then $I = [d(\alpha), d(\beta)]$ is a nontrivial interval, which is a subset of $d[[\alpha, \beta]]$, since d is Darboux. Let $S = [\alpha, \beta] \cap d^{-1}(I)$ and notice that $h = d \upharpoonright S$ maps S onto I. So, by (F*), we can find an $A_\alpha^\beta \in [S]^{\mathfrak{c}}$ such that $d \upharpoonright A_\alpha^\beta$ is uniformly continuous.

It is easy to see that $\mathcal{A}_0$ satisfies (b) and (c); so it is enough to decrease its elements to get condition (a), while assuring that they still have cardinality $\mathfrak{c}$. This can be done by applying Lemma 6.3.2. ∎

Theorem 6.3.5 *Assume that* (A) *and* (F*) *hold. Then for every Darboux function* $d: \mathbb{R} \to \mathbb{R}$ *there exist a complete metric* ρ *on* $\mathcal{C}(\mathbb{R})$ *and a dense* G_δ *subset* G *of* $\langle \mathcal{C}(\mathbb{R}), \rho \rangle$ *of nowhere constant functions such that* $d + g$ *is Darboux for every* $g \in G$. *In particular,* (A)&(F*) *implies* (G).

PROOF. Take a Darboux function $d: \mathbb{R} \to \mathbb{R}$ and choose a countable dense set $D \subset \mathbb{R}$ satisfying condition (∗) from Lemma 6.3.4 such that $d \upharpoonright D$ is dense in d. Let $\mathcal{A} \subset [\mathbb{R}]^{\mathfrak{c}}$ be the family from Lemma 6.3.4 and let $\{A_n : n < \omega\}$ be an enumeration of $\mathcal{A} \cup \{\{d\} : d \in D \setminus \bigcup \mathcal{A}\}$. Let ρ_0 be the uniform convergence metric on $\mathcal{C}(\mathbb{R})$, that is,

$$\rho_0(f, g) = \min\{1, \sup\{|f(x) - g(x)| : x \in \mathbb{R}\}\},$$

and, for $f, g \in \mathcal{C}(\mathbb{R})$, let

$$\rho_1(f, g) = 2^{-\min\{n < \omega : f \upharpoonright A_n \neq g \upharpoonright A_n\}}.$$

(If $\{n < \omega : f \upharpoonright A_n \neq g \upharpoonright A_n\} = \emptyset$, we assume that $\rho_1(f, g) = 0$.) Then ρ_1 is a pseudometric on $\mathcal{C}(\mathbb{R})$. Consider $\mathcal{C}(\mathbb{R})$ with the following metric ρ:

$$\rho(f, g) = \max\{\rho_0(f, g), \rho_1(f, g)\},$$

and notice that $\langle \mathcal{C}(\mathbb{R}), \rho \rangle$ forms a complete metric space. We will prove that if G is the set of all nowhere constant continuous functions $g: \mathbb{R} \to \mathbb{R}$ for which $d + g$ satisfies (6.1) of Lemma 6.3.1, then G contains a dense G_δ subset of $\langle \mathcal{C}(\mathbb{R}), \rho \rangle$. For this we will show that for every $\alpha < \beta$ from D the following two kinds of sets contain dense open subsets of $\langle \mathcal{C}(\mathbb{R}), \rho \rangle$:

$$H_\alpha^\beta = \{g \in \mathcal{C}(\mathbb{R}): g \text{ is not constant on } (\alpha, \beta)\}$$

and

$$G_\alpha^\beta = \{g \in \mathcal{C}(\mathbb{R}): [(d+g)(\alpha), (d+g)(\beta)] \subset (d+g)[[\alpha, \beta]]\}.$$

Then

$$\bigcap \{H_\alpha^\beta \cap G_\alpha^\beta : \alpha, \beta \in D \ \& \ \alpha < \beta\} \subset G$$

will contain a dense G_δ subset of $\langle \mathcal{C}(\mathbb{R}), \rho \rangle$.

To see that H_α^β contains a dense open subset, first note that, by (a) and (c) of Lemma 6.3.4, the elements of $\mathcal{A}$ are nowhere dense. Next, take an $f \in \mathcal{C}(\mathbb{R})$ and fix an $\varepsilon > 0$. Let $B(f, \varepsilon)$ be the open ρ-ball centered at f and of radius ε. We will find $g \in B(f, \varepsilon)$ and $\delta > 0$ such that $B(g, \delta) \subset B(f, \varepsilon) \cap H_\alpha^\beta$. So take an $n < \omega$ such that $2^{-n} < \varepsilon/4$, choose a nonempty open interval $J \subset (\alpha, \beta) \setminus \bigcup_{i=1}^n A_i$, and pick different $x, y \in J$. It is easy to find $g \in \mathcal{C}(\mathbb{R})$ such that $g(x) \neq g(y)$, $g \upharpoonright (\mathbb{R} \setminus J) = f \upharpoonright (\mathbb{R} \setminus J)$, and $\rho_0(f, g) < 2^{-n}$. Then, by the choice of J we also have $\rho_1(f, g) < 2^{-n}$, so $g \in B(f, \varepsilon/4)$. Now, if $\delta = \min\{|g(x) - g(y)|/4, \varepsilon/4\}$, then we have $B(g, \delta) \subset B(f, \varepsilon) \cap H_\alpha^\beta$.

To see that each G_α^β contains a dense open subset, take $f \in \mathcal{C}(\mathbb{R})$ and $\varepsilon > 0$. As previously we will find a $g \in B(f, \varepsilon)$ and a $\delta > 0$ such that $B(g, \delta) \subset B(f, \varepsilon) \cap G_\alpha^\beta$. Find $\alpha = x_0 < x_1 < \cdots < x_m = \beta$, $x_i \in D$, such that the variation of f on each interval $[x_i, x_{i+1}]$ is less than $\varepsilon/8$. Also, since $d \upharpoonright [x_i, x_{i+1}]$ is Darboux, we can partition each $[x_i, x_{i+1}]$ even further to get also that $|d(x_i) - d(x_{i+1})| < \varepsilon/8$ for all $i < m$.

Pick an $n < \omega$ such that $2^{-n} < \varepsilon/8$ and $\{x_i : i \leq m\} \subset \bigcup_{i=1}^n A_i$. For every $i < m$ choose a $k_i > n$ such that $d \upharpoonright A_{k_i} \subset (x_i, x_{i+1}) \times [d(x_i), d(x_{i+1})]$. By (A), for every $i < m$ we can also pick a uniformly continuous function h_i from A_{k_i} onto $[(d+f)(x_i), (d+f)(x_{i+1})]$. Now, for each $i < m$, define $h \upharpoonright A_{k_i} = (h_i - d - f) \upharpoonright A_{k_i}$ and note that

$$h[A_{k_i}] \subset h_i[A_{k_i}] - d[A_{k_i}] - f[A_{k_i}] \subset [-\varepsilon/2, \varepsilon/2]$$

since $[(d+f)(x_i), (d+f)(x_{i+1})] \subset [d(x_i) + f(x_i) - \varepsilon/4, d(x_i) + f(x_i) + \varepsilon/4]$, $d[A_{k_i}] \subset [d(x_i), d(x_{i+1})] \subset [d(x_i) - \varepsilon/8, d(x_i) + \varepsilon/8]$, and $f[A_{k_i}]$ is a subset of $(f(x_i) - \varepsilon/8, f(x_i) + \varepsilon/8)$. Also define h as 0 on $\bigcup_{i=1}^n A_i$ and extend it to a uniformly continuous function from $\mathbb{R}$ into $[-\varepsilon/2, \varepsilon/2]$. Put $g = f + h$ and note that $\rho(f,g) \leq \varepsilon/2$. Also let $k = \max\{n, k_0, \ldots, k_{m-1}\}$ and pick $\delta \in (0, 2^{-k})$. We claim that $B(g, \delta) \subset B(f, \varepsilon) \cap G_\alpha^\beta$.

Indeed, it is easy to see that $B(g, \delta) \subset B(f, \varepsilon)$. To see that $B(g, \delta) \subset G_\alpha^\beta$, take a $g_0 \in B(g, \delta)$. By the choice of δ, h, and g for every $i \leq m$ we have

$$(d + g_0)(x_i) = (d + g)(x_i) = (d + f + h)(x_i) = (d + f)(x_i)$$

and, for $A = \bigcup_{j<m} A_{k_j}$,

$$g_0 \upharpoonright A = g \upharpoonright A = (f + h) \upharpoonright A = (f + (h_i - d - f)) \upharpoonright A = (h_i - d) \upharpoonright A.$$

So,

$$
\begin{aligned}
[(d + g_0)(\alpha), (d + g_0)(\beta)] &= [(d + f)(x_0), (d + f)(x_m)] \\
&\subset \bigcup_{j<m} [(d + f)(x_j), (d + f)(x_{j+1})] \\
&= \bigcup_{j<m} h_j[A_{k_j}] \\
&= \bigcup_{j<m} (d + (h_j - d))[A_{k_j}] \\
&= \bigcup_{j<m} (d + g_0)[A_{k_j}] \\
&\subset (d + g_0)[[\alpha, \beta]],
\end{aligned}
$$

proving that $g_0 \in G_\alpha^\beta$. This finishes the proof of the theorem. ∎

Corollary 6.3.6 $\text{CPA}_{\text{cube}}^{\text{sec}}$ *implies* (G).

Let us also notice that in (G) we cannot require that function g is "$\mathcal{C}^1$". This follows from the following fact, which, for the functions from the class $\mathcal{C}^1$, was first noticed by J. Steprāns [123, p. 118]. Since Steprāns leaves his statement without any comments concerning its proof, we include here a missing argument.

Proposition 6.3.7 *There exists, in ZFC, a Darboux function* $d: \mathbb{R} \to \mathbb{R}$ *such that* $d + f$ *is not Darboux for every nonconstant "$\mathcal{C}^1$" function* f.

PROOF. Let $\mathbb{R} = \{x_\xi : \xi < \mathfrak{c}\}$ and let $\{f_\xi : \xi < \mathfrak{c}\}$ be an enumeration of all nonconstant "$\mathcal{C}^1$" functions. Notice that for every $\xi < \mathfrak{c}$ there exists a nonempty open interval I_ξ such that $f_\xi \upharpoonright I_\xi$ is strictly monotone. (Just take an $x \in \mathbb{R}$ such that $f_\xi'(x) \neq 0$, which exists since f_ξ is nonconstant. Then I_ξ is chosen as a neighborhood of x on which f_ξ' is nonzero.)

The function d we construct will be strongly Darboux in a sense that $d^{-1}(r)$ is dense in $\mathbb{R}$ for every $r \in \mathbb{R}$. For such a d, in order to show that $d + f_\xi$ is not Darboux, it is enough to show that $(d + f_\xi)^{-1}(y_\xi)$ is not dense in $\mathbb{R}$ for some $y_\xi \in \mathbb{R}$. (See, e.g., [25, prop. 7.2.4].)

By induction we construct a sequence $\{\langle Q_\xi, d_\xi, y_\xi \rangle : \xi < \mathfrak{c}\}$ such that for every $\xi < \mathfrak{c}$ we have:

(i) The sets $\{Q_\eta : \eta \leq \xi\}$ are countable pairwise disjoint and $x_\xi \in \bigcup_{\eta \leq \xi} Q_\eta$.

(ii) $d_\xi : Q_\xi \to \mathbb{R}$ and $d_\xi^{-1}(x_\xi)$ is dense in $\mathbb{R}$.

(iii) $y_\zeta \notin (d_\eta + f_\zeta)[Q_\eta \cap I_\zeta]$ for every $\zeta, \eta \leq \xi$.

Notice that, if we have such a sequence, then, by (i), $\mathbb{R} = \bigcup_{\xi < \mathfrak{c}} Q_\xi$; so $d = \bigcup_{\xi < \mu} d_\xi : \mathbb{R} \to \mathbb{R}$. It is strongly Darboux by condition (ii) while, by (iii), for every $\zeta < \mathfrak{c}$ the set $(d + f_\zeta)^{-1}(y_\zeta)$ is not dense in $\mathbb{R}$, since it misses I_ζ. Thus, d is as desired.

To make an inductive ξ-th step, first choose a countable dense subset Q_ξ^0 of $\mathbb{R}$ disjoint with

$$\bigcup_{\eta < \xi} Q_\eta \cup \bigcup_{\zeta < \xi} \left(I_\zeta \cap f_\zeta^{-1}(y_\zeta - x_\xi) \right).$$

This can be done since, by the choice of of the intervals I_ζ's, each of the sets $I_\zeta \cap f_\zeta^{-1}(y_\zeta - x_\xi)$ has at most one element.

Define $d_\xi \restriction Q_\xi^0$ as constantly equal to x_ξ. This guarantees (ii), while condition (iii) is satisfied for the part defined so far: For every $\zeta < \xi$ and $x \in I_\zeta \cap Q_\xi^0$ we have $d_\xi(x) + f_\zeta(x) = f_\zeta(x) + x_\xi \neq y_\zeta$, since otherwise we would have $x \in I_\zeta \cap f_\zeta^{-1}(y_\zeta - x_\xi)$, contradicting the choice of Q_ξ^0.

If $x_\xi \in \bigcup_{\eta < \xi} Q_\eta$, we define $Q_\xi = Q_\xi^0$. Otherwise we put $Q_\xi = Q_\xi^0 \cup \{x_\xi\}$, and, if $d_\xi(x_\xi)$ is not defined yet (i.e., if $x_\xi \notin Q_\xi^0$), we define $d_\xi(x_\xi)$ by choosing $d_\xi(x_\xi) \notin \{y_\zeta - f_\zeta(x_\xi) : \zeta < \xi\}$. This guarantee that (i) is satisfied while (iii) is preserved.

Finally, we choose y_ξ to have

$$y_\xi \in \mathbb{R} \setminus \bigcup_{\eta \leq \xi} (d_\eta + f_\xi)[Q_\eta].$$

This will guarantee that (iii) also holds for $\zeta = \xi$. ∎

Let us also note that the following question, for the case of D^1, due to J. Steprāns [123, question 5.1], remains open.

Problem 6.3.8 Does there exist, in ZFC, a Darboux function $d : \mathbb{R} \to \mathbb{R}$ such that $d + f$ is not Darboux for every nowhere constant "D^1" function f? What if we restrict the choice of f to D^1?

6.4 Remark on a form of $\text{CPA}^{\text{sec}}_{\text{cube}}$

Remark 6.4.1 Note that if $X = \mathfrak{C}$ and h is any autohomeomorphism of $\mathfrak{C}$, then the conclusion of Fact 6.2.1 may hold only for a one-dimensional perfect cube $E \subset \mathfrak{C}^1$.

PROOF. Indeed, for every perfect cube $E \subset \mathfrak{C}^\eta = \mathfrak{C}^1 \times \mathfrak{C}^{\eta \setminus 1}$ with $\eta > 1$, the set $\{p \in E : h(f(p)) = \pi_0(p)\}$ is closed with vertical sections having at most one point. So, it is meager in E and it cannot meet all perfect subcubes of E. Since in the proof of Fact 6.2.1 was independent of the dimension of perfect cubes, this implies that the axiom $\text{CPA}^{\text{sec}}_{\text{cube}}$ would have been false if we did not include one-dimensional perfect cubes $\mathfrak{C}^1$ in the definition of $\mathcal{F}^{\text{sec}}_{\text{cube}}$. ∎

Notice also that, if $\text{CPA}[X]$ stands for CPA for a fixed Polish space X, then, similarly as in Remark 1.8.3, we can also prove

Remark 6.4.2 For any Polish space X axiom $\text{CPA}[X]$ implies CPA.

7

CPA in the Sacks model

7.1 Notations and basic forcing facts

We will use here forcing terminology and notation similar to that from [4]. The terminology specific to the Sacks forcing comes from [7], [95], or [123].

Let $\mathbb{P} = \langle \mathrm{Perf}(\mathfrak{C}), \subset \rangle$. Recall that perfect set (Sacks) forcing $\mathbb{S}$ is usually defined as the set of all trees $T(P) = \{x \upharpoonright n \in 2^{<\omega} : x \in P \ \& \ n < \omega\}$, where $P \in \mathbb{P}$, and is ordered by inclusion; that is, $s \in \mathbb{S}$ is stronger than $t \in \mathbb{S}$, $s \leq t$, if $s \subset t$. It is important to realize that

$$P \subset Q \quad \text{if and only if} \quad T(P) \subset T(Q),$$

so $T : \mathbb{P} \to \mathbb{S}$ establishes isomorphism between forcings $\mathbb{P}$ and $\mathbb{S}$. Also, if for $s \in \mathbb{S}$ we define $\lim(s) = \{x \in 2^\omega : \forall n < \omega \ (x \upharpoonright n \in s)\}$, then $\lim : \mathbb{S} \to \mathbb{P}$ is the inverse of T. A perfect set forcing is usually represented as $\mathbb{S}$ rather than in its more natural form $\mathbb{P}$ since the conditions in $\mathbb{S}$ are absolute, unlike those in $\mathbb{P}$. However, in light of our axiom, it is important to think of this forcing in terms of $\mathbb{P}$. For $s \in 2^{<\omega}$ and $q \in \mathbb{S}$ we put

$$q^s = \{t \in q : s \cup t \text{ is a function}\}.$$

Note that $q^s \in \mathbb{S}$ if and only if $s \in q$. Condition $q^s \in \mathbb{S}$ is a subtree of all elements of q consistent with s.

In what follows we will show that CPA holds in the generic extension $V[G]$ of a model V of ZFC+CH, where G is a V-generic filter over $\mathbb{S}_{\omega_2}$, the countable support iteration of length ω_2 of the Sacks forcing. Here, for $0 < \alpha \leq \omega_2$, the countable support iteration $\mathbb{S}_\alpha$ of length α of forcing $\mathbb{S}$ will be defined, by induction, as the family of all functions p defined on α such that for every $0 < \beta < \alpha$

155

- $p(\beta)$ is a $\mathbb{S}_\beta$-name for a subset of $2^{<\omega}$, $p \restriction \beta \in \mathbb{S}_\beta$;
- $p \restriction \beta \Vdash p(\beta) \in \mathbb{S}$; and
- there is a countable set $\mathrm{supp}(p) \subset \alpha$, called the *support of p* such that $p(\gamma)$ is a standard $\mathbb{S}_\gamma$-name for $2^{<\omega}$ for all $\gamma \in \alpha \setminus \mathrm{supp}(p)$.

For $p \in \mathbb{S}_\alpha$ and $\sigma \colon F \to 2^k$, where $k < \omega$ and $F \in [\alpha]^{<\omega}$, we define a function $p|\sigma$ on α inductively as follows. For $\beta < \alpha$ we put

- $(p|\sigma)(\beta) = p(\beta)$ if either $\beta \notin F$ or $(p|\sigma) \restriction \beta \notin \mathbb{S}_\beta$; otherwise
- $(p|\sigma)(\beta) = \tau$, where τ is a $\mathbb{S}_\beta$-name such that $(p|\sigma) \restriction \beta \Vdash \tau = [p(\beta)]^{\sigma(\beta)}$.

Since, in general, $q^{\sigma(\beta)}$ need not to belong to $\mathbb{S}$, it may easily happen that $p|\sigma \notin \mathbb{S}_\alpha$. If $p|\sigma \in \mathbb{S}_\alpha$, we say that σ is *consistent* with p. For $F \in [\alpha]^{<\omega}$ and $k < \omega$ we say that $p \in \mathbb{S}_\alpha$ is $\langle F, k \rangle$-*determined* if for every $\beta \in F$ and $\sigma \colon F \to 2^k$ consistent with p the condition $(p|\sigma) \restriction \beta$ decides already the value of $p(\beta) \cap 2^k$; that is, if for every $s \in 2^k$

$$\text{either} \quad (p|\sigma) \restriction \beta \Vdash s \in p(\beta) \quad \text{or} \quad (p|\sigma) \restriction \beta \Vdash s \notin p(\beta). \tag{7.1}$$

Note that in this case (7.1) also holds for every $s \in 2^{\leq k}$, since $p \restriction \beta$ forces that $p(\beta)$ is a tree.

We say that $p \in \mathbb{S}_\alpha$ is *determined* provided for every $m < \omega$ and $F_0 \in [\alpha]^{<\omega}$ there exist $m < k < \omega$ and $F_0 \subset F \in [\alpha]^{<\omega}$ such that p is $\langle F, k \rangle$-determined. In [95], A. Miller notices that the result of a fusion of a sequence is a determined condition. In particular, if

$$\mathbb{S}_\alpha^D = \{p \in \mathbb{S}_\alpha \colon p \text{ is determined}\},$$

then $\mathbb{S}_\alpha^D$ is dense in $\mathbb{S}_\alpha$.

Recall also that for a countable $A \subset \omega_2$ we defined $\Phi_{\mathrm{prism}}(A)$ as the family of all projection-keeping homeomorphisms $f \colon \mathfrak{C}^A \to \mathfrak{C}^A$. We also defined $\mathbb{P}_A$ as $\{\mathrm{range}(f) \colon f \in \Phi_{\mathrm{prism}}(A)\}$. However, in this chapter, for $A \subset \alpha \leq \omega_2$, we will be more interested in the following order-isomorphic copy of $\mathbb{P}_A$:

$$\mathbb{P}_A^\alpha = \{[\mathrm{range}(f)]_\alpha \colon f \in \Phi_{\mathrm{prism}}(A)\},$$

where $[E]_\alpha = \{g \in \mathfrak{C}^\alpha \colon g \restriction A \in E\}$ for every $E \subset \mathfrak{C}^A$. Also, we put

$$\mathbb{P}_\alpha = \bigcup \{\mathbb{P}_A^\alpha \colon A \in [\alpha]^{\leq \omega}\}$$

and order it by inclusion. (Thus, for a countable α we have two different definitions of $\mathbb{P}_\alpha$. However, it is not difficult to see that they describe the same family.)

It is known that forcing $\mathbb{P}_\alpha$ is equivalent to $\mathbb{S}_\alpha$. This fact is stated explicitly by V. Kanovei in [73], though it was also used, in a less explicit form, in earlier papers of A. Miller [95] and J. Steprāns [123]. However, none of these papers contains a proof of this fact. Since this fact is of major importance for the CPA axiom, we include below its proof.

Remark 7.1.1 In [73], the author considers forcings $\mathbb{P}_X$ for partial ordered sets X that are not necessary well ordered, so they have no standard iteration equivalences $\mathbb{S}_X$. Thus, he proves the basic properties of the forcings $\mathbb{P}_X$, like the preservation of ω_1, directly from their definition. It could be interesting to follow this path and prove the basic facts on $\mathbb{P}_\alpha$ we need straight from the definition, instead of deducing them from the known properties of $\mathbb{S}_\alpha$. This would allow us to avoid all together the complicated matter of iterated forcing, with conditions being the names with respect to the initial segments of the iteration. In fact, in the proof of the consistency of CPA we will use the equivalence of forcings $\mathbb{P}_\alpha$ and $\mathbb{S}_\alpha$ only to get the following properties, where V is a model for ZFC+CH and G is a V-generic filter over $\mathbb{P}_{\omega_2}$.

(1) Cardinal numbers in V and $V[G]$ are the same.
(2) In $V[G]$ we have $2^\omega = \omega_2$.
(3) If $0 < \gamma < \omega_2$, then $G_\gamma \stackrel{\text{def}}{=} G \upharpoonright \gamma$ is a V-generic filter over $\mathbb{P}_\gamma$.
(4) If $0 < \gamma \le \omega_2$ has uncountable cofinality and $r \in 2^\omega \cap V[G_\gamma]$, then there exists a $\beta < \gamma$ such that $r \in V[G_\beta]$.

Fact 7.1.2 $\mathbb{S}_\alpha$ *is order isomorphic to* $\mathbb{S}_\alpha^D$ *for every* $0 < \alpha \le \omega_2$. *In particular, the forcings* $\mathbb{S}_\alpha$ *and* $\mathbb{P}_\alpha$ *are equivalent.*

PROOF. Fix an $\alpha \le \omega_2$, $\alpha > 0$. For $F \in [\alpha]^{<\omega}$, $k < \omega$, and $g \in \mathfrak{C}^\alpha$, let $g|_{\langle F,k\rangle} \in (2^k)^F$ be defined by

$$(g|_{\langle F,k\rangle})(\xi) = g(\xi) \upharpoonright k \text{ for every } \xi \in F.$$

For $p \in \mathbb{S}_\alpha^D$, let

$$j(p) = \{g \in \mathfrak{C}^\alpha : g|_{\langle F,k\rangle} \text{ is consistent with } p \text{ for every } F \in [\alpha]^{<\omega}, k < \omega\}.$$

We will show that j establishes an order isomorphism between $\mathbb{S}_\alpha^D$ and $\mathbb{P}_\alpha$.

It is clear that for $p, q \in \mathbb{S}_\alpha^D$ if $p \le q$, then $j(p) \subseteq j(q)$. Thus, j is order-preserving.

To see that j is one to one, let $p, q \in \mathbb{S}_\alpha^D$ be such that $p \not\le q$. It is enough to show that $j(p) \not\subseteq j(q)$. Since $p \not\le q$, there exists a $\beta < \alpha$ with the property that $p \upharpoonright \beta \le q \upharpoonright \beta$, but $p \upharpoonright \beta$ does not force $p(\beta) \subset q(\beta)$. So, there are $r_0 \in \mathbb{P}_\beta$ below $p \upharpoonright \beta$ and $s \in 2^m$ with $m < \omega$ such that

$r_0 \Vdash s \in p(\beta) \setminus q(\beta)$. Choose $m \le k_0 \le k_1 < \omega$ and $\beta \in F_0 \subset F_1 \in [\alpha]^{<\omega}$ such that p is $\langle F_0, k_0 \rangle$-determined and q is $\langle F_1, k_1 \rangle$-determined. Since

$$\text{the family } \left\{ p|\sigma : \sigma \in \left(2^{k_1} \right)^{F_1} \text{ is consistent with } p \right\} \text{ is dense below } p, \quad (7.2)$$

there exists a $\sigma \in \left(2^{k_1} \right)^{F_1}$ such that $(p|\sigma) \upharpoonright \beta = (p \upharpoonright \beta)|(\sigma \upharpoonright \beta)$ is consistent with r_0. Let $r \in \mathbb{P}_\beta$ be their common extension. So, $r \Vdash s \in p(\beta) \setminus q(\beta)$ and $r \le (p|\sigma) \upharpoonright \beta \le (q|\sigma) \upharpoonright \beta$. Since condition $(p|\sigma) \upharpoonright \beta$ already decides whether s belongs to $p(\beta)$, we must have $(p|\sigma) \upharpoonright \beta \Vdash s \in p(\beta)$. Similarly, $(q|\sigma) \upharpoonright \beta \Vdash s \notin q(\beta)$. From $(p|\sigma) \upharpoonright \beta \Vdash s \in p(\beta)$ we conclude that there exists a $\sigma_1 \in \left(2^{k_1} \right)^{F_1}$ extending $\sigma \upharpoonright \beta$ such that $\sigma_1(\beta) \supset s$ and σ_1 is consistent with p. Now, using the fact that p is determined, by an easy induction we can find a $g \in j(p)$ such that $g|_{\langle F_1, k_1 \rangle} = \sigma_1$. But σ_1 cannot be consistent with q, since this would contradict $(q|\sigma) \upharpoonright \beta \Vdash s \notin q(\beta)$. So, $g \in j(p) \setminus j(q)$.

Next we will show that $j(p) \in S_\alpha$. For this first put $A = \mathrm{supp}(p)$ and define $j_A(p)$ as

$$\left\{ g \in \mathfrak{C}^A : g|_{\langle F, k \rangle} \text{ is consistent with } p \text{ for every } F \in [A]^{<\omega} \text{ and } k < \omega \right\}.$$

By the definition of supp it is easy to see that $[j_A(p)]_\alpha = j(p)$. So, it is enough to show that $j_A(p) \in \mathbb{P}_A$. To avoid some indexing complications we assume that A is an ordinal number. Then we can assume as well that $A = \alpha$, which implies that $\alpha < \omega_1$. So, it is enough to find a projection-keeping homeomorphism f from $\mathfrak{C}^\alpha$ onto $j(p)$.

To define f we need some preliminary facts. For a tree $T \in \mathbb{P}$ define a function $g_T : 2^{<\omega} \to T$, where we construct g_T on 2^n by induction on $n < \omega$. We put $g_T(\emptyset) = \emptyset$ and for $s \in 2^n$ and $i < 2$ we define $g_T(s\hat{\ }i) = t\hat{\ }i$, where t is the smallest element of T such that $g_T(s) \subset t$ and $t\hat{\ }0, t\hat{\ }1 \in T$. Notice that $|g_T(s)| \ge |s|$ for every $T \in \mathbb{P}$ and $s \in 2^{<\omega}$. In particular, for every $s \in 2^{<\omega}$, $k < \omega$, and $T, T' \in \mathbb{P}$,

$$\text{if } T \cap 2^k = T' \cap 2^k, \text{ then } g_T(s) \upharpoonright k = g_{T'}(s) \upharpoonright k. \quad (7.3)$$

Notice also that if $P = \lim T$, then a function $h_T : \mathfrak{C} \to P$ defined by $h_T(x) = \bigcup_{n<\omega} g_T(x \upharpoonright n)$ establishes a homeomorphism between $\mathfrak{C}$ and P.

For $x \in \mathfrak{C}^\alpha$, the values of $f(x)(\beta)$ will be defined by induction on $\beta < \alpha$ and will depend only on $x \upharpoonright \beta+1$. So assume that $f(x) \upharpoonright \beta$ has been already defined, let $P_{x \upharpoonright \beta} = \{ h(\beta) : f(x) \upharpoonright \beta \subset h \in j(p) \}$, put $T_{x \upharpoonright \beta} = T(P_{x \upharpoonright \beta})$, and define

$$f(x)(\beta) = h_{T_{x \upharpoonright \beta}}(x(\beta)) = \bigcup_{n<\omega} g_{T_{x \upharpoonright \beta}}(x(\beta) \upharpoonright n).$$

Clearly function f is a projection-keeping bijection from $\mathfrak{C}^\alpha$ onto $j(p)$. Thus, it is enough to show that f is continuous. For this, it is enough to prove that $f_\beta = f \upharpoonright \beta$ is continuous for every $\beta \leq \alpha$. So, assume that for some $\gamma \leq \alpha$ all functions f_ξ for $\xi < \gamma$ are continuous. We need to prove that f_γ is continuous. If γ is a limit ordinal, it is obvious. So, assume that $\gamma = \beta + 1$. We need to show that the mapping $\mathfrak{C}^\alpha \ni x \mapsto f(x)(\beta) \in \mathfrak{C}$ is continuous. So, fix an $m < \omega$. We will show that $f(x)(\beta) \upharpoonright m = f(x')(\beta) \upharpoonright m$ for x' close enough to x. For this find $\beta \in F \in [\alpha]^{<\omega}$ and $m \leq k < \omega$ such that p is $\langle F, k\rangle$-determined and let $x, x' \in \mathfrak{C}^\alpha$ be such that $\sigma_0 = x|_{\langle F,k\rangle} = x'|_{\langle F,k\rangle}$. Then

$$T_{x\upharpoonright\beta} \cap 2^k = \left\{\sigma(\beta) \colon \sigma_0 \upharpoonright \beta \subset \sigma \in \left(2^k\right)^F \ \& \ \sigma \text{ consistent with } p\right\} = T_{x'\upharpoonright\beta} \cap 2^k,$$

and, by (7.3), we obtain that

$$f(x)(\beta) \upharpoonright k = g_{T_{x\upharpoonright\beta}}(x(\beta) \upharpoonright k) \upharpoonright k = g_{T_{x'\upharpoonright\beta}}(x'(\beta) \upharpoonright k) \upharpoonright k = f(x')(\beta) \upharpoonright k.$$

So, f is as desired.

To finish the proof it is enough to show that j is onto $\mathbb{P}_\alpha$. For this, by induction on $0 < \beta \leq \alpha$, we will define $p_\beta \colon \mathbb{P}_\alpha \to \mathbb{S}_\beta$ such that $p_\beta(P) \subset p_\gamma(P)$ for every $0 < \beta \leq \gamma \leq \alpha$ and $P \in \mathbb{P}_\alpha$. We will aim for $p_\alpha = j^{-1}$.

For $P \in \mathbb{P}_\alpha$ we define $p_1(P) = \{x(0) \upharpoonright n \in 2^n \colon x \in P \ \& \ n < \omega\}$. So, assume that for some $1 < \gamma \leq \alpha$ functions $\{p_\beta \colon \beta < \alpha\}$ are already defined. If γ is a limit ordinal, we define $p_\gamma(P) = \bigcup_{\beta<\gamma} p_\beta(P)$. If γ is a successor ordinal, say $\gamma = \beta + 1$, we define $p_\gamma(P) \upharpoonright \beta = p_\beta(P)$ and

$$p_\gamma(P)(\beta) = \{\langle p_\beta(Q), s\rangle \colon Q \in \mathbb{P}_\alpha \ \& \ Q \subset P \ \& \ (\forall f \in Q) \ s \subset f(\beta)\}.$$

Note that if $P = [P_0]_\alpha$ for some $P_0 \in \mathbb{P}_A$, with $A \in [\alpha]^{\leq\omega}$, then $p_\gamma(\beta)$ is a name for $2^{<\omega}$ unless $\beta \in A$. Thus, support of each $p_\beta(P)$ is at most countable as required.

Finally, to show that $j(p_\alpha(P)) = P$, where $P \in \mathbb{P}_\alpha$, notice that for every $F \in [\alpha]^{<\omega}$, $k < \omega$, and $\sigma \colon F \to 2^k$ we have:

$$(\exists g \in P) \ g|_{\langle F,k\rangle} = \sigma \quad \Leftrightarrow \quad \sigma \text{ is consistent with } p_\alpha(P)$$
$$\Leftrightarrow \quad (\exists g \in j(p_\alpha(P))) \ g|_{\langle F,k\rangle} = \sigma.$$

Thus, $j(p_\alpha(P)) = P$. ∎

7.2 Consistency of CPA

Theorem 7.2.1 CPA *holds in the iterated perfect set model. In particular,* it is consistent with ZFC set theory. Moreover, 2^{ω_1} can be equal to ω_2 or can be bigger than ω_2.

PROOF. Start with a model V of ZFC+CH and let $V[G]$ be a generic extension of V with respect to forcing $\mathbb{P}_{\omega_2}$. By Fact 7.1.2, forcing $\mathbb{P}_{\omega_2}$ is equivalent to $\mathbb{S}_{\omega_2}$, so it preserves cardinals and $\mathfrak{c} = \omega_2$ in $V[G]$. Moreover, it preserves the value of 2^{ω_1} from V. For $\alpha \leq \omega_2$, let $G_\alpha = G \upharpoonright \alpha$. Then G_α is V-generic over $\mathbb{P}_\alpha$.

By Remark 6.4.2 it is enough to prove only CPA[X] for $X = \mathfrak{C}$. In this case the sequence $\{x_\alpha : \alpha < \omega_2\}$ will be equal to $\{c_\alpha : \alpha < \omega_2\}$.

Let $\{c_\alpha : \alpha < \omega_2\}$ be an enumeration, in $V[G]$, of $\mathfrak{C}$ such that for every α-th element $\omega_1\alpha$ of Γ, $\alpha > 0$, we have:

- $\{c_\xi : \xi < \omega_1\,\alpha\} = \mathfrak{C} \cap V[G_\alpha]$;
- $c_{\omega_1\alpha}$ is the Sacks generic real in $V[G_{\alpha+1}]$ over $V[G_\alpha]$.

Moreover, for each $\alpha \in \Gamma$, let $\{f_\xi^\alpha : \xi < \omega_1\}$ be an enumeration of a family $\mathcal{F}_{\text{prism}}^*(X) \cap V[G_\alpha]$.[1] We will show that these sequences satisfy CPA in $V[G]$.

So let S be a strategy for Player II. Thus, S is a function from a subset of $D = \bigcup_{\xi < \omega_1} \left(\mathcal{F}_{\text{prism}}^*(X) \times \mathcal{F}_{\text{prism}}^*(X) \right)^\xi \times \mathcal{F}_{\text{prism}}^*(X)$ into $\mathcal{F}_{\text{prism}}^*(X)$. Since $\mathbb{P}_{\omega_2}$ is ω_2-cc and satisfies axiom A, there is a closed unbounded subset Γ_S of Γ such that for every $\alpha \in \Gamma_S$

$$\omega_1\alpha = \alpha \ \& \ S \cap V[G_\alpha] = S \cap [(D \times \mathcal{F}_{\text{prism}}^*(X)) \cap V[G_\alpha]] \in V[G_\alpha]. \ (7.4)$$

Since the quotient forcing $\mathbb{P}_{\omega_2}/\mathbb{P}_\alpha$ is equivalent to $\mathbb{P}_{\omega_2}$, we can assume that $\alpha = 0$, that is, that $V[G_\alpha]$ is our ground model V.

Let $\langle\langle f_\xi, g_\xi \rangle : \xi < \omega_1 \rangle$ be a game played according to S, where $f_\xi = f_\xi^0$ for every $\xi < \omega_1$. Then $\mathcal{G} = \{g_\xi : \xi < \omega_1\} \in V$ is $\mathcal{F}_{\text{prism}}^*$-dense. It is enough to show that for $c = c_0$ we have

$$X \setminus V \subset \bigcup_{g_\xi \in \mathcal{F}_{\text{prism}}} g_\xi \left[\{ p \in \text{dom}(g_\xi) : \pi_0(p) = c \} \right].$$

So, take an $r \in X \setminus V$. Then there exists a $\mathbb{P}_{\omega_2}$-name τ for r such that

$$\mathbb{P}_{\omega_2} \Vdash \tau \in X \setminus V.$$

We can also choose τ such that it is a $\mathbb{P}_A$-name for some $A \in [\omega_2]^{\leq\omega}$ with $0 \in A$. Then all the information on r is coded by $G_A = G \upharpoonright A$. Therefore

---

[1] Formally no $f \in \mathcal{F}_{\text{prism}}^*(X) \cap V[G_{\omega_2}]$ belongs to $V[G_\alpha]$ with $\alpha < \omega_2$. However, in this proof the expression $f \in \mathcal{F}_{\text{prism}}^*(X) \cap V[G_\alpha]$ will be understood as "Code(f) belongs to $V[G_\alpha]$," where for dom(f) $= P \subset \mathfrak{C}^\alpha$ with $P = \text{range}(g)$, $g \in \Phi_{\text{prism}}(\alpha)$, and $D_\alpha \in V$ being a fixed countable dense subset of $\mathfrak{C}^\alpha$ we define Code(f) $= f \upharpoonright g[D_\alpha]$.

$r \in V[\{c_{\omega_1 \xi} : \xi \in A\}]$. Assume that A has an order type α. Clearly $\alpha < \omega_1$ and $\mathbb{P}_A$ is isomorphic to $\mathbb{P}_\alpha$. Applying this isomorphism we can assume that τ is a $\mathbb{P}_\alpha$-name for r and $\mathbb{P}_\alpha \parallel \tau \in X \setminus V$. Picking the smallest α with this property, we can also assume that for every $\beta < \alpha$ we have

$$\mathbb{P}_\alpha \parallel \tau \in X \setminus V[G_\beta].$$

Now, for any such name τ and any $R \in \mathbb{P}_\alpha$ there exist $P \in \mathbb{P}_\alpha$, $P \subset R$, and a continuous injection $f : P \to X$ (so $f \in \mathcal{F}_{\text{prism}}(X) \cap V$) that "reads τ continuously" in the sense that

$$Q \parallel \tau \in f[Q] \tag{7.5}$$

for every $Q \subset P$, $Q \in \mathbb{P}_\alpha$. (See [123, lem. 3.1] or [95, lem. 6, p. 580]. This can be also deduced from our Lemma 3.2.5.) So, the set

$$D = \{Q \in \mathbb{P}_\alpha : (\exists \xi < \omega_1) \, Q = \text{dom}(g_\xi) \, \& \, Q \parallel \tau \in g_\xi[Q]\} \in V$$

is dense in $\mathbb{P}_\alpha$. (For $R \in \mathbb{P}_\alpha$, take f as in (7.5), find $\xi < \omega_1$ with $f = f_\xi$, and notice that $Q = \text{dom}(g_\xi)$ justifies the density of D.)

Take $Q \in D \cap G_\alpha$ and $\xi < \omega_1$ such that $Q = \text{dom}(g_\xi)$. Then there is a $z \in Q$ such that $\pi_0(z) = c$ and $g_\xi(z) = r$. This finishes the proof. ∎

Notation

References

[1]   Agronsky, S., Bruckner, A.M., Laczkovich, M., and Preiss, D. Convexity conditions and intersections with smooth functions, *Trans. Amer. Math. Soc.* **289** (1985), 659–677.

[2]   Balcerzak, M., Ciesielski, K., and Natkaniec, T. Sierpiński–Zygmund functions that are Darboux, almost continuous, or have a perfect road, *Arch. Math. Logic* **37** (1997), 29–35. (Preprint* available.[1])

[3]   Banach, S. Sur une classe de fonctions continues, *Fund. Math.* **8** (1926), 166–172.

[4]   Bartoszyński, T., and Judah, H. *Set Theory. On the Structure of the Real Line*, A K Peters Ltd, Wellesley, MA, 1995.

[5]   Bartoszyński, T., and Recław, I. Not every γ-set is strongly meager, in *Set Theory* (Boise, ID, 1992–1994), 25–29, Contemp. Math. **192**, Amer. Math. Soc., Providence, RI, 1996.

[6]   Baumgartner, J. Sacks forcing and the total failure of Martin's axiom, *Topology Appl.* **19**(3) (1985), 211–225.

[7]   Baumgartner J., and Laver, R. Iterated perfect-set forcing, *Ann. Math. Logic* **17** (1979), 271–288.

[8]   Berarducci, A., and Dikranjan, D. Uniformly approachable functions and UA spaces, *Rend. Ist. Matematico Univ. di Trieste* **25** (1993), 23–56.

[9]   Blass, A. Combinatorial cardinal characteristics of the continuum, to appear in *Handbook of Set Theory*, ed. M. Foreman, M. Magidor, and A. Kanamori.

[10]   Blumberg, H. New properties of all real functions, *Trans. Amer. Math. Soc.* **24** (1922), 113–128.

[11]   Brendle, J. An e-mail to J. Pawlikowski, June 11, 2000.

[12]   Brendle, J., Larson, P., and Todorcevic, S. Rectangular axioms, perfect set properties and decomposition, preprint of November, 2002.

[13]   Brodskiĭ, M.L. On some properties of sets of positive measure, *Uspehi Matem. Nauk (N.S.)* **4, No. 3 (31)** (1949), 136–138.

---

[1] Preprints marked by * are available in electronic form from the *Set Theoretic Analysis Web Page:* http://www.math.wvu.edu/homepages/kcies/STA/STA.html

[14]   Brown, J.B. Differentiable restrictions of real functions, *Proc. Amer. Math. Soc.* **108** (1990), 391–398.

[15]   Brown, J.B. Restriction theorems in real analysis, *Real Anal. Exchange* **20**(1) (1994–95), 510–526.

[16]   Brown, J.B. Intersections of continuous, Lipschitz, Hölder class, and smooth functions, *Proc. Amer. Math. Soc.* **123** (1995), 1157–1165.

[17]   Bruckner, A.M. *Differentiation of Real Functions*, CMR Series **5**, Amer. Math. Soc., Providence, RI, 1994.

[18]   Bukovský, L., Kholshchevnikova, N.N., and Repický, M. Thin sets of harmonic analysis and infinite combinatorics, *Real Anal. Exchange* **20**(2) (1994–95), 454–509.

[19]   Burges, J.P. A selector principle for Σ_1^1 equivalence relations, *Michigan Math. J.* **24** (1977), 65–76.

[20]   Burke, M.R., and Ciesielski, K. Sets on which measurable functions are determined by their range, *Canad. J. Math.* **49** (1997), 1089–1116. (Preprint* available.)

[21]   Burke, M.R., and Ciesielski, K. Sets of range uniqueness for classes of continuous functions, *Proc. Amer. Math. Soc.* **127** (1999), 3295–3304. (Preprint* available.)

[22]   Ceder, J., and Ganguly, D.K. On projections of big planar sets, *Real Anal. Exchange* **9** (1983–84), 206–214.

[23]   Cichoń, J., Jasiński, A., Kamburelis, A., and Szczepaniak, P. On translations of subsets of the real line, *Proc. Amer. Math. Soc.* **130**(6) (2002), 1833–1842. (Preprint* available.)

[24]   Cichoń, J., Morayne, M., Pawlikowski, J., and Solecki, S. Decomposing Baire functions, *J. Symbolic Logic* **56**(4) (1991), 1273–1283.

[25]   Ciesielski, K. *Set Theory for the Working Mathematician*, London Math. Soc. Stud. Texts **39**, Cambridge Univ. Press, Cambridge, 1997.

[26]   Ciesielski, K. Set theoretic real analysis, *J. Appl. Anal.* **3**(2) (1997), 143–190. (Preprint* available.)

[27]   Ciesielski, K. Some additive Darboux-like functions, *J. Appl. Anal.* **4**(1) (1998), 43–51. (Preprint* available.)

[28]   Ciesielski, K. Decomposing symmetrically continuous functions and Sierpiński-Zygmund functions into continuous functions, *Proc. Amer. Math. Soc.* **127** (1999), 3615–3622. (Preprint* available.)

[29]   Ciesielski, K., and Jastrzębski, J. Darboux-like functions within the classes of Baire one, Baire two, and additive functions, *Topology Appl.* **103** (2000), 203–219. (Preprint* available.)

[30]   Ciesielski, K., and Millán, A. Separately nowhere constant functions; n-cube and α-prism densities, *J. Appl. Anal.*, to appear. (Preprint* available.)

[31]   Ciesielski, K., Millán, A., and Pawlikowski, J. Uncountable γ-sets under axiom $\text{CPA}_{\text{cube}}^{\text{game}}$, *Fund. Math.* **176**(1) (2003), 143–155. (Preprint* available.)

[32]   Ciesielski, K., and Natkaniec, T. On Sierpiński-Zygmund bijections and their inverses, *Topology Proc.* **22** (1997), 155–164. (Preprint* available.)

[33] Ciesielski, K., and Natkaniec, T. A big symmetric planar set with small category projections, *Fund. Math.* **178**(3) (2003), 237–253. (Preprint* available.)

[34] Ciesielski, K., and Pawlikowski, J. On sums of Darboux and nowhere constant continuous functions, *Proc. Amer. Math. Soc.* **130**(7) (2002), 2007–2013. (Preprint* available.)

[35] Ciesielski, K., and Pawlikowski, J. On the cofinalities of Boolean algebras and the ideal of null sets, *Algebra Universalis* **47**(2) (2002), 139–143. (Preprint* available.)

[36] Ciesielski, K., and Pawlikowski, J. Small combinatorial cardinal characteristics and theorems of Egorov and Blumberg, *Real Anal. Exchange* **26**(2) (2000–2001), 905–911. (Preprint* available.)

[37] Ciesielski, K., and Pawlikowski, J. Crowded and selective ultrafilters under the Covering Property Axiom, *J. Appl. Anal.* **9(1)** (2003), 19–55. (Preprint* available.)

[38] Ciesielski, K., and Pawlikowski, J. Small coverings with smooth functions under the Covering Property Axiom, *Canad. J. Math.*, to appear. (Preprint* available.)

[39] Ciesielski, K., and Pawlikowski, J. Covering Property Axiom CPA$_{\mathrm{cube}}$ and its consequences, *Fund. Math.* **176**(1) (2003), 63–75. (Preprint* available.)

[40] Ciesielski, K., and Pawlikowski, J. Nice Hamel bases under the Covering Property Axiom, *Acta Math. Hungar.*, to appear. (Preprint* available.)

[41] Ciesielski, K., and Pawlikowski, J. Uncountable intersections of open sets under CPA$_{\mathrm{prism}}$, *Proc. Amer. Math. Soc.*, to appear. (Preprint* available.)

[42] Ciesielski, K., and Pawlikowski, J. On additive almost continuous functions under CPA$_{\mathrm{prism}}^{\mathrm{game}}$, *J. Appl. Anal.*, to appear. (Preprint* available.)

[43] Ciesielski, K., and Pawlikowski, J. Continuous images of big sets and additivity of s_0 under CPA$_{\mathrm{prism}}$, *Real Anal. Exchange*, to appear. (Preprint* available.)

[44] Ciesielski, K., and Shelah, S. Model with no magic set, *J. Symbolic Logic* **64(4)** (1999), 1467–1490. (Preprint* available.)

[45] Ciesielski, K., and Wojciechowski, J. Sums of connectivity functions on $\mathbb{R}^n$, *Proc. London Math. Soc.* **76**(2) (1998), 406–426. (Preprint* available.)

[46] Coplakova, E., and Hart, K.P. Crowded rational ultrafilters, *Topology Appl.* **97** (1999), 79–84.

[47] Darji, U. On completely Ramsey sets, *Colloq. Math.* **64**(2) (1993), 163–171.

[48] Davies, R.O. Second category E with each proj($\mathbb{R}^2 \setminus E^2$) dense, *Real Anal. Exchange* **10** (1984–85), 231–232.

[49] van Douwen, E.K. The integers and topology, in *Handbook of Set-Theoretic Topology*, ed. K. Kunen and J.E. Vaughan, North-Holland, Amsterdam (1984), 111–167.

[50] van Douwen, E.K. Better closed ultrafilters on $\mathbb{Q}$, *Topology Appl.* **47** (1992), 173–177.

[51] van Douwen, E.K., Monk, J.D., and Rubin, M. Some questions about Boolean algebras, *Algebra Universalis* **11** (1980), 220–243.

[52] Eggleston, H.G. Two measure properties of Cartesian product sets, *Quart. J. Math. Oxford (2)* **5** (1954), 108–115.

[53] Engelking, R. *General Topology*, Polish Sci. Publ. PWN, Warszawa, 1977.

[54] Erdős, P. On some properties of Hamel bases, *Colloq. Math.* **10** (1963), 267–269.

[55] Farah, I. Semiselective coideals, *Mathematika* **45** (1998), 79–103.

[56] Federer, H. *Geometric Measure Theory*, Springer-Verlag, New York, 1969.

[57] Filipów, R., and Recław, I. On the difference property of Borel measurable and (s)-measurable functions, *Acta Math. Hungar.* **96**(1) (2002), 21–25.

[58] Foran, J. Continuous functions: A survey, *Real Anal. Exchange* **2** (1977), 85–103.

[59] Fuchino, S., and Plewik, Sz. On a theorem of E. Helly, *Proc. Amer. Math. Soc.* **127(2)** (1999), 491–497.

[60] Galvin, F. Partition theorems for the real line, *Notices Amer. Math. Soc.* **15** (1968), 660.

[61] Galvin, F. Errata to "Partition theorems for the real line," *Notices Amer. Math. Soc.* **16** (1969), 1095.

[62] Galvin, F., and Miller, A.W. γ-sets and other singular sets of real numbers, *Topology Appl.* **17** (1984), 145–155.

[63] Galvin, F., and Prikry, K. Borel sets and Ramsey's theorem, *J. Symbolic Logic* **38** (1973), 193–198.

[64] Gerlits, J., and Nagy, Zs. Some properties of $C(X)$, I, *Topology Appl.* **14** (1982), 151–161.

[65] Grigorieff, S. Combinatorics on ideals and forcing, *Ann. Math. Logic* **3**(4) (1971), 363–394.

[66] Hart, K.P. Ultrafilters of character ω_1, *J. Symbolic Logic* **54**(1) (1989), 1–15.

[67] Hrušák, M. Private communication (e-mail to K. Ciesielski), March 2000.

[68] Hulanicki, A. Invariant extensions of the Lebesgue measure, *Fund. Math.* **51** (1962), 111–115.

[69] Jech, T. *Set Theory*, Academic Press, New York, 1978.

[70] Jordan, F. Generalizing the Blumberg theorem, *Real Anal. Exchange* **27**(2) (2001–2002), 423–439. (Preprint* available.)

[71] Judah, H., Miller, A.W., and Shelah, S. Sacks forcing, Laver forcing, and Martin's axiom, *Arch. Math. Logic* **31**(3) (1992), 145–161.

[72] Just, W. and Koszmider, P. Remarks on cofinalities and homomorphism types of Boolean algebras, *Algebra Universalis* **28**(1) (1991), 138–149.

[73] Kanovei, V. Non-Glimm–Effros equivalence relations at second projective level, *Fund. Math.* **154** (1997), 1–35.

[74] Kechris, A.S. *Classical Descriptive Set Theory*, Springer-Verlag, Berlin, 1995.

[75] Kellum, K.R. Sums and limits of almost continuous functions, *Colloq. Math.* **31** (1974), 125–128.

[76] Kellum, K.R. Almost Continuity and connectivity - sometimes it's as easy as to prove a stronger result, *Real Anal. Exchange* **8** (1982-83), 244–252.

[77] Kharazishvili, A.B. *Strange Functions in Real Analysis*, Pure and Applied Mathematics **229**, Marcel Dekker, New York, 2000.

[78] Kirchheim, B., and Natkaniec, T. On universally bad Darboux functions, *Real Anal. Exchange* **16** (1990–91), 481–486.

[79]   Koppelberg, S. Boolean algebras as unions of chains of subalgebras, *Algebra Universalis* **7** (1977), 195–204.

[80]   Kunen, K. Some points in $\beta\mathbf{N}$, *Math. Proc. Cambridge Philos. Soc.* **80**(3) (1976), 385–398.

[81]   Kunen, K. *Set Theory*, North-Holland, Amsterdam, 1983.

[82]   Laczkovich, M. Functions with measurable differences, *Acta Mathematica Academiae Scientarum Hungaricae* **35**(1-2) (1980), 217–235.

[83]   Laczkovich, M. Two constructions of Sierpiński and some cardinal invariants of ideals, *Real Anal. Exchange* **24**(2) (1998–99), 663–676.

[84]   Laver, R. On the consistency of Borel's conjecture, *Acta Math.* **137** (1976), 151–169.

[85]   Laver, R. Products of infinitely many perfect trees, *J. London Math. Soc.* **29** (1984), 385–396.

[86]   Loomis, L.H. On the representation of the σ-complete Boolean algebra, *Bull. Amer. Math. Soc.* **53** (1947), 757–760.

[87]   Luzin, N. Sur un problème de M. Baire, *Hebdomadaires Seances Acad. Sci. Paris* **158** (1914), 1258–1261.

[88]   Mahlo, P. Über Teilmengen des Kontinuums von dessen Machtigkeit, *Sitzungsber. Sächs. Akad. Wiss. Leipzig Math.-Natu. Kl.* **65** (1913), 283–315.

[89]   Martin, D.A., and Solovay, R.M. Internal Cohen extensions, *Ann. Math. Logic* **2**(2) (1970), 143–178.

[90]   Mauldin, R.D. *The Scottish Book*, Birkhäuser, Boston, 1981.

[91]   Mazurkiewicz, S. Sur les suites de fonctions continues, *Fund. Math.* **18** (1932), 114–117.

[92]   Millán, A. A crowded Q-point under $\mathrm{CPA}^{\mathrm{game}}_{\mathrm{prism}}$, preprint*.

[93]   Millán, A. $\mathrm{CPA}^{\mathrm{game}}_{\mathrm{prism}}$ and ultrafilters on $\mathbb{Q}$, preprint*.

[94]   Miller, A.W. Covering 2^ω with ω_1 disjoint closed sets, *The Kleene Symposium*, North-Holland, Amsterdam (1980), 415–421.

[95]   Miller, A.W. Mapping a set of reals onto the reals, *J. Symbolic Logic* **48** (1983), 575–584.

[96]   Miller, A.W. Special Subsets of the Real Line, in *Handbook of Set-Theoretic Topology*, ed. K. Kunen and J.E. Vaughan, North-Holland, Amsterdam (1984), 201–233.

[97]   Miller, A.W. Additivity of measure implies dominating reals, *Proc. Amer. Math. Soc.* **91** (1984), 111–117.

[98]   Miller, H. On a property of Hamel bases, *Boll. Un. Mat. Ital A(7)* **3** (1989), 39–43.

[99]   Morayne, M. On continuity of symmetric restrictions of Borel functions, *Proc. Amer. Math. Soc.* **98** (1985), 440–442.

[100]   Muthuvel, K. Some results concerning Hamel bases, *Real Anal. Exchange* **18**(2) (1992-93), 571–574.

[101]   Mycielski, J. Independent sets in topological algebras, *Fund. Math.* **55** (1964), 139–147.

[102]   Mycielski, J. Algebraic independence and measure, *Fund. Math.* **61** (1967), 165–169.

[103] Natkaniec, T. On category projections of cartesian product $A \times A$, *Real Anal. Exchange* **10** (1984–85), 233–234.

[104] Natkaniec, T. On projections of planar sets, *Real Anal. Exchange* **11** (1985–86), 411–416.

[105] Natkaniec, T. Almost Continuity, *Real Anal. Exchange* **17** (1991-92), 462–520.

[106] Natkaniec, T. The density topology can be not extraresolvable, *Real Anal. Exchange*, to appear. (Preprint* available.)

[107] Nowik, A. Additive properties and uniformly completely Ramsey sets, *Colloq. Math.* **82** (1999), 191–199.

[108] Nowik, A. Possibly there is no uniformly completely Ramsey null set of size 2^ω, *Colloq. Math.* **93** (2002), 251–258. (Preprint* available.)

[109] Olevskiĭ, A. Ulam-Zahorski problem on free interpolation by smooth functions, *Trans. Amer. Math. Soc.* **342** (1994), 713–727.

[110] Recław, I. Every Lusin set is undetermined in the point-open game, *Fund. Math.* **144** (1994), 43–54.

[111] Repický, M. Perfect sets and collapsing continuum, *Comment. Math. Univ. Carolin.* **44**(2) (2003), 315–327.

[112] Rosłanowski, A, and Shelah, S. Measured Creatures, preprint.

[113] Sacks, G. Forcing with perfect sets, in *Axiomatic Set Theory*, ed. D. Scott, *Proc. Symp. Pure Math.* **13**(1), Amer. Math. Soc. (1971), 331–355.

[114] Saks, S. *Theory of the Integral*, 2nd ed., Monografie Mat., vol. 7, PWN, Warsaw, 1937.

[115] Shelah, S. Possibly every real function is continuous on a non–meagre set, *Publ. Inst. Mat. (Beograd) (N.S.)* **57**(71) (1995), 47–60.

[116] Sierpiński, W. Sur un ensemble non dénombrable donte toute image continue est de 1-re catégorie, *Bull. Intern. Acad. Polon. Sci. A* 1928, 455–458. Reprinted in *Oeuvres Choisies*, vol. II, 671–674.

[117] Sierpiński, W. Sur un ensemble non dénombrable donte toute image continue est de mesure null, *Fund. Math.* **11** (1928), 302–304. Reprinted in *Oeuvres Choisies*, vol. II, 702–704.

[118] Sierpiński, W. Remarque sur les suites infinies de fonctions (Solution d'un problème de M. S. Saks), *Fund. Math.* **18** (1932), 110–113.

[119] Sierpiński, W. Sur les translations des ensembles linéaires, *Fund. Math.* **19** (1932), 22–28. Reprinted in *Oeuvres Choisies*, vol. III, 95–100.

[120] Sierpiński, W. *Hypothèse du continu*, Monografie Matematyczne, Tom IV, Warsaw 1934.

[121] Sikorski, R. *Boolean Algebras*, 3rd ed., Springer-Verlag, New York, 1969.

[122] Simon, P. Sacks forcing collapses $\mathfrak{c}$ to $\mathfrak{b}$, *Comment. Math. Univ. Carolin.* **34**(4) (1993), 707–710.

[123] Steprāns, J. Sums of Darboux and continuous functions, *Fund. Math.* **146** (1995), 107–120.

[124] Steprāns, J. Decomposing Euclidean space with a small number of smooth sets, *Trans. Amer. Math. Soc.* **351** (1999), 1461–1480. (Preprint* available.)

[125] Todorcevic, S. *Partition problems in topology*, Contemp. Math. **84**, Amer. Math. Soc., Providence, RI, 1989.

[126] Todorchevich, S., and Farah, I. *Some Applications of the Method of Forcing*, Yenisei Series in Pure and Applied Mathematics, Yenisei, Moscow; Lycée, Troitsk, 1995.

[127] Ulam, S. *A Collection of Mathematical Problems*, Interscience Tracts in Pure and Applied Mathematics **8**, Interscience Publishers, New York-London, 1960.

[128] Vaughan, J.E. Small uncountable cardinals and topology, in *Open Problems in Topology*. ed. J. van Mill and G. M. Reed, North-Holland, Amsterdam (1990), 195–216.

[129] Whitney, H. Analytic extensions of differentiable functions defined in closed sets, *Trans. Amer. Math. Soc.* **36** (1934), 63–89.

[130] Zahorski, Z. Sur l'ensamble des points singuliére d'une fonction d'une variable réele admettand des dérivées des tous orders, *Fund. Math.* **34** (1947), 183–245.

[131] Zapletal, J. *Descriptive Set Theory and Definable Forcing*, Mem. Amer. Math. Soc. **167**, 2004.

Index